Advances in Experimental Medicine and Biology

Volume 1143

More information about this series at http://www.springer.com/series/5584

Haojian Zhang • Shaoguang Li
Editors

Leukemia Stem Cells in Hematologic Malignancies

Springer

Editors
Haojian Zhang
Medical Research Institute
Wuhan University
Wuhan, Hubei, China

Shaoguang Li
Department of Medicine
University of Massachusetts Medical School
Worcester, Massachusetts, USA

ISSN 0065-2598 ISSN 2214-8019 (electronic)
Advances in Experimental Medicine and Biology
ISBN 978-981-13-7344-2 ISBN 978-981-13-7342-8 (eBook)
https://doi.org/10.1007/978-981-13-7342-8

This Springer imprint is published by the registered company Springer Nature Singapore Pte Ltd.
The registered company address is: 152 Beach Road, #21-01/04 Gateway East, Singapore 189721, Singapore

Preface

In many types of cancer, a curative therapy requires eradication of cancer-initiating (stem) cells, and the success relies on a full understanding of survival mechanisms for these stem cells. In this approach, it is equally important to minimize an unwanted effect on normal stem cell counterparts. To realize these goals, a key initial step is to identify and functionally characterize critical target genes and study their associated molecular pathways in cancer stem cells. Relative to solid tumors, a significant number of blood cancers are believed to be derived from leukemic stem cells (LSCs) and responsible for disease progression, relapse, and drug resistance. Therefore, new therapeutic strategies need to be developed by focusing on a complete eradication of LSCs. In this book, we review our current understanding of the biology of LSCs and emphasize the necessity and importance of targeting LSCs in the treatment of hematopoietic malignancies.

Wuhan, Hubei, China | Haojian Zhang
Worcester, Massachusetts, USA | Shaoguang Li

Contents

Chapter 1
Signaling Pathways in Leukemic Stem Cells

Lindsay M. Gurska, Kristina Ames, and Kira Gritsman

Contents

Abstract Hematopoietic stem cells (HSCs) and leukemic stem cells (LSCs) utilize many of the same signaling pathways for their maintenance and survival. In this review, we will focus on several signaling pathways whose roles have been extensively studied in both HSCs and LSCs. Our main focus will be on the PI3K/AKT/mTOR pathway and several of its regulators and downstream effectors. We will also discuss several other signaling pathways of particular importance in LSCs, including the WNT/β-catenin pathway, the NOTCH pathway, and the TGFβ pathway. For

L. M. Gurska · K. Ames
Department of Cell Biology, Albert Einstein College of Medicine, Bronx, New York, USA

Department of Medicine, Albert Einstein College of Medicine, Bronx, New York, USA

K. Gritsman (✉)
Department of Cell Biology, Albert Einstein College of Medicine, Bronx, New York, USA

Department of Medicine, Albert Einstein College of Medicine, Bronx, New York, USA

Department of Medical Oncology, Montefiore Hospital, Bronx, New York, USA
e-mail: kira.gritsman@einstein.yu.edu

H. Zhang, S. Li (eds.), *Leukemia Stem Cells in Hematologic Malignancies*, Advances in Experimental Medicine and Biology 1143,
https://doi.org/10.1007/978-981-13-7342-8_1

each of these pathways, we will emphasize differences in how these pathways operate in LSCs, compared to their function in HSCs, to highlight opportunities for the specific therapeutic targeting of LSCs. We will also highlight areas of crosstalk between multiple signaling pathways that may affect LSC function.

Keywords Hematopoietic stem cell · Leukemic stem cell · Signaling · PI3K · Akt · mTOR · FOXO · WNT · β-Catenin · NOTCH · TGFβ

Abbreviations

AML	Acute myeloid leukemia
APC	Adenomatous polyposis coli
BCR-ABL	Fusion of breakpoint cluster region (BCR) and Abelson murine leukemia viral oncogene homolog 1 (ABL)
Cbf-1	Centromere-binding protein 1
CLL	Chronic lymphoblastic leukemia
CML	Chronic myeloid leukemia
Co-Smad	Common-partner Smad
DEPTOR	DEP domain-containing mTOR-interacting protein
GPCRs	G-protein-coupled receptors
GSK-3β	Glycogen synthase kinase-3β
Hes-1	Hairy and enhancer of split-1
HSC	Hematopoietic stem cell
HSPC	Hematopoietic stem and progenitor cell
IKK	IκB kinase
I-Smad	Inhibitory Smad
JNK	Jun N-terminal kinase
LEF	Lymphoid enhancer factor
Lrp5/6	Lipoprotein receptor-related protein 5 or 6
LSC	Leukemic stem cell
LSK	Lineage negative, Sca1+, cKit-
MAPK	Mitogen-activated protein kinase
MEK	Mitogen-activated protein kinase
MLL	Mixed lineage leukemia
mLST8	Mammalian lethal with sec-13 protein 8
mSin1	Mammalian stress-activated map kinase-interacting protein 1
mTOR	Mechanistic target of rapamycin
mTORC1	mTOR complex 1
mTORC2	mTOR complex 2
NICD	Notch intracellular domain
PDK1	Pyruvate dehydrogenase kinase 1
PH	Pleckstrin homology

PI3K	Phosphatidylinositol 3-kinase
PIP2	Phosphatidylinositol (3,4)-diphosphate
PIP3	Phosphatidylinositol (3,4,5)-triphosphate
PORCN	Porcupine
PRAS40	Proline-rich Akt substrate 40 kDa
Protor1/2	Protein observed with Rictor 1 and 2
PTEN	Phosphatase and tensin homolog
Raptor	Regulatory-associated protein of mammalian target of rapamycin
R-Smad	Receptor-mediated Smad
RTK	Receptor tyrosine kinase
SH2	Src-homology 2
SHIP	SH2-containing inositol 5′-phosphatase
SIRT1	NAD-dependent deacetylase sirtuin-1
T-ALL	T-cell acute lymphoblastic leukemia
TCF	T-cell factor
TGFβ	Transforming growth factor-beta
TKI	Tyrosine kinase inhibitor
TSC1	Tuberous sclerosis 1
TSC2	Tuberous sclerosis 2
TβRI	TGFβ-type I receptor
TβRII	TGFβ-type II receptor

1.1 Introduction

Hematopoiesis is a hierarchical process, with hematopoietic stem cells (HSCs) residing at the top of the hierarchy. These cells have the ability to both self-renew and differentiate to generate all mature blood cells. To retain homeostasis under normal conditions, as well as to initiate the required responses under stress conditions, hematopoiesis needs to be a tightly regulated process. To maintain homeostasis, HSCs need to respond to extracellular cues, such as growth factors and chemokines within the bone marrow niche, in order to make the decision whether to maintain self-renewal or to enter the cell cycle and differentiate. These growth factors and chemokines are the ligands for key signaling networks that either instruct or facilitate these critical decisions that orchestrate HSC function. Examples of signaling pathways required for HSC regulation include the PI3K, WNT/β-catenin, NOTCH, and TGFβ pathways.

Without proper regulation by such signaling pathways, HSCs and downstream progenitors can acquire unlimited self-renewal potential at the expense of differentiation, as well as increased proliferation and survival. The acquisition of the first genetic or epigenetic alteration in an HSC can give that cell a clonal advantage over its counterparts and may lead to the generation of preleukemic stem cells (pre-LSCs) (reviewed in Corces-Zimmerman et al. [1]) (Fig. 1.1). Pre-LSCs are still capable of multilineage differentiation but have increased self-renewal capacity. They can then acquire additional genetic alterations that will promote full transformation to leukemic stem cells (LSCs) (Fig. 1.1). Unlike pre-LSCs, LSCs cannot contribute to normal adult blood cells. Instead, they differentiate only into leukemic blasts, which form the bulk of the leukemia. Some studies have shown that LSCs can also arise directly from more committed myeloid progenitors, through genetic alterations that reprogram these progenitors to express a self-renewal program usually associated with HSCs [2–4] (Fig. 1.1). Phenotypically, LSCs may share some of the same cell surface markers as HSCs, but they are defined functionally by their ability to generate leukemia in the bone marrow transplantation setting. As mutations in components of signaling pathways are prominent in leukemias, it is important to understand how these signaling pathways regulate LSC function.

In this review, we will focus on several signaling pathways whose roles have been extensively studied in both HSCs and LSCs and their effects on myeloid LSC self-renewal, cycling, quiescence, and differentiation capacity. Our main focus will be on the PI3K/AKT/mTOR pathway and some of its regulators and downstream effectors, whose roles in HSC and LSC function have been explored. We will also discuss other signaling pathways of particular importance in LSCs, including the WNT/β-catenin pathway, the NOTCH pathway, and the TGFβ pathway. For each of these pathways, we will emphasize differences in how these pathways operate in LSCs, compared to their function in HSCs, to highlight potential opportunities for the specific therapeutic targeting of LSCs. We will also highlight areas of crosstalk between multiple signaling pathways that may affect LSC function.

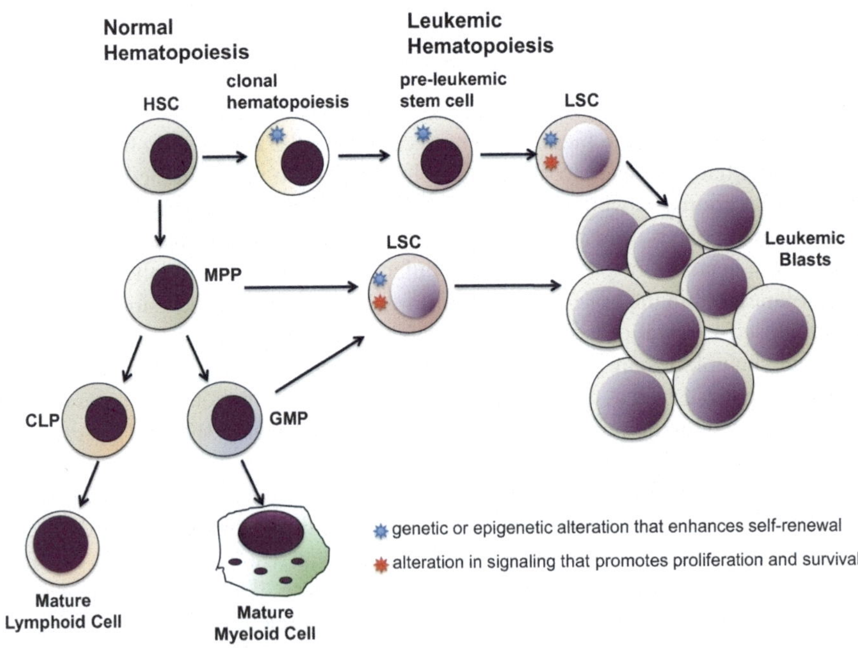

Fig. 1.1 Normal and leukemic hematopoiesis

1.2 The PI3K/AKT/mTOR Signaling Pathway

Many hematopoietic growth factors and cytokines signal through the PI3K/AKT pathway. Phosphatidylinositol 3-kinase (PI3K) can be activated by growth factors through receptor tyrosine kinase (RTK) receptors, by chemokines through G-protein-coupled receptors (GPCRs) or, by RAS family proteins (Fig. 1.2). The PI3K family consists of three distinct classes, and the Class I isoforms have been the most extensively studied in hematopoiesis, so we will focus on Class I PI3K for this discussion. Class I PI3Ks are heterodimeric lipid kinases, which are composed of a p110-kDa catalytic subunit and a regulatory subunit (reviewed in Vanhaesebroeck et al. [5]). Depending on the type of regulatory subunit, Class I PI3Ks are classified as either Class 1A or Class1B enzymes. Mammals have three Class 1A p110 isoforms (p110α, p110β, and p110δ), which are encoded by three separate genes (*PIK3CA*, *PIK3CB*, and *PIK3CD*, respectively), and one Class 1B isoform p110γ, encoded by the *PIK3CG* gene. The regulatory subunits are encoded by different genes and diversified by alternative promoter usage as follows: *PIK3R1* encodes p85α and the splice variants p55α and p50α, while *PIK3R2* encodes the p85β subunit, *PIK3R3* encodes p55γ, *PIK3R5* gene encodes p101, and *PIK3R6* encodes the p84/p87regulatory subunit. Class 1A p85 regulatory subunits (p85, p55, p50) harbor two SRC homology 2 domains (nSH2, cSH2) and an intervening p110-binding

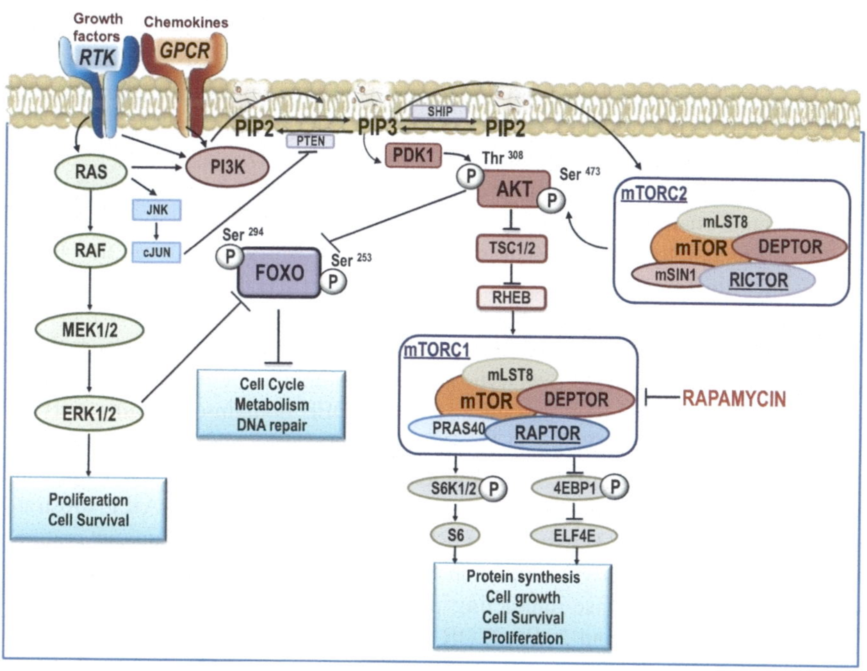

Fig. 1.2 The PI3K/AKT/mTOR and RAS/MEK/ERK signaling cascades and their main downstream effectors

region (iSH2) and constitutively interact with the p110α, p110β, and p110δ catalytic subunits [6]. The Class 1A catalytic isoforms can all bind to the same p85 regulatory subunits, so they can functionally compensate for one another (reviewed in [5]). In contrast, the Class1B catalytic subunit p110γ does not have a p85-binding domain and is almost exclusively activated by GPCRs. The Class I PI3Ks p110α, p110δ, and p110γ also harbor a RAS-binding domain and all except p110β are thought to be RAS effectors [5]. In contrast, p110β uses its RBD to bind to RAC and RHO GTPase family members and also interacts with Rab5 GTPase [7–9]. Upon stimulation, Class I PI3Ks produce the lipid second messenger phosphatidylinositol (3,4,5)-triphosphate (PIP3) from phosphatidylinositol (3,4)-diphosphate (PIP2), and this process can be antagonized by phosphatase and tensin homolog (PTEN) or Src-homology 2 (SH2)-containing inositol 5′-phosphatase (SHIP), both of which dephosphorylate PIP3 to PIP2. PIP3 recruits the inactive serine/threonine-protein kinase AKT and pyruvate dehydrogenase kinase 1 (PDK1) from the cytosol through their pleckstrin homology (PH) domains, where PDK1 then phosphorylates AKT at Thr308. For complete activation, AKT must also be phosphorylated by the mTOR complex 2 (mTORC2) at Ser473 [10]. Intriguingly, activation of PI3K/AKT in tumors can be frequently accompanied by JNK activation, and this activation seems to be PI3K-dependent, since it is promoted by loss of *PTEN*, with constitutive activation of JNK and the transcription factor c-JUN [11]. Furthermore, c-JUN promotes cellular survival by negatively regulating the expression of *PTEN* through direct binding to a variant AP-1 site on the *PTEN* promoter, thus activating the AKT pathway [12]. AKT has multiple downstream effectors, which regulate diverse cell processes, including cellular metabolism, glucose homeostasis, inflammation, apoptosis, cell cycle regulation, protein synthesis, and autophagy [5]. Here we will focus on those AKT effectors that have been shown to play a role in HSCs and LSCs: mechanistic target of rapamycin (mTOR) and FOXO.

mTOR is the major intracellular component that senses and reacts to dynamic environmental changes in response to nutrient and growth factor fluctuation to coordinate cell metabolism and growth. mTOR is a serine/threonine kinase that forms two distinct functional complexes, mTOR complex 1 (mTORC1) and mTOR complex 2 (mTORC2) (reviewed in Zoncu et al. [13]). mTORC1 has six known protein components, while mTORC2 has seven components [14]. These two complexes share five proteins: the mTOR catalytic subunit, mammalian lethal with sec-13 protein 8 (mLST8, also known as GbL) [15, 16], DEP domain-containing mTOR-interacting protein (DEPTOR) [17], and the TTI1/TEL2 complex counts as 2 proteins [18]. In addition, mTORC1 contains regulatory-associated protein of mammalian target of rapamycin (RAPTOR) [19, 20] and proline-rich AKT substrate 40 kDa (PRAS40) [21–23], while mTORC2 contains the following unique components: rapamycin-insensitive companion of mTOR (RICTOR) [24, 25], mammalian stress-activated map kinase-interacting protein 1 (mSIN1) [24, 26], and protein observed with Rictor 1 and 2 (PROTOR1/2) [27, 28]. mTORC1 integrates signals from nutrients, growth factors, energy, and stress, while mTORC2 is primarily regulated by growth factors [13]. The small GTPase RAS homolog enriched in brain

(RHEB) stimulates mTORC1 activity when it is GTP-bound. AKT activates mTORC1 indirectly, through phosphorylation and subsequent inactivation of tuberous sclerosis 2 (TSC2) in the tuberous sclerosis 1 (TSC1)/TSC2 complex. TSC1-TSC2 acts as a GTPase-activating protein (GAP) for RHEB, thereby inactivating RHEB. Therefore, AKT activation ultimately leads to the stimulation of mTORC1. AKT also increases mTORC1 activity through phosphorylation of proline-rich AKT substrate PRAS40 [13].

The FOXO transcription factors are evolutionarily conserved, and their roles have been implicated in a multitude of fundamental biological processes, including metabolism, stem cell maintenance, lifespan regulation, and tumorigenesis [29, 30]. The mammalian FOXOs include FOXO1, FOXO3, FOXO4, and FOXO6. The FOXOs are potent transcriptional activators that bind to the consensus motif TTGTTTAC (termed DBE for DAF-16a family member binding element) to regulate the transcription of target genes [31–33]. Multiple studies have shown complex regulation of FOXOs by phosphorylation, which determines their subcellular localization and governs their transcriptional activity (reviewed in Tothova et al. [34)]). In HSC and progenitor cells, upon growth factor stimulation, AKT phosphorylates FOXO 1, FOXO 3, and FOXO 4 at three conserved residues (in FOXO3 – Thr32, Ser253, Ser315), which inactivates FOXOs by creating a binding motif for the 14-3-3 chaperone proteins. This then facilitates translocation of the FOXOs from the nucleus and to the cytoplasm while preventing nuclear reentry by masking the nuclear localization sequence [31, 35, 36]. It is interesting to note that FOXO phosphorylation by AKT can be overridden by stress stimuli that activate Jun N-terminal kinase (JNK), which phosphorylates FOXOs at distinct sites from that of AKT to facilitate their nuclear import [37–40]. Interestingly, Yang et al. showed that ERK is also able to downregulate FOXO3a activity by phosphorylation of FOXO3a at Ser 294, Ser 344, and Ser 425, consequently promoting cell proliferation and tumorigenesis in a breast cancer model [41]. In addition to phosphorylation, FOXOs are posttranslationally regulated by acetylation/deacetylation, methylation, ubiquitination, and redox modulation. This fine-tuned regulation allows FOXOs to induce specific gene-expression programs in response to many environmental stimuli, such as insulin, growth factors, nutrients, and oxidative stress (reviewed in [42, 43]).

1.3 The PI3K Isoforms in HSCs and in Leukemia

Of the Class I catalytic isoforms, *Pik3ca* and *Pik3cb* are ubiquitously expressed, and their germline deletion in mice results in embryonic lethality (reviewed in Vanhaesebroeck et al. [5]). In contrast, *Pik3cd* and *Pik3cg* are enriched in leukocytes and are not required for development. In fact, mice with germline deletion of *Pik3cd* or *Pik3cg* are viable and have normal blood counts. While p110α and p110δ isoforms primarily transduce RTK signals, p110β and p110γ play a less prominent role in RTK signaling and are functionally redundant in signaling downstream of GPCRs [9]. P110α is essential for growth factor signaling and oncogenic signaling in fibroblasts [44]. However, functional redundancy between p110α and p110β in transducing RTK signals has also been described in some contexts [45, 46]. Surprisingly, postnatal conditional deletion of p110α in the hematopoietic system results only in mild anemia, without affecting HSC self-renewal, while conditional deletion of p110β in hematopoietic cells does not affect blood counts (Table 1.1) [47, 48]. In fetal liver HSCs, compound deletion of the regulatory subunits p85α and p85β decreases HSC frequency and multilineage repopulating capacity [49]. This suggests that there is increased redundancy between the Class IA isoforms in transducing growth factor signals in HSCs.

However, there is evidence that the same level of redundancy may not be present in the context of hematologic malignancies. For example, a requirement for p85α, but not of p85β, regulatory subunit was shown to be necessary for myeloproliferative disease driven by an activation loop mutant of proto-oncogene receptor tyrosine kinase (KIT), which encodes the receptor for stem cell factor [50] and has been described in AML [51–53]. An isoform-selective role for the catalytic isoform p110α was also demonstrated in RAS-mutated myeloid leukemia, in a mouse model of juvenile myelomonocytic leukemia (JMML) driven by oncogenic Kras, and in a xenograft model of human RAS-mutated AML [47]. Furthermore, this study showed that combined pharmacologic inhibition of p110α and mitogen-activated protein kinase (MEK) could be an effective therapeutic strategy for RAS-mutated myeloid malignancies [47]. In contrast, in a mouse leukemia model driven by loss of *Pten* in HSCs, of all the Class I PI3K isoforms, p110β was the most important in promoting myeloid leukemogenesis, though a *Rac*-dependent mechanism [48]. Importantly, deletion of p110β also was able to restore normal HSC function, while depleting leukemia-initiating cells, as was demonstrated in transplantation assays. This is consistent with studies in other tumor models driven by *Pten* loss, such as prostate cancer, which also showed preferential dependency upon p110β [54]. Furthermore, this study showed that a p110β-selective inhibitor could prolong survival in leukemia driven by *Pten* loss [48]. All of these studies demonstrate the concept of isoform-selective PI3K dependencies in myeloid leukemia, suggesting that it should be possible to achieve an antitumor effect with a good therapeutic window using selective PI3K isoform inhibition. In fact, several isoform-selective PI3K inhibitors, such as idelalisib, or dual-isoform inhibitors, such as duvelisib, are already in clinical use for indolent lymphoid malignancies, with reasonable toxicity profiles. However, these studies also illustrate that the specific PI3K isoform dependencies may vary depending on the oncogenic driver, as in other tumor types.

1.4 AKT in HSCs and Leukemia

Constitutive activation of PI3K/AKT signaling is observed in many cancers, including AML. Interestingly, unlike in solid tumors, mutations in members of the PI3K pathway are rarely detected in AML. However, increased AKT phosphorylation has been observed in up to 72% of AML cases [55]. In many cases, AKT activation in

Table 1.1 HSC and LSC phenotypes resulting from genetic alterations in the PI3K pathway

Gene	Alteration	Phenotype	References
Pik3r1 and Pik3r2	KO	Decreased repopulating ability of fetal liver HSCs	Haneline et al. [49]
Akt	myr-AKT BMT model	Constitutively active AKT depletes HSCs and induces myeloproliferation, AML, and T-ALL	Kharas et al. [58]
Akt1/ Akt2	KO	Compound deletion of Akt1 and Akt2 impairs fetal HSC repopulation	Juntilla et al. [59]
Pik3ca	cKO (Mx1-Cre)	No effect on HSC self-renewal	Gritsman et al. [47]
		Increased survival in $KRAS^{G12D}$JMML model	
Pten	cKO	Pten deletion using Mx1-Cre perturbed HSC function, myeloproliferation, AML, and T-ALL when Pten is deleted in adult HSCs	Yilmaz et al. [63], Zhang et al. [64], and Magee et al. [73]
		Accelerated disease progression in CML model (using BCR-ABL-iCre retrovirus)	Peng et al. [68]
Pik3cb	cKO (Mx1-Cre)	p110β deletion prevents myeloproliferative neoplasia in context of Pten deletion and improves HSC function	Yuzugullu et al. [48]
Ship	KO	Myeloproliferative phenotype, hypersensitivity to growth factors	Helgason et al. [70]
Pdk1	cKO (Vav1-iCre)	Impaired repopulating potential and differentiation of fetal liver HSCs	Wang et al. [75]
	cKO (Mx1-Cre)	Prolonged survival in MLL-AF9 AML model	Hu et al. [76]
TSC1	cKO (Mx1-Cre)	Increased HSC cycling and impaired self-renewal	Chen et al. [79]
mTOR	cKO(Mx1-Cre)	Loss of HSC quiescence	Guo et al. [78]
Rictor	cKO (Mx1-Cre/ Vav1-iCre)	No effect on HSC function	Kalaitzidis et al. [74] and Magee et al. [77]
		Improved survival in Pten−/− model of leukemia	
Raptor	cKO (Mx1-Cre)	Impaired hematopoietic reconstitution	Kalaitzidis et al. [74]
		Improved survival in Pten−/− model of leukemia	
Foxo	KO or cKO (Mx1-Cre)	Increased HSC cycling, reduced repopulation activity, myeloproliferative phenotype, increased ROS levels	Tothova et al. [88], Yalcin et al. [89], Rimmele et al. [90]. and Liang et al. [91]
		Foxo3a is important for maintenance of CML LICs	Naka et al. [96]
	cKO (Mx1-Cre)	*Foxo1/3/4* deletion induces differentiation of AML cells	Sykes et al. [97]
S6k1	KO	*S6k1* KO impairs HSC self-renewal and improves survival in MLL-AF9 AML model	Ghosh et al. [86]

AML is thought to be secondary to mutations or chromosomal rearrangements causing constitutive activation of upstream RTKs, such as FLT3 or c-KIT. Inactivation of PTEN, either through genetic or epigenetic mechanisms, may also play a role in AKT activation. However, the prognostic significance of AKT phosphorylation is controversial. Phosphorylation of the AKT at Thr308 was shown to confer a poor prognosis in AML [56], whereas phosphorylation of AKT at Ser473 correlates with a favorable response to chemotherapy [57]. To determine whether AKT activation is a true mechanism of transformation, or simply a marker of the transformed state, Kharas et al. generated a murine retroviral bone marrow transplantation model in which they expressed myristoylated AKT in mouse bone marrow cells [58]. The transplant recipients developed a myeloproliferative phenotype, as well as transplantable AML or T-ALL, demonstrating that AKT activation is an important driver of transformation in leukemia. Interestingly, this study also showed that PI3K/AKT pathway activation is deleterious to normal HSC function, due to increased HSC cycling and depletion of the stem cell pool. Both the T-ALL disease phenotype and the HSC maintenance phenotype could be rescued with the mTORC1 inhibitor rapamycin, highlighting the important role of mTORC1 in HSCs and in T-ALL, but not in AML. Interestingly, this study did not show any change in ROS levels in HSCs expressing myristoylated AKT [58].

In a different study, Juntilla et al. have shown that AKT1 and AKT2 play essential redundant roles in fetal HSC maintenance. *Akt1/2* double knockout fetal HSCs are unable to enter cell cycle and persist in the G0 phase, causing long-term functional defects due to increased quiescence. This increased quiescence of AKT1/2-deficient HSCs was associated with decreased ROS levels, suggesting that AKT1 and AKT2 are important for regulating ROS levels in hematopoiesis [59]. This work confirmed the important roles of AKT in the regulation of HSC homeostasis.

1.5 The Phosphatases PTEN and SHIP in HSCs and LSCs

PTEN is commonly deleted or otherwise inactivated in diverse cancers (reviewed in [60]), including hematologic malignancies [61, 62]. The phenotype observed with conditional deletion of *Pten* in the hematopoietic system using Mx1-Cre resembles the phenotype observed with activation of AKT in hematopoietic cells [63, 64], also reviewed in [65]). *Pten* deletion in hematopoietic cells led to the development of MPN, AML, and T-ALL and resulted in enhanced LSC function while compromising normal HSC function [63]. Zhang and colleagues also demonstrated that *Pten* inactivation in murine bone marrow causes short-term expansion, followed by the subsequent long-term decline of HSCs, primarily due to enhanced HSC activation, followed by exhaustion of the HSC pool [64]. Although *Pten*-deficient HSCs engraft normally in recipient mice, their ability to sustain hematopoietic reconstitution is impaired due to cell cycle dysregulation and decreased bone marrow retention [64]. Both phenotypes, the HSC exhaustion and leukemia initiation, are mediated by mTORC1, because they can be rescued by rapamycin [63]. However, it was later shown by Guo et al. that in established T-ALL, the effects of *Pten* deletion in LSCs couldn't be completely reversed by rapamycin [66]. This likely reflects the different roles of mTORC1 in leukemia initiation, as opposed to leukemia maintenance. Interestingly, Lee et al. showed that the effects of rapamycin on *Pten*-deficient HSCs and LSCs can be partially rescued by increased expression of cyclin-dependent kinase inhibitor 2A, p16(Ink4a) and p53 in HSCs. However, while both p53 and p16(Ink4a) promoted the depletion in *Pten−/−* HSCs, only p53 could suppress leukemogenesis [67]. In a CML model driven by BCR-ABL expression, Peng et al. demonstrated that *Pten* deletion accelerates disease development, while *Pten* overexpression delayed disease progression by suppressing LSC function [68]. This study further supports the role of PTEN as a tumor suppressor in myeloid LSCs. All of these studies highlight the distinct effects of PTEN deletion in HSCs and in LSCs, again suggesting that there is a therapeutic window for targeting elevated AKT signaling in LSCs.

In another study, Li et al. showed that PTEN protein levels are similar among different HSC populations and multipotent progenitors (MPPs), but that there is increased PTEN phosphorylation (p-PTEN) during the transition from long-term to short-term HSCs, and further to MPPs, where p-PTEN levels are inversely correlated with the self-renewal capacity and long-term repopulation capacity of HSPCs [69]. Serial transplantation experiments demonstrated that unphosphorylated Pten augments the self-renewal of wild-type long-term repopulating cells and promotes myeloid over lymphoid differentiation. Moreover, unphosphorylated Pten restricts HSCs to a quiescent state in the bone marrow niche by dual inhibition of PI3K/AKT and SRC signaling. In contrast, p-PTEN represses PI3K/AKT, but promotes the niche contact-stimulated signaling that regulates the proliferation and lodging of hematopoietic stem and progenitor cells (HSPCs) in the bone marrow niche, whereas unphosphorylated PTEN represses both types of signaling [69]. Thus, this study highlights the dual cell-autonomous and non-cell-autonomous regulation of HSCs by PTEN.

Conditional deletion of another PIP3 phosphatase, *Ship*, also led to the development of a myeloproliferative disease, and bone marrow progenitor cells from Ship−/− mice were hypersensitive to hematopoietic growth factors [70, 71]. Luo et al. evaluated the potential tumor suppressor role of *SHIP* in myeloid leukemogenesis, using AML patient samples and myeloid leukemia cell lines [72]. Strikingly, they discovered in one AML patient a dominant negative mutation in the *SHIP* gene, $SHIP^{V684E}$, which replaces Val with Glu at the phosphatase active site, reducing its catalytic activity.

Leukemia cells bearing this mutation have enhanced AKT phosphorylation following IL-3 stimulation. These cells also have a growth advantage, and are resistant to apoptosis upon serum deprivation, and in response to the chemotherapy drug etoposide [72]. These results suggest that mutated SHIP is a potential mechanism for the development of chemotherapy resistance through the PI3K/AKT signaling pathway.

Magee et al. addressed the effects of patient age on activation of the PI3K/AKT pathway in HSCs [73]. They asked why *PTEN* is sometimes mutated in adult, but rarely in early childhood leukemias, and showed that there is a different effect of *Pten* deletion on the neonatal vs. adult HSCs. They revealed that PTEN is required by adult HSCs to dampen AKT signaling and prevents leukemia development in adults. However, neonatal HSCs do not activate AKT signaling as a result of *Pten* deletion. Moreover, they showed that in adult mice, *Rictor* deletion in *Pten−/−* HSCs prevents leukemogenesis and HSC depletion, which for the first time implicates mTORC2 in HSC maintenance in the context of *Pten* inactivation. On the contrary, in normal HSCs, deletion of *Rictor* has little effect. Together with the fact that in neonatal HSCs, *Pten* deletion does not activate PI3K/AKT pathway and does not promote HSC proliferation, depletion, or leukemogenesis, this demonstrates the difference in critical tumor suppressor mechanisms and in the roles of PI3K/AKT signaling between adult and neonatal HSCs [73]. This study highlights the importance of the developmental changes in key signaling pathways that confer temporal alterations upon stem cell self-renewal and tumor suppressor mechanisms.

Kalaitizidis et al. confirmed the observation of the importance of RICTOR in the context of *Pten* loss in adult mice in their study of *Raptor* and *Rictor* deletion in the hematopoietic system (74). They found that deletion of *Raptor* in *Pten*-lox-lox;Mx1-Cre; mice significantly improved survival, but without much improvement in HSC function in transplantation assays. This is consistent with prior studies showing that rapamycin can prevent leukemia initiation in the context of hematopoietic *Pten* deletion [63]. However, in contrast to *Raptor* deletion, rapamycin was also able to rescue HSC function in the context of *Pten* loss, suggesting that RAPTOR may have some rapamycin-insensitive functions that are required for HSC maintenance [74]. Interestingly, conditional deletion of only *Rictor*, but not *Raptor*, after disease onset could significantly improve HSC function in the context of *Pten* loss [74]. Overall, these studies underscore the importance of both mTORC1 and mTORC2 in leukemic transformation induced by *Pten* loss, while only mTORC1 appears to play an important role in the maintenance of normal HSCs. Therefore, components of mTORC2 should be explored further as potential therapeutic targets for myeloid malignancies.

1.6 PDK1 in HSCs and Leukemia

It has been reported that pyruvate dehydrogenase kinase 1 (PDK1), which contributes to AKT activation by phosphorylating AKT on Thr308, also plays important roles in the hematopoietic system. Wang et al. revealed an indispensable role for PDK1 in fetal liver HSCs. Using the Vav-iCre system, they showed that Pdk1 is crucial for HSC expansion, and *Pdk1* deletion severely impairs the repopulating potential and differentiation capacity of fetal liver HSCs [75]. Interestingly, another study showed that conditional deletion of *Pdk1* in the MLL-AF9 murine AML model prolonged survival due to increased apoptosis of LSCs, which was accompanied by reduced expression of *Stat5*, which is often activated in leukemia [76]. These data highlight the essential role of PDK1 during fetal hematopoiesis and in

LSC maintenance. It would be interesting to further delineate the differences between the function of PDK1 in adult HSCs and in LSCs, given the differences in PI3K/AKT signaling previously reported between neonatal and adult HSCs [77].

1.7 mTOR in HSCs and Leukemia

mTOR has been shown to play a central role in both embryonic and adult hematopoiesis. Guo et al. showed that conditional mTOR deletion using Mx1-Cre results in loss of HSC quiescence, with the consequent exhaustion of HSCs, as well as impaired HSC engraftment and repopulation (78). Interestingly, conditional deletion of *Tsc1*, and consequent mTOR activation, also causes HSC cycling and premature depletion, due to increased mitochondrial biogenesis followed by elevated levels of reactive oxygen species (ROS) [79]. This suggests that TSC1 is an essential regulator of HSC quiescence. In a follow-up study, Chen et al. also showed that mTOR activity in HSCs increases with age, and that *Tsc1* deletion mimics the characteristics of aged HSCs in young mice [80]. Furthermore, they showed that decreasing mTOR signaling with rapamycin restores HSC function in old mice, thus highlighting a fundamental function of mTOR signaling in HSC aging [80]. However, activation of mTOR signaling by conditional deletion of *Tsc1* is insufficient to cause hematologic malignancies in mice [79]. Interestingly, Kalaitzidis et al. also showed that Raptor deficiency leads to HSC expansion in *Pten*-wild-type HSCs and promotes the transition of short-term HSCs from the G0 to G1 phase [74]. Furthermore, Raptor-deficient BM cells cannot reconstitute hematopoiesis in lethally irradiated recipient mice [74]. Although under homeostatic conditions mTORC1 is not required for HSC maintenance, its loss leads to HSC and progenitor mobilization. A similar phenotype was reported by Hoshii et al. for *Raptor* deletion in the hematopoietic system; they observed myeloid expansion, but no effects on progenitor proliferation or survival [81]. Both groups also showed that deletion of Raptor, but not of Rictor, caused an increase in AKT activation in hematopoietic stem and progenitor cells (HSPCs). This is consistent with the previously described effects of prolonged rapamycin treatment on AKT activation, which can occur through the unopposed activity of mTORC2, leading to phosphorylation of AKT on Ser473 [82]. Given the known effects of AKT activation on HSC function, this could potentially explain the HSC cycling phenotype in Raptor−/− mice [58, 74].

Surprisingly, the GTP-binding protein RHEB, which is known to activate mTORC1 activity in other cell types, was found to only moderately activate mTORC1 in HSCs and progenitors [83]. Peng et al. showed that *Rheb* deletion did not significantly affect HSC maintenance in mice, in contrast to *Raptor* deletion, signifying that under steady-state conditions, RHEB activity is expendable for mTORC1 activity and HSC self-renewal [83].

Dysregulation of the mTORC1-S6K1 signaling pathway occurs frequently in AML patients. Primary AML cells derived from patients show increased activation of S6K1 [84]. The role of mTORC1-S6K1 signaling and its role in regulation of

HSCs and LSCs were recently reviewed in detail by Ghosh and Kapur, where they highlight recent progress in identifying the roles of different components of the pathway and pharmacological targeting of mTORC1 and S6K1 [85]. Ghosh et al. showed that S6K1 deficiency in hematopoietic cells results in the reduced quiescence of L-HSCs, and their impaired self-renewal, concomitant with the reduced expression of cyclin-dependent kinase inhibitor 1A (Cdkn1a) [86]. They showed that loss of S6K1 increases the susceptibility of mice to repeated stress. Moreover, the authors reported that the leukemic potential of cells driven by the mixed lineage leukemia (MLL) fusion oncogene MLL-AF9 is similarly dependent on S6K1. Inhibition of S6K1 activity reduced the proliferation of MLL-rearranged human AML cells, and loss of S6K1 impaired in vivo LSC function in MLL-AF9 AML model, by modulating AKT and 4E-BP1 phosphorylation [86]. This data demonstrated that S6K1 loss decreases mTOR activation, consequently leading to reduced phosphorylation of AKT and 4E-BP1 in HSCs. Thus, S6K1 is a potential therapeutic target in AML LSCs. Hoshii reported similar findings with *Raptor* deletion in MLL-AF9 LSCs and also showed that loss of Raptor led to differentiation of LSCs [81]. Also, consistent with the results of Kalaitzidis et al. in the Pten−/− leukemia model, they reported that loss of *Raptor* failed to rescue normal hematopoiesis in the MLL-AF9 AML model. However, Hoshii et al. also reported the intriguing finding that *Raptor* deletion compromised the leukemia-initiating cells in MLL-AF9 AML mice, without compromising the self-renewal of LSCs, arguing that these are two different cell populations with differential dependence on RAPTOR [81]. This finding brings into question the usefulness of targeting mTORC1 alone in AML.

1.8 The FOXO Transcription Factors in HSCs and Leukemia

In the hematopoietic system, FOXO1 and FOXO3 are thought to be the main players, with FOXO3 being more active than FOXO1, and they have been shown to be key regulators of both HSC and LSC maintenance. It was shown that FOXOs regulate the intracellular levels of reactive oxygen species (ROS), where loss of FOXO activity elevates ROS levels, leading to increased HSC cycling and apoptosis [87, 88]. Yalcin et al. also showed that loss of FOXO3 leads to an increased number of hematopoietic progenitors with hypersensitivity to cytokines and led to a myeloproliferative phenotype. These progenitors had increased ROS levels, increased AKT/mTOR signaling, and a relative deficiency of LNK, a negative regulator of cytokine receptor signaling [89]. More recently, it was shown by Rimmele et al. that in *Foxo3(−/−)* HSCs and HSPCs, mitochondrial metabolism is suppressed, affecting HSC long-term repopulation activity [90]. This repression was associated with altered expression of mitochondrial and metabolic genes, which was independent of ROS levels or mTOR signaling [90].

Remarkably, accumulating evidence in the field raises the possibility that in stem cells, including in HSCs, AKT is not the only regulator of FOXO. It has been demonstrated that nuclear FOXO localization and activity in HSPCs are dependent on other upstream regulators, such as P38 mitogen-activated protein kinases (p38MAPK) or NAD-dependent deacetylase sirtuin-1 (SIRT1) [87, 91]. In fact, at steady state, FOXO seems to be primarily regulated independently from PI3K/AKT signaling. As Miyamoto et al. have shown using their *Foxo3a* conditional knockout mouse model, *Foxo3a−/−* HSCs have normal proliferation and differentiation in the bone marrow, but lose their self-renewal capacity due to loss of HSC quiescence, associated with increased p38-MAPK phosphorylation [87]. However, during myelosuppressive stress and aging, FOXO3a in HSCs is regulated via AKT and ERK signaling [92].

It has been reported that high levels of FOXO3a mRNA and phosphorylation of FOXO3a are adverse prognostic factors in AML [93, 94]. Interestingly, Chapuis et al. showed that in AML samples, the localization of FOXO3a is able to escape AKT control and is also independent of the ERK/MAPK signaling pathway, but it is dependent on FOXO3a Ser644 phosphorylation by IκB kinase (IKK) [95]. They proposed an interesting possibility that IKK activity retains FOXO3a in the cytoplasm, thereby inactivating FOXO3a and affecting the proliferation and survival of AML cells. Although FOXOs are primarily known as tumor suppressors, the roles of FOXOs in myeloid malignancies are controversial. High level of FOXO3 expression is associated with an adverse prognosis in AML [93], and genetic ablation of *Foxo3* has been shown to reduce disease burden in a murine model of CML [96]. Strikingly, Sykes et al. demonstrated the opposite roles for AKT/FOXOs in AML [97]. There is a generally accepted concept that activated AKT signaling, which should lead to FOXO inhibition, is associated with hematopoietic transformation, but they observed that in 40% of AML patient samples, regardless of genetic subtype, FOXOs are active. In these patients, instead of decreased FOXO activity, FOXO is active and inversely correlated with c-Jun N-terminal kinase of (c-JUN)/JNK signaling. Moreover, leukemic cells resistant to FOXO inhibition responded to JNK inhibition, which revealed a molecular role for AKT/FOXO and JNK/c-JUN in maintaining the differentiation blockade, opening an exciting avenue for the inhibition of leukemias with a range of genetic lesions [97].

1.9 Integration of PI3K/AKT with the MAPK Pathway in HSCs and LSCs

Baumgartner et al. recently showed that during emergency hematopoiesis, the MEK/ERK and PI3K pathways are synchronously activated in HSCs [98]. Moreover, MEK1 phosphorylation by activated ERK provides a feedback mechanism to counterbalance AKT/mTORC1 activation, presenting HSCs with the rate-limiting feedback mechanism that controls their fitness and reentry into quiescence (Fig. 1.2)

[98]. In myeloid leukemogenesis, oncogenic NRAS^{G12D} activates both MAPK and PI3K signaling and can promote the expansion of preleukemic clones in some contexts [99–102]. It is interesting to consider that disruption of this feedback loop that controls the duration of PI3K signaling and the production of PIP3 could exhaust the pool of fast-proliferating HSCs induced by chemotherapy, while at the same time this could benefit the expansion of a pool of self-renewing preleukemic clones.

Many of the growth factors and cytokines that orchestrate hematopoiesis and HSC maintenance signal through the PI3K/AKT canonical signaling pathway, and dysregulation of this pathway can lead to hematopoietic transformation. It is becoming clear that tight regulation of PI3K/AKT pathway activation is critical for the maintenance of HSC homeostasis, at least in adults. There is also sufficient evidence that activation of AKT in HSCs, either directly or indirectly through inactivation of PTEN or SHIP, can lead to hematopoietic transformation. However, modulation of the downstream components of AKT generally is not sufficient to promote full transformation, though it can cause significant alterations in HSC homeostasis. This highlights the importance of understanding the roles of the individual components of the PI3K/AKT pathway in hematopoietic transformation, as well as their crosstalk with parallel pathways, such as with the MAPK pathway. Exciting studies have already provided evidence that the FOXOs can act as a "crosstalk hub" between several signaling pathways, including the PI3K/AKT, MEK/ERK, and JNK/c-JUN pathways, suggesting that any mutation that alters FOXO3 function in HSCs might contribute to malignant transformation. Further research in this area should focus not only on further delineating the roles of PI3K/AKT pathway members in LSCs but also on the interplay between PI3K and other signaling pathways.

1.10 The WNT/β-Catenin Signaling Pathway in HSCs and LSCs

The WNT signaling pathway has been well characterized for its stimulatory role in adult HSC maintenance and self-renewal [103, 104] (Fig. 1.3, Table 1.2). The canonical WNT signaling pathway is activated when WNT ligands within the bone marrow stroma interact with the FRIZZLED-LRP5/6 receptor complex, which is expressed on the HSC surface (reviewed in Staal et al. [105]). This leads to inhibition of the multi-protein destruction complex composed of GSK-3β, AXIN1 or AXIN2, and APC, which normally phosphorylates β-catenin, leading to its proteasomal degradation. β-Catenin is then able to accumulate in cytoplasm, leading to its translocation into the nucleus, where it activates the TCF/LEF transcription factor family to activate the expression of genes essential for expanding the HSC pool [105].

Despite the conserved role of WNT signaling in regulating HSC self-renewal, there has been conflicting evidence regarding the effects of β-catenin signaling on

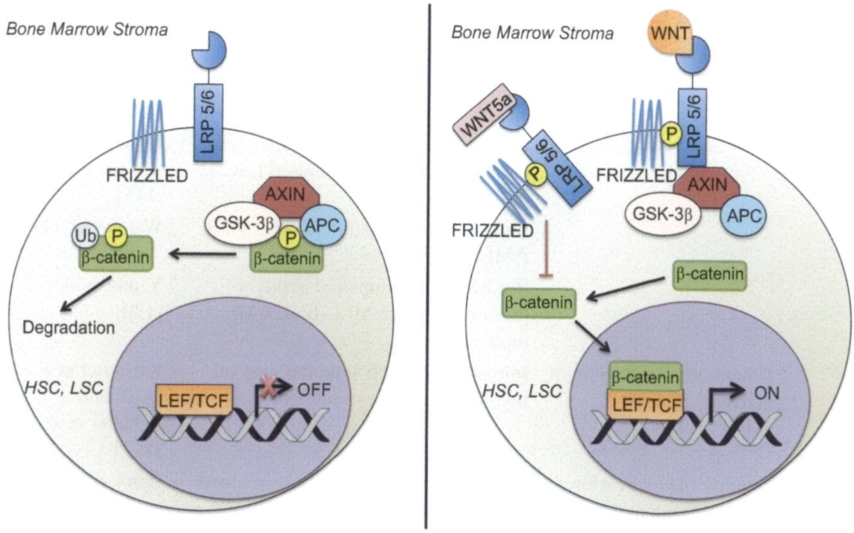

Fig. 1.3 Summary of WNT/β-catenin pathway activation in HSCs and LSCs

the HSC population. Knockout studies have demonstrated that WNT signaling is dispensable for HSC function. For example, deletion of β-catenin in bone marrow progenitors did not impair self-renewal or differentiation [106]. In support of this, deletion of *Porcupine* (*Porcn*), an O-acyltransferase essential for all WNT ligands, in HSCs did not impair their ability to self-renew, proliferate, or differentiate [107]. A recent review has covered the concept that the proper dosage of WNT signaling must be achieved to moderate HSC self-renewal versus differentiation decisions [108]. Using transgenic mouse lines expressing different doses of the negative regulator APC, they elegantly demonstrated that slightly enhanced levels of WNT signaling can enhance HSC reconstitution, while much higher levels of WNT signaling can result in the failure to engraft and reconstitute lethally irradiated mice [109]. This supports earlier work, which showed through conditional expression of a stable constitutively active form of β-catenin that HSCs lose their ability to repopulate and also lose the capacity for myeloid lineage commitment [110]. Scheller et al. also observed a similar HSC exhaustion phenotype when they expressed a different constitutively active mutant of β-catenin in hematopoietic cells [111]. Notably, using the SCL-Cre-ER mouse model, β-catenin activation more specifically in HSCs resulted in HSC apoptosis [112]. However, when β-catenin activation was coupled with *Pten* deletion, there was an expansion of LT-HSCs at the expense of differentiation, suggesting that the WNT and PI3K/AKT pathways cooperate in regulating LT-HSC proliferation, apoptosis, and differentiation [112]. Altogether, it is clear that the HSCs require some level of activation of the WNT signaling pathway for self-renewal, survival, and proliferation and that this level must be tightly regulated.

Table 1.2 HSC and LSC phenotypes resulting from genetic alterations in the WNT/β-catenin pathway

Gene	Alteration	Phenotype	References
Ctnnb1	cKO (Mx1-Cre)	No effects on HSC function	Cobas et al. [106]
		Delays initiation of disease in mouse model of CML, compromised LSC self-renewal	Zhao et al. [114] and Hu et al. [115]
		Impaired LSC function in MLL-AF9 AML	Wang et al. [119]
		Pre-LSCs have impaired ability to generate LSCs in MLL-ENL AML model	Yeung et al. [120]
	Conditional expression of activated mutant (Mx1-Cre)	Impaired HSC repopulation, loss of myeloid lineage commitment	Kirstetter et al. [110] and Scheller et al. [111]
	Conditional expression of activated mutant (Scl-Cre-ER)	HSC apoptosis, which can be rescued by activation of the PI3K pathway	Perry et al. [112]
Apc	cKO (hypomorphic allele)	Mildly increased WNT signaling enhances HSC reconstitution in vivo, but higher levels of WNT signaling impair engraftment and HSC reconstitution	Luis et al. [109]
Lef/Tcf	*Lef* cKO (Vav1-iCre)	Impaired LSC self-renewal in CML model	Yu et al. [118]
	Tcf germline KO	Normal HSC self-renewal	
Wnt5a	Overexpression	Inhibits HSC expansion by inducing a quiescent state at the expense of canonical WNT signaling	Nemeth et al. [124]
	Heterozygous deletion	Loss of one *Wnt5a* allele leads to spontaneous CML or B cell lymphoma development	Liang et al. [125]
	Haploinsufficiency in the bone marrow niche	Impaired HSC engraftment	Schreck et al. [127]
		Impaired leukemia engraftment in CML model	

The WNT signaling pathway has also been shown to play important roles in LSCs, as this population also requires the WNT signaling pathway for self-renewal and survival. In human CML LSCs, it has been shown that β-catenin is localized in the nucleus and that in vitro LSC activity could be reduced by expression of AXIN [113]. Transplantation of BCR-ABL LSCs depleted of β-catenin into mice results in a decreased disease burden and impaired serial transplantation [114, 115]. Importantly, it has been shown that progression of CML from the chronic phase to blast crisis occurs through the stimulation of β-catenin activity through activation of the PI3K pathway by BCR-ABL [116]. Hu et al. also showed that inhibiting BCR-ABL or the PI3K pathway results in decreased levels of β-catenin.

Subsequently, in vivo inhibition of this pathway delays CML development and impairs the tumor-initiating ability of CML LSCs in secondary transplantation, implying that the crosstalk between the PI3K pathway and WNT pathway is important for LSC maintenance in CML [116].

Importantly, β-catenin also plays a role in BCR-ABL tyrosine kinase inhibitor (TKI) resistance, as it has been reported that β-catenin is stabilized in CML cells to promote TKI resistance through a BCR-ABL kinase-independent resistance mechanism [117]. It has been reported that CML cells treated with imatinib maintain their β-catenin expression and have subsequent activation of downstream targets, promoting their survival and colony formation ability [117]. In addition, it has been shown that CML LSCs require the transcription factors TCF1 and LEF1 to propagate CML in secondary recipients, and combining imatinib with induced deletion of *Tcf1* and *Lef1* in vivo diminishes the LSC population [118]. Importantly, HSCs are less sensitive to loss of *Tcf1* and *Lef1*, suggesting that targeting TCF1 and LEF1 in combination with a TKI could be a novel therapeutic strategy to eliminate CML LSCs, without much toxicity to HSCs [118].

In AML LSCs, inhibiting β-catenin in either the HOXA9/MEIS1 or MLL-AF9 mouse models prevents AML initiation from either the HSC or GMP populations, suggesting that WNT signaling is essential for LSC function [119]. In addition, conditional deletion of *Ctnnb1*, the gene that encodes β-catenin, in MLL-rearranged AML mouse models reduces the number of LSCs and overall incidence of leukemia development [120]. In both CML and AML LSC populations, β-catenin- and WNT pathway-associated genes are increased in comparison to normal HSCs [113, 120], further supporting the important roles of the WNT signaling pathway in LSCs. More recently, it has been shown that GPR84, a positive regulator of β-catenin, is also upregulated in LSCs in comparison with normal HSCs, and suppression of GPR84 reduced MLL-rearranged LSC functionality in vitro and in vivo [121]. WNT signaling has also been studied in FLT3-mutant AML models, as FLT3 signaling can increase nuclear β-catenin levels [122]. High levels of β-catenin were found in bone marrow resident leukemic cells in patients with FLT3-mutated AML. Notably, the combination of a β-catenin antagonist and a FLT3 TKI impaired the engraftment of FLT3-mutated leukemic cells into immunodeficient mice, suggesting that β-catenin inhibitors can synergize with FLT3 inhibitors in FLT3-mutated AML, due to the effects of this drug combination on LSC function [122].

Noncanonical WNT signaling has also been shown to play a role in myeloid LSCs. The noncanonical WNT ligand WNT5a can signal through both the planar cell polarity and WNT/calcium noncanonical pathways, both of which are extensively reviewed in Komiya and Habas [123]. In short, WNT5a signaling through the WNT/calcium pathway results in increased intracellular calcium levels and subsequent inhibition of β-catenin [123]. This has been shown in HSCs, as WNT5a can inhibit the activation of canonical WNT signaling, and this inhibition results in an increase of quiescent HSCs [124]. WNT5a has implications in AML, as WNT5a levels are low or absent in primary AML samples, and WNT5a hemizygous mice develop myeloid leukemias [125]. In addition, high levels of methylation of the WNT5a gene locus correlate with lower expression of WNT5a, and this high methylation status correlates with poor prognosis of a subset of AML patients [126]. More recently, the role of WNT5a in the niche has been explored. It was shown that both HSCs and BCR-ABL LSCs are impaired in engrafting into a WNT5a-deficient niche [127]. Furthermore, transplanting LSCs exposed to a WNT5a-deficient niche reduces the incidence of leukemia in mice, as well as the ability to propagate leukemia to secondary recipient mice [127]. A more recent study has confirmed a role for both canonical and noncanonical WNT signaling in leukemia, showing in vivo that the canonical WNT signaling pathway is upregulated in the leukemic HSPC compartment at the expense of a downregulated WNT5a signaling axis [128]. Altogether, it is evident that myeloid LSCs, like HSCs, rely on WNT signaling for functionality, but that different levels of WNT/β-catenin signaling may be required to promote HSC and LSC function.

1.11 The NOTCH Pathway in HSCs and LSCs

The NOTCH pathway is best characterized for its crucial role in embryonic development and in HSC emergence during embryogenesis. However, the efforts to elucidate the roles of NOTCH signaling in adult hematopoiesis and HSC maintenance have expanded in an effort to understand how dysregulated NOTCH signaling affects the leukemic state (Fig. 1.4, Table 1.3). Within this signaling network is a family of four NOTCH receptors expressed on the cell surface that can interact with one of five ligands (Jagged 1, Jagged 2, Delta 1, Delta 2, and Delta 3), which are presented on the surface of a neighboring cell (reviewed in Pajcini et al. [129]) . Ligand binding results in cleavage of the intracellular domain of the NOTCH receptor by a γ-secretase complex to allow this portion of the receptor to translocate into the nucleus. In the nucleus, the intracellular domain of NOTCH (NICD) can interact with co-activator proteins such as CBF-1 and MAM1, resulting in transcriptional activation of NOTCH target genes, such as the HES family [129].

NOTCH signaling has been shown to play a stimulatory role in adult HSCs, and promotes HSC expansion. Notably, constitutive activation of NOTCH1 signaling via expression of NICD in hematopoietic cells results in immortalized, cytokine-dependent HSCs capable of differentiating into both myeloid and lymphoid progeny [130]. This result was confirmed by additional work showing that expression of NICD in vivo resulted in HSC expansion and self-renewal at the expense of differentiation [131]. Several groups have shown that NOTCH pathway activation also leads to expansion of the hematopoietic progenitor compartment. For example, Karanu et al. showed that adding soluble forms of the human ligands Delta 1 and Delta 4 results in increased proliferation of hematopoietic progenitors in vivo, and transplantation of Jagged 1 or Delta 1 into immunodeficient mice expands HSPCs that are capable of pluripotent hematopoietic reconstitution [132, 133]. Supporting this work, it was shown in transgenic *Notch* reporter mice that NOTCH signaling is upregulated in HSCs, but that this signaling is lost upon lineage commitment [134]. Moreover, inhibiting NOTCH signaling was able to accelerate the differentiation of

HSCs in vitro, and this inhibition resulted in the depletion of HSCs in vivo [134]. Thus, activation of the NOTCH pathway, either through overexpression of NICD or by exogenous ligands, promotes HSC expansion and self-renewal.

Surprisingly, examination of the role of NOTCH through genetic deletion has suggested that NOTCH signaling is not required for adult HSC function. It has been shown that genetic deletion of *Notch1* does not affect the reconstitution capacity of HSCs, though it does lead to a block in T-cell development [135]. Studying the effects of *Notch2* deletion in the hematopoietic context has also shown that HSC function is NOTCH-independent [136]. However, *Notch2*-deficient mice have impaired hematopoietic recovery after 5-fluorouracil (5-FU), suggesting a requirement for NOTCH signaling for HSC function during stress hematopoiesis [136]. In addition, NOTCH may regulate HSC function through the niche. For example, osteoblasts within the niche express high levels of JAGGED 1, correlating with an increase in HSC number in vivo [137]. In addition, endothelial cells within the niche promote the proliferation and survival of LT-HSCs via expression of NOTCH

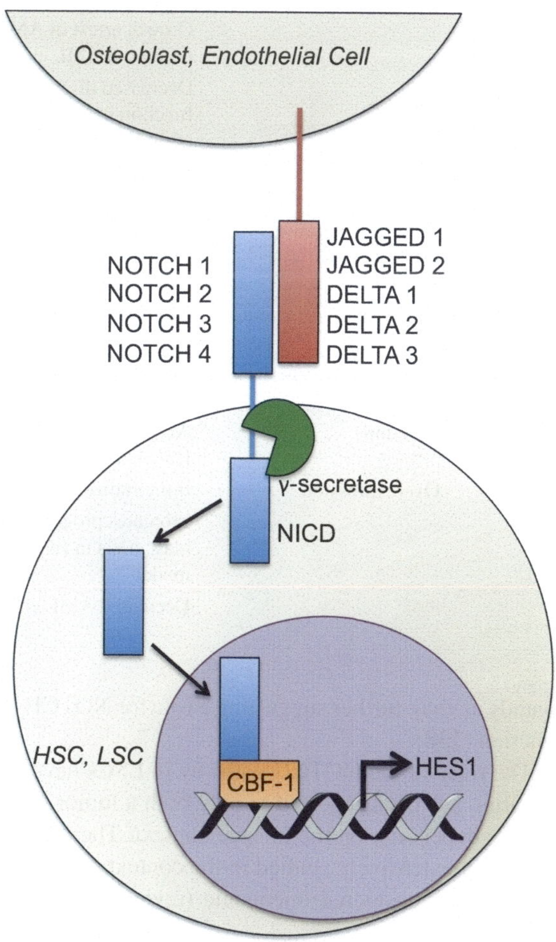

Fig. 1.4 NOTCH signaling in HSCs and LSCs

Table 1.3 HSC and LSC phenotypes resulting from genetic alterations in the NOTCH pathway

Gene	Alteration	Phenotype	References
Jagged 1/ Delta 1	Soluble ligand administration	HSC expansion and enhanced hematopoietic reconstitution	Karanu et al. [132, 133]
	Osteoblast ablation	Expands LT-HSC population with impaired self-renewal capacity	Bowers et al. [149]
		Accelerated leukemia in CML murine model	
Notch 1	Deletion	HSC reconstitution intact, although blockage of T-cell development	Radtke et al. [135]
Notch 2	Deletion	HSC reconstitution impaired only under 5-FU stress conditions	Varnum-Finney et al. [136]
Notch 1/2	Overexpression or cKI of constitutive active ICD	Immortalization of HSCs capable of both myeloid and lymphoid differentiation	Varnum-Finney et al. [130]
		HSC expansion and increased self-renewal at the expense of differentiation	Stier et al. [131]
		Growth arrest of AML cell lines, decreased AML engraftment	Kannan et al. [143]
		Decreased disease burden and LSC function in MLL-AF9 murine model of AML	Lobry et al. [139]
		Inhibits proliferation of CML K562 cell line	Yin et al. [148]
Maml	Dominant negative MAML expression	HSC expansion and increased self-renewal at the expense of differentiation	Kannan et al. [143]
Nicastrin	cKO (Mx1-Cre; Vav-iCre)	CMML-like disease	Klinakis et al. [141]
Cbf-1	Dominant negative	Retroviral expression accelerated HSC myeloid differentiation	Duncan et al. [134]
Hes1	Deletion	Accelerates development of leukemia in MLL-AF9 model	Kato et al. [144]
	Overexpression	Immortalizes CMPs and GMPs	Nakahara et al. [146]
		Promotes progression of CML to blast crisis in BCR-ABL murine model	
		Decreased AML engraftment	Kannan et al. [143]

ligands in vivo, further suggesting a role for NOTCH signaling dependence for HSC function [138].

The roles of the NOTCH pathway in LSCs have been controversial, as NOTCH signaling has been reported to play both a tumor suppressor and oncogenic role in LSCs – depending on the lineage context. The roles of NOTCH signaling in LSCs have been extensively studied in the context of T-ALL and CLL, in which NOTCH signaling is considered oncogenic (reviewed in Lobry et al. [139]). Work is still being done to fully elucidate the role of the NOTCH pathway in myeloid LSCs. However, several groups have begun to unravel that NOTCH signaling may play a tumor suppressor role in the myeloid context [140–143]. NOTCH signaling has an accepted role in promoting myeloid differentiation by upregulation of the transcription factor PU.1 [140]. It was shown that *Notch 1* expression is downregulated in AML cell lines and in primary patient blasts, and this downregulation correlates with a decrease in PU.1-mediated differentiation, suggesting that NOTCH signaling may be responsible for maintaining the immature myeloid compartment seen in leukemia [140]. Furthermore, data suggests that induced activation of the NOTCH signaling axis in AML in vivo results in cell growth arrest, differentiation, and apoptosis [142]. Consistent with this, activation of NOTCH signaling in AML by NOTCH ICN1 expression in vivo decreases AML cell proliferation, and this phenotype is reversed through the inhibition of NOTCH signaling using a dominant-negative MAML construct [143]. Additionally, inactivation of NOTCH signaling in HSCs in vivo can promote a chronic myelomonocytic leukemia (CMML)-like phenotype, further supporting a tumor-suppressive function [141]. Additionally, recent work has shown that the NOTCH target HES1 binds to the FLT3 promoter to suppress its transcription, and loss of HES1 results in AML development in vivo [144]. Importantly, in MLL-AF9 leukemic cells, FLT3 was significantly upregulated upon HES1 deletion [144]. However, the role of NOTCH signaling as a tumor suppressor pathway in AML LSCs was recently challenged by results revealing that crosstalk between the WNT and NOTCH signaling pathways in osteoblasts of the bone marrow results in activation of NOTCH signaling between osteoblasts and HSPCs, actually promoting leukemic transformation [145]. Thus, more work is still needed to fully elucidate the roles of NOTCH signaling in AML LSCs.

There is also evidence to suggest that the NOTCH signaling pathway plays a role in CML LSCs, although the specific effects of NOTCH in this context also require further exploration. For example, induced expression of *Hes1* was shown to cooperate with BCR-ABL to induce blast crisis CML in vivo [146]. Importantly, HES1 is highly expressed in CML patients in blast crisis, suggesting its potential role in the progression from chronic phase to blast phase in CML [146]. In addition, genetic deletion of the transcription factor *Twist1* in the microenvironment activates the NOTCH signaling pathway via upregulation of *Jagged 2*, and this activation is required for MLL-AF9-induced AML [147]. In contrast, it was shown that overexpression of NICD was sufficient to inhibit the proliferation of the CML blast crisis cell line K562 [148]. A tumor-suppressive role for NOTCH has also been demonstrated within the CML niche. Specifically, ablation of osteoblasts in a CML transgenic mouse model results in accelerated development of CML [149]. Since it was shown that JAGGED 1 levels were increased in CML osteoblasts, this group went on to show that the addition of JAGGED 1 to CML LSCs reduced their cycling, thus supporting the hypothesis that NOTCH signaling promotes an antileukemic environment [149]. However, while it is clear that NOTCH signaling is important in normal HSC and myeloid LSC function, understanding and deciphering the contexts in which it promotes or suppresses the leukemic state still remains an active area of research.

1.12 The TGFβ Pathway in HSCs and LSCs

The TGFβ pathway is known to play an important role in controlling HSC function, but its potential role in myeloid LSCs has only been explored within the last decade (Fig. 1.5, Table 1.4). Recently reviewed by Vaidya and Kale [150], the TGFβ family consists of three ligands – TGFβ1, TGFβ2, and TGFβ3 – all of which signal through type I and type II serine/threonine kinase membrane receptors [150]. The primary type I receptor for TGFβ is ALK5/TβRI and the sole type II receptor for TGFβ is TβRII. In short, TGFβ is found in its dormant form within the extracellular matrix, but upon cleavage of its pro-domain, it is able to bind to a TβRII homodimer. TβRII is then able to multimerize with a TβRI homodimer, resulting in subsequent phosphorylation of TβRI. After the required receptor transphosphorylation, the

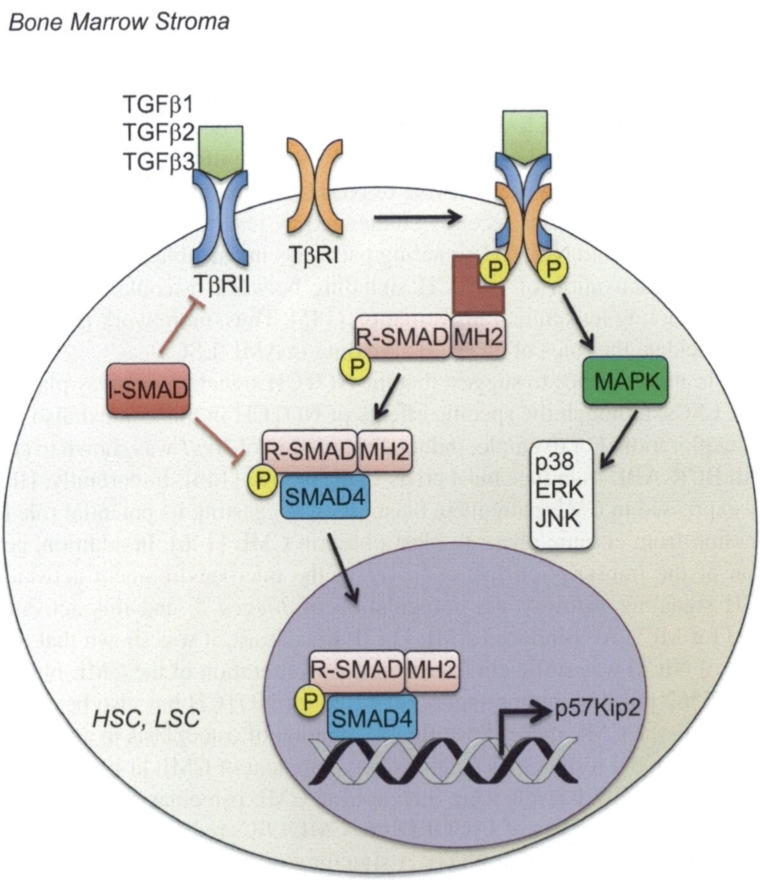

Fig. 1.5 Summary of TGFβ pathway activation in HSCs and LSCs

Table 1.4 HSC and LSC phenotypes resulting from genetic alterations in the TGFβ pathway

Gene	Alteration	Phenotype	References
TBRI	cKO (Mx1-Cre)	No effect on HSC self-renewal, differentiation, or long-term repopulation capacity	Larsson et al. [162]
	Pharmacological inhibition	Inhibitors Ly364947 and SD208 suppress CML LSC colony formation	Naka et al. [96]
		Combination of TGFβ inhibition and imatinib decreases CML LSC function	
		Inhibitor SB431542 sensitizes CML cell lines to imatinib treatment by reducing LYN kinase turnover	Smith et al. [175]
		Combination of TGFβ-neutralizing antibody 1D11 and CXCR4 inhibitor in FLT3-mutated AML model decreases leukemic burden and prolongs survival	Tabe et al. [176]
Tgfβ1	KO	Homozygous mice are embryonic lethal with increased inflammatory response	Kulkarni et al. [160]
Tgfβ1	cKO in megakaryocytes (Pf4-Cre)	HSC activation and proliferation	Zhao et al. [152]
TBRII	KO	Homozygous mice are embryonic lethal with increased inflammatory response	Oshima et al. [161]
TBRII	cKO (Mx1-Cre)	Increased HSC cycling and reduces long-term repopulation capacity	Yamazaki et al. [163]
SMAD 4	cKO (Mx1-Cre)	Impaired HSC self-renewal and reconstitution capacity	Karlsson et al. [153]
		Deletion of SMAD 4 in *HOXA9*- and *NUP98-HOXA9*-induced AML model increases LSC frequency	Quere et al. [174]
SMAD 7	Retroviral overexpression	Increased HSC self-renewal, but no impact on reconstitution or differentiation capacities	Blank et al. [164]
p57Kip2	cKO (Mx1-Cre)	Impaired HSC self-renewal and reconstitution capacity	Matsumoto et al. [166]

canonical downstream TGFβ pathway involving the SMAD transcription factors is activated. The SMAD family consists of five receptor-mediated SMADs (R-SMADs 1, 2, 3, 5, and 8), one common-partner SMAD (Co-SMAD 4), and two inhibitory SMADs (I-SMADs 6 and 7). The R-SMADs contain MH2 domains that can bind via adaptor proteins to activated TβRI receptors. This interaction results in the phosphorylation of the R-SMAD, which is now able to dissociate and bind to SMAD 4. This heterodimeric complex is able to translocate into the nucleus to regulate the transcription of TGFβ target genes. The inhibitory SMADs turn off TGFβ signaling by either dephosphorylating or recruiting E3 ligases to TβRI or by competitively binding to SMAD 4. TGFβ can also signal through a SMAD-independent pathway, specifically by activating the MAPK pathway and subsequent ERK, JNK, and/or p38 activity [150]. TGFβ is expressed by both hematopoietic and bone marrow stromal cells [151], as well as by other niche components, such as megakaryocytes [152]. It has also been shown that both TβRI and TβRII are expressed on hematopoietic cells, and they can induce SMAD activity [153]. Thus, HSCs can both express TGFβ and respond to TGFβ signaling.

It has been well documented that TGFβ is an inhibitor of HSC proliferation and an inducer of HSC quiescence [154–158]. In addition, it was reported that *Tgfβ1*-null mice have enhanced myelopoiesis [159]. Due to this profound regulatory role of the TGFβ pathway, many groups have examined the effect of deleting its components in HSCs. It is evident that TGFβ plays a crucial role in embryonic development, as both *Tgfβ1* and *Tgfβ2* knockout mice have an embryonic lethal phenotype [160, 161]. While conditional deletion of *TβRI* in vivo does not impact HSC self-renewal or long-term repopulating capacity [162], deletion of *TβRII* results in impaired HSC long-term repopulation [163]. Consistent with this finding, conditional deletion of *Smad 4* also results in impaired HSC self-renewal and reconstitution capacity [153]. In an attempt to remove the major source of TGFβ from HSCs, Zhao et al. depleted megakaryocytes, which resulted in the loss of HSC quiescence and subsequent HSC proliferation [152]. Importantly, these HSCs had reduced SMAD 2 and SMAD 3 phosphorylation, supporting the hypothesis that megakaryocyte ablation is sufficient to reduce TGFβ signaling in HSCs. This loss of HSC quiescence phenotype could be rescued by injecting TGFβ1 into megakaryocyte-depleted mice. Furthermore, conditional deletion of *TGFβ1* in megakaryocytes induced the same HSC proliferation phenotype as megakaryocyte ablation [152].

Several groups have investigated the mechanisms by which TGFβ signaling controls HSC quiescence and self-renewal. Most evidence suggests a SMAD-dependent mechanism, as seen in one study in which the inhibitory SMAD 7 was overexpressed in murine HSCs, leading to an increase in their self-renewal capacity [164]. Furthermore, the TGFβ-SMAD-p57Kip2 axis has been shown to be essential for inducing HSC quiescence. Expression of p57Kip2 increases upon TGFβ stimulation in primary human hematopoietic cells, and this upregulation is essential to drive cell cycle arrest [165]. It was later shown that quiescent murine CD34- HSCs exhibit both increased SMAD 2/SMAD 3 and p57Kip2 activity as opposed to their cycling CD34+ counterparts [163], further suggesting that SMAD-dependent TGFβ signaling is required to maintain a dormant HSC population. In support of this, conditional deletion of p57 was shown to impair HSC self-renewal and reduce the size of the LSK population [166]. Additionally, Musashi 2 (MSI2), an RNA-binding protein that is important for HSC function and myeloid malignancies, was shown to maintain HSC function and regulate HSC self-renewal fate decisions by regulating TGFβ-SMAD 2/3-p57Kip2 [167]. *Msi2*-deficient HSCs have reduced reconstitution capability due to enhanced myeloid differentiation at the expense of HSC maintenance, and this is accompanied by decreased TGFβ-SMAD 2/3-p57Kip2 signaling [167]. The TGFβ-SMAD-p57 axis has also been implemented in restoring HSC quiescence after stress-induced cycling. After 5-FU treatment, murine HSCs exhibit a transient increase in TGFβ expression that correlates with the amount of time required for HSCs to reenter a quiescent state [168]. In summary, there is clear evidence that the TGFβ-SMAD-p57Kip2 signaling axis is required for HSC quiescence.

Furthermore, recent evidence suggests that the tyrosine phosphatase SHP-1 is essential for activating the TGFβ pathway in HSCs to induce quiescence, raising the possibility that the TGFβ-SMAD-p57Kip2 axis requires upstream SHP-1 activity. Specifically, it was shown that SHP-1 physically interacts with TβRI to potentiate downstream signaling, and SHP-1-depleted HSCs lose their quiescence and subsequently enter cycling until exhaustion [169]. As expected, these *Shp-1*-depleted HSCs have impaired long-term self-renewal, and, importantly, they are incapable of becoming quiescent upon TGFβ stimulation [169]. While more work is required to delineate all the possible mechanisms by which TGFβ controls HSC function, it is evident that a TGFβ-induced SMAD-dependent mechanism mediates HSC quiescence and cycling.

Despite the apparent role that TGFβ plays in regulating HSC function, very little information is known about how the TGFβ pathway regulates LSCs. As expected by its accepted role in limiting HSC proliferation, there is evidence that TGFβ acts as a tumor suppressor. The cyclin-dependent kinase p27 is critical for regulating the cell cycle of hematopoietic cells, and one growth factor that regulates p27 expression is TGFβ [170]. It was shown that BCR-ABL-expressing cells have decreased expression of p27 and are incapable of responding to TGFβ stimulation, suggesting that the BCR-ABL fusion protein functions to turn off TGFβ signaling to maintain a proliferative advantage [170]. It was previously reported that loss of the transcription factor JunB in HSCs is adequate to transform HSCs and induce myeloid malignancies in vivo [171, 172]. One possible mechanism for this phenotype is that loss of JunB renders HSCs incapable of responding to TGFβ stimulation due to dysregulation of the JunB downstream target gene *Hes1*, suggesting a role for the antiproliferative effects induced by the TGFβ pathway in HSCs in protection against myeloid malignancies [173].

The SMADs have also been shown to play tumor suppressor roles. For example, it was shown that HOXA9, a key transcriptional regulator of HSC self-renewal that is commonly dysregulated in AML, can be stabilized in the cytoplasm by SMAD 4, thereby losing its ability to translocate into the nucleus and mediate events required for leukemic transformation [174]. Importantly, transplanting SMAD 4-deficient HSCs transduced with the oncogenic NUP98-HOXA9 fusion into mice resulted in rapid onset of a more pronounced myeloproliferative disease, and secondary transplantation of these cells displayed a competitive advantage compared to wild-type cells, with faster development of leukemia. Notably, reintroducing the SMAD 4 domain required for HOXA9 cytoplasmic stabilization induced apoptosis of the HOXA9-transduced cells in vitro and limited their ability to initiate leukemia in vivo [174]. In summary, it is plausible that TGFβ plays a protective role in myeloid LSCs by regulating downstream parallel pathways that are essential for regulating proliferation.

However, other studies have suggested that TGFβ plays an oncogenic role in LSCs. In one study, TGFβ was shown to be a critical regulator of AKT activity, and

AKT-dependent suppression of active nuclear FOXO3a is required for eliminating CML LSCs [96]. Importantly, the combination of inhibiting TGFβ signaling and imatinib was sufficient to impair CML transformation, and this was shown to be due to decreased nuclear FOXO levels [96]. It has also been suggested that TGFβ can regulate the activity of LYN kinase to protect CML cells from imatinib, and inhibiting the TGFβ pathway can suppress LYN kinase activity and subsequently sensitize CML cells to imatinib treatment [175]. In AML, examining the role of TGFβ in the microenvironment has shown that the hypoxic bone marrow microenvironment reinforced by leukemic cells stimulates the TGFβ pathway, which in turn induces the expression of CXCR4 – a critical factor involved in promoting the survival of resident chemoresistant LSCs [176]. Furthermore, the inhibition of both TGFβ and CXCR4 in a FLT3-mutated AML model was sufficient to decrease the leukemic burden and prolong survival [176]. Lastly, it has recently been suggested that several different isoforms of TβRII play a role in AML. The TβRII splice variant TβRII-B is expressed in normal hematopoietic cells and functions to promote apoptosis and differentiation of AML cells [177]. However, the canonical TβRII receptor, which is the isoform primarily expressed in AML cells, inhibits the ability of TβRII-B to mediate its tumor suppressor role in vivo, overall blocking the differentiation of the leukemic cells [177]. This further complicates the potential role that the TGFβ pathway is playing in LSC function, displaying the necessity for further exploring the TGFβ pathway in the leukemic context.

1.13 Concluding Remarks

It is evident that proper signaling through the PI3K, WNT/β-catenin, NOTCH, and TGFβ pathways is fundamental for regulating the fate decisions and proliferation of HSCs. With the frontline treatment of AML still being chemotherapy and stem cell transplantation for most patients, and given the resistance observed in CML patients to targeted therapies like imatinib, it is critical that new treatment options become available to fully eradicate LSCs. Key advances in considering new therapeutic avenues include the emerging field of studying the characteristics of the LSC, as well as the acceptance of a pre-LSC that has the ability to contribute to normal hematopoiesis, while accumulating key mutations that eventually push it toward the threshold of transforming to a LSC. Mutations in components of each of these signaling pathways are common in myeloid leukemia, further validating studying these signaling pathways in the context of regulating the function of the LSC.

A classic "two-hit" model for the accumulation of mutations in AML to generate a fully functional LSC posits that two distinct types of mutations are required for full transformation (Fig. 1.1) [178]. First, a mutation in a gene critical for regulating HSC differentiation and self-renewal decisions is required. Traditionally, genes within this class were thought to be transcription factors, but it is now understood that alterations in multiple other types of genes, including splicing factors and epigenetic regulators, could also serve a similar function. This type of mutation will confer properties of self-renewal to the LSC at the expense of differentiation. Second, the LSC also requires an activating mutation in a signaling pathway essential for regulating HSC proliferation. This mutation will confer a proliferative and survival advantage to the LSC. This requirement for second mutations, the so-called "two-hit hypothesis", has been supported by multiple studies. For example, it was shown that signaling mutations alone, such as BCR-ABL and FLT3-ITD, are not sufficient to impair LSC differentiation and confer self-renewal, suggesting that the proliferative advantage alone is not adequate to generate a fully functional LSC [179]. On the other hand, mutations and/or rearrangements in transcription factors, such as AML1-ETO or CBFB-MYH11, are sufficient to confer self-renewal properties to LSCs, but insufficient to trigger leukemogenesis without additional mutations [177, 180, 181]. Given this, since the PI3K, WNT/β-catenin, NOTCH, and TGFβ signaling pathways play essential roles in regulating HSC cycling and self-renewal decisions, mutations in components of either of these pathways can directly impact the functionality of the LSC.

Epigenetic mutations also play an essential role in leukemia development at the stem cell level. However, their potential cooperation with signaling pathway mutations is often overlooked. Mutations in enzymes that affect DNA methylation or chromatin modifications, including TET2, DNMT3a, IDH1/2, and ASXL1/2, are considered to be "initiating" mutations, which are acquired at or before the pre-LSC stage (reviewed in Corces-Zimmerman et al. [1]). It is clear how such mutations may cooperate with mutations in transcription factors in leukemogenesis, since these epigenetic modifiers function in establishing the genomic landscape required for the recruitment of transcription factors. However, little has been explored regarding how epigenetic mutations cooperate with signaling mutations in LSCs. The "two-hit" model of leukemogenesis still helps us to understand the requirements for leukemic transformation. However, perhaps it would be helpful to think of these two, or more, classes of mutations not as simply additive, but as potentially synergistic in promoting hematopoietic transformation. The specific nonrandom patterns of co-occurrence of genetic alterations in AML, such as AML1-ETO with C-KIT mutations, or MLL rearrangements with RAS mutations [182], could provide clues about such potential synergistic relationships that should be further explored. For example, in other tumors, evidence for the direct regulation of epigenetic modifying enzymes by PI3K/AKT signaling is emerging (reviewed in Spangle et al. [183]). Thus, it is necessary to explore the interactions between signaling pathways and epigenetics in the context of LSCs, ultimately with the goal of refining the standard of care for myeloid leukemia patients to prevent relapse.

Lastly, given that the PI3K, WNT/β-catenin, NOTCH, and TGFβ pathways all play important roles in regulating HSC self-renewal and proliferation, it is not surprising that there would be crosstalk between these pathways. One example of crosstalk in other cell types includes the NOTCH and PI3K pathways, in which NOTCH signaling requires activation of the PI3K/AKT pathway to induce megakaryocyte differentiation [184]. A second example is between the PI3K and WNT pathways, as it has been shown that β-catenin is a direct substrate of AKT [185]. Lastly, it has been demonstrated extensively that PI3K/AKT signaling is a

SMAD-independent pathway that can be activated by TGFβ. Exploring this crosstalk in the context of LSCs may help to elucidate the contradictory role of TGFβ in LSCs. Although such crosstalk between signaling pathways has been studied in hematopoietic cells and in other tissues, it may be different in LSCs, and it is crucial to decipher these targetable differences. In summary, the PI3K, WNT/β-catenin, NOTCH, and TGFβ pathways are all important signaling pathways in HSCs, but with distinct roles in LSCs in some contexts. Further characterization of these pathways in the context of LSCs is important for understanding leukemogenesis and for developing new treatment paradigms with a safe therapeutic window.

Acknowledgments We thank all members of the Gritsman lab for helpful discussions. We apologize to any authors whose work was not cited due to space limitations. This work was supported by the NIH/NCI R01 #R01CA196973 (KG), the American Society of Hematology (KG), the V Foundation for Cancer Research (KG), the Sinsheimer Foundation (KG), the Training Program in Cellular and Molecular Biology and Genetics at Albert Einstein College of Medicine #5T32GM007491-44 (LG), and the IRACDA-BETTR Postdoctoral Training Grant at Albert Einstein College of Medicine #2K12GM102779 (KA).

Conflicts of Interest The authors declare that no conflict of interest exists.

References

1. Corces-Zimmerman MR, Hong WJ, Weissman IL, Medeiros BC, Majeti R (2014) Preleukemic mutations in human acute myeloid leukemia affect epigenetic regulators and persist in remission. Proc Natl Acad Sci U S A 111(7):2548–2553
2. Huntly BJ, Shigematsu H, Deguchi K, Lee BH, Mizuno S, Duclos N et al (2004) MOZ-TIF2, but not BCR-ABL, confers properties of leukemic stem cells to committed murine hematopoietic progenitors. Cancer Cell 6(6):587–596
3. Goardon N, Marchi E, Atzberger A, Quek L, Schuh A, Soneji S et al (2011) Coexistence of LMPP-like and GMP-like leukemia stem cells in acute myeloid leukemia. Cancer Cell 19(1):138–152
4. Krivtsov AV, Twomey D, Feng Z, Stubbs MC, Wang Y, Faber J et al (2006) Transformation from committed progenitor to leukaemia stem cell initiated by MLL-AF9. Nature 442(7104):818–822
5. Vanhaesebroeck B, Guillermet-Guibert J, Graupera M, Bilanges B (2010) The emerging mechanisms of isoform-specific PI3K signalling. Nat Rev Mol Cell Biol 11(5):329–341
6. Vadas O, Burke JE, Zhang X, Berndt A, Williams RL (2011) Structural basis for activation and inhibition of class I phosphoinositide 3-kinases. Sci Signal 4(195):re2
7. Christoforidis S, Miaczynska M, Ashman K, Wilm M, Zhao L, Yip SC et al (1999) Phosphatidylinositol-3-OH kinases are Rab5 effectors. Nat Cell Biol 1(4):249–252
8. Fritsch R, de Krijger I, Fritsch K, George R, Reason B, Kumar MS et al (2013) RAS and RHO families of GTPases directly regulate distinct phosphoinositide 3-kinase isoforms. Cell 153(5):1050–1063
9. Guillermet-Guibert J, Bjorklof K, Salpekar A, Gonella C, Ramadani F, Bilancio A et al (2008) The p110beta isoform of phosphoinositide 3-kinase signals downstream of G protein-coupled receptors and is functionally redundant with p110gamma. Proc Natl Acad Sci U S A 105(24):8292–8297

10. Alessi DR, Andjelkovic M, Caudwell B, Cron P, Morrice N, Cohen P et al (1996) Mechanism of activation of protein kinase B by insulin and IGF-1. EMBO J 15(23):6541–6551
11. Vivanco I, Palaskas N, Tran C, Finn SP, Getz G, Kennedy NJ et al (2007) Identification of the JNK signaling pathway as a functional target of the tumor suppressor PTEN. Cancer Cell 11(6):555–569
12. Hettinger K, Vikhanskaya F, Poh MK, Lee MK, de Belle I, Zhang JT et al (2007) c-Jun promotes cellular survival by suppression of PTEN. Cell Death Differ 14(2):218–229
13. Zoncu R, Efeyan A, Sabatini DM (2011) mTOR: from growth signal integration to cancer, diabetes and ageing. Nat Rev Mol Cell Biol 12(1):21–35
14. Laplante M, Sabatini DM (2012) mTOR signaling in growth control and disease. Cell 149(2):274–293
15. Jacinto E, Loewith R, Schmidt A, Lin S, Ruegg MA, Hall A et al (2004) Mammalian TOR complex 2 controls the actin cytoskeleton and is rapamycin insensitive. Nat Cell Biol 6(11):1122–1128
16. Kim E, Goraksha-Hicks P, Li L, Neufeld TP, Guan KL (2008) Regulation of TORC1 by Rag GTPases in nutrient response. Nat Cell Biol 10(8):935–945
17. Peterson TR, Laplante M, Thoreen CC, Sancak Y, Kang SA, Kuehl WM et al (2009) DEPTOR is an mTOR inhibitor frequently overexpressed in multiple myeloma cells and required for their survival. Cell 137(5):873–886
18. Kaizuka T, Hara T, Oshiro N, Kikkawa U, Yonezawa K, Takehana K et al (2010) Tti1 and Tel2 are critical factors in mammalian target of rapamycin complex assembly. J Biol Chem 285(26):20109–20116
19. Hara K, Maruki Y, Long X, Yoshino K, Oshiro N, Hidayat S et al (2002) Raptor, a binding partner of target of rapamycin (TOR), mediates TOR action. Cell 110(2):177–189
20. Kim DH, Sarbassov DD, Ali SM, King JE, Latek RR, Erdjument-Bromage H et al (2002) mTOR interacts with raptor to form a nutrient-sensitive complex that signals to the cell growth machinery. Cell 110(2):163–175
21. Sancak Y, Thoreen CC, Peterson TR, Lindquist RA, Kang SA, Spooner E et al (2007) PRAS40 is an insulin-regulated inhibitor of the mTORC1 protein kinase. Mol Cell 25(6):903–915
22. Thedieck K, Polak P, Kim ML, Molle KD, Cohen A, Jeno P et al (2007) PRAS40 and PRR5-like protein are new mTOR interactors that regulate apoptosis. PLoS One 2(11):e1217
23. Vander Haar E, Lee SI, Bandhakavi S, Griffin TJ, Kim DH (2007) Insulin signalling to mTOR mediated by the Akt/PKB substrate PRAS40. Nat Cell Biol 9(3):316–323
24. Jacinto E, Facchinetti V, Liu D, Soto N, Wei S, Jung SY et al (2006) SIN1/MIP1 maintains rictor-mTOR complex integrity and regulates Akt phosphorylation and substrate specificity. Cell 127(1):125–137
25. Sarbassov DD, Ali SM, Kim DH, Guertin DA, Latek RR, Erdjument-Bromage H et al (2004) Rictor, a novel binding partner of mTOR, defines a rapamycin-insensitive and raptor-independent pathway that regulates the cytoskeleton. Curr Biol CB 14(14):1296–1302
26. Frias MA, Thoreen CC, Jaffe JD, Schroder W, Sculley T, Carr SA et al (2006) mSin1 is necessary for Akt/PKB phosphorylation, and its isoforms define three distinct mTORC2s. Curr Biol CB 16(18):1865–1870
27. Pearce LR, Huang X, Boudeau J, Pawlowski R, Wullschleger S, Deak M et al (2007) Identification of Protor as a novel Rictor-binding component of mTOR complex-2. Biochem J 405:513–522
28. Pearce LR, Sommer EM, Sakamoto K, Wullschleger S, Alessi DR (2011) Protor-1 is required for efficient mTORC2-mediated activation of SGK1 in the kidney. Biochem J 436:169–179
29. Greer EL, Brunet A (2008) FOXO transcription factors in ageing and cancer. Acta Physiol 192(1):19–28
30. Eijkelenboom A, Mokry M, Smits LM, Nieuwenhuis EE, Burgering BM (2013) FOXO3 selectively amplifies enhancer activity to establish target gene regulation. Cell Rep 5(6):1664–1678

31. Brunet A, Bonni A, Zigmond MJ, Lin MZ, Juo P, Hu LS et al (1999) Akt promotes cell survival by phosphorylating and inhibiting a Forkhead transcription factor. Cell 96(6):857–868
32. Kops GJ, Burgering BM (1999) Forkhead transcription factors: new insights into protein kinase B (c-akt) signaling. J Mol Med 77(9):656–665
33. Furuyama T, Nakazawa T, Nakano I, Mori N (2000) Identification of the differential distribution patterns of mRNAs and consensus binding sequences for mouse DAF-16 homologues. Biochem J 349(Pt 2):629–634
34. Tothova Z, Gilliland DG (2007) FoxO transcription factors and stem cell homeostasis: insights from the hematopoietic system. Cell Stem Cell 1(2):140–152
35. Brunet A, Kanai F, Stehn J, Xu J, Sarbassova D, Frangioni JV et al (2002) 14-3-3 transits to the nucleus and participates in dynamic nucleocytoplasmic transport. J Cell Biol 156(5):817–828
36. Rena G, Prescott AR, Guo SD, Cohen P, Unterman TG (2001) Roles of the forkhead in rhabdomyosarcoma (FKHR) phosphorylation sites in regulating 14-3-3 binding, transactivation and nuclear targeting. Biochem J 354:605–612
37. Wang MC, Bohmann D, Jasper H (2005) JNK extends life span and limits growth by antagonizing cellular and organism-wide responses to insulin signaling. Cell 121(1):115–125
38. Lehtinen MK, Yuan Z, Boag PR, Yang Y, Villen J, Becker EB et al (2006) A conserved MST-FOXO signaling pathway mediates oxidative-stress responses and extends life span. Cell 125(5):987–1001
39. Essers MA, Weijzen S, de Vries-Smits AM, Saarloos I, de Ruiter ND, Bos JL et al (2004) FOXO transcription factor activation by oxidative stress mediated by the small GTPase Ral and JNK. EMBO J 23(24):4802–4812
40. Brunet A, Sweeney LB, Sturgill JF, Chua KF, Greer PL, Lin Y et al (2004) Stress-dependent regulation of FOXO transcription factors by the SIRT1 deacetylase. Science 303(5666):2011–2015
41. Yang JY, Zong CS, Xia W, Yamaguchi H, Ding Q, Xie X et al (2008) ERK promotes tumorigenesis by inhibiting FOXO3a via MDM2-mediated degradation. Nat Cell Biol 10(2):138–148
42. Calnan DR, Brunet A (2008) The FoxO code. Oncogene 27(16):2276–2288
43. Salih DA, Brunet A (2008) FoxO transcription factors in the maintenance of cellular homeostasis during aging. Curr Opin Cell Biol 20(2):126–136
44. Zhao JJ, Cheng H, Jia S, Wang L, Gjoerup OV, Mikami A et al (2006) The p110alpha isoform of PI3K is essential for proper growth factor signaling and oncogenic transformation. Proc Natl Acad Sci U S A 103(44):16296–16300
45. Marques M, Kumar A, Cortes I, Gonzalez-Garcia A, Hernandez C, Moreno-Ortiz MC et al (2008) Phosphoinositide 3-kinases p110 alpha and p110 beta regulate cell cycle entry, exhibiting distinct activation kinetics in G(1) phase. Mol Cell Biol 28(8):2803–2814
46. Utermark T, Rao T, Cheng H, Wang Q, Lee SH, Wang ZC et al (2012) The p110alpha and p110beta isoforms of PI3K play divergent roles in mammary gland development and tumorigenesis. Genes Dev 26(14):1573–1586
47. Gritsman K, Yuzugullu H, Von T, Yan H, Clayton L, Fritsch C et al (2014) Hematopoiesis and RAS-driven myeloid leukemia differentially require PI3K isoform p110alpha. J Clin Invest 124(4):1794–1809
48. Yuzugullu H, Baitsch L, Von T, Steiner A, Tong H, Ni J et al (2015) A PI3K p110beta-Rac signalling loop mediates Pten-loss-induced perturbation of haematopoiesis and leukaemogenesis. Nat Commun 6:8501
49. Haneline LS, White H, Yang FC, Chen S, Orschell C, Kapur R et al (2006) Genetic reduction of class IA PI-3 kinase activity alters fetal hematopoiesis and competitive repopulating ability of hematopoietic stem cells in vivo. Blood 107(4):1375–1382
50. Munugalavadla V, Sims EC, Chan RJ, Lenz SD, Kapur R (2008) Requirement for p85alpha regulatory subunit of class IA PI3K in myeloproliferative disease driven by an activation loop mutant of KIT. Exp Hematol 36(3):301–308

51. Cairoli R, Beghini A, Grillo G, Nadali G, Elice F, Ripamonti CB et al (2006) Prognostic impact of c-KIT mutations in core binding factor leukemias: an Italian retrospective study. Blood 107(9):3463–3468
52. Cairoli R, Grillo G, Beghini A, Tedeschi A, Ripamonti CB, Larizza L et al (2003) C-Kit point mutations in core binding factor leukemias: correlation with white blood cell count and the white blood cell index. Leukemia 17(2):471–472
53. Beghini A, Peterlongo P, Ripamonti CB, Larizza L, Cairoli R, Morra E et al (2000) C-kit mutations in core binding factor leukemias. Blood 95(2):726–727
54. Jia S, Liu Z, Zhang S, Liu P, Zhang L, Lee SH et al (2008) Essential roles of PI(3)K-p110beta in cell growth, metabolism and tumorigenesis. Nature 454(7205):776–779
55. Min YH, Eom JI, Cheong JW, Maeng HO, Kim JY, Jeung HK et al (2003) Constitutive phosphorylation of Akt/PKB protein in acute myeloid leukemia: its significance as a prognostic variable. Leukemia 17(5):995–997
56. Gallay N, Dos Santos C, Cuzin L, Bousquet M, Simmonet Gouy V, Chaussade C et al (2009) The level of AKT phosphorylation on threonine 308 but not on serine 473 is associated with high-risk cytogenetics and predicts poor overall survival in acute myeloid leukaemia. Leukemia 23(6):1029–1038
57. Tamburini J, Elie C, Bardet V, Chapuis N, Park S, Broet P et al (2007) Constitutive phosphoinositide 3-kinase/Akt activation represents a favorable prognostic factor in de novo acute myelogenous leukemia patients. Blood 110(3):1025–1028
58. Kharas MG, Okabe R, Ganis JJ, Gozo M, Khandan T, Paktinat M et al (2010) Constitutively active AKT depletes hematopoietic stem cells and induces leukemia in mice. Blood 115(7):1406–1415
59. Juntilla MM, Patil VD, Calamito M, Joshi RP, Birnbaum MJ, Koretzky GA (2010) AKT1 and AKT2 maintain hematopoietic stem cell function by regulating reactive oxygen species. Blood 115(20):4030–4038
60. Di Cristofano A, Pandolfi PP (2000) The multiple roles of PTEN in tumor suppression. Cell 100(4):387–390
61. Aggerholm A, Gronbaek K, Guldberg P, Hokland P (2000) Mutational analysis of the tumour suppressor gene MMAC1/PTEN in malignant myeloid disorders. Eur J Haematol 65(2):109–113
62. Dahia PLM, Aguiar RCT, Alberta J, Kum JB, Caron S, Sill H et al (1999) PTEN is inversely correlated with the cell survival factor Akt/PKB and is inactivated via multiple mechanisms in haematological malignancies. Hum Mol Genet 8(2):185–193
63. Yilmaz OH, Valdez R, Theisen BK, Guo W, Ferguson DO, Wu H et al (2006) Pten dependence distinguishes haematopoietic stem cells from leukaemia-initiating cells. Nature 441(7092):475–482
64. Zhang J, Grindley JC, Yin T, Jayasinghe S, He XC, Ross JT et al (2006) PTEN maintains haematopoietic stem cells and acts in lineage choice and leukaemia prevention. Nature 441(7092):518–522
65. Fragoso R, Barata JT (2014) PTEN and leukemia stem cells. Adv Biol Regul 56:22–29
66. Guo W, Schubbert S, Chen JY, Valamehr B, Mosessian S, Shi H et al (2011) Suppression of leukemia development caused by PTEN loss. Proc Natl Acad Sci U S A 108(4):1409–1414
67. Lee JY, Nakada D, Yilmaz OH, Tothova Z, Joseph NM, Lim MS et al (2010) mTOR activation induces tumor suppressors that inhibit leukemogenesis and deplete hematopoietic stem cells after Pten deletion. Cell Stem Cell 7(5):593–605
68. Peng C, Chen Y, Yang Z, Zhang H, Osterby L, Rosmarin AG et al (2010) PTEN is a tumor suppressor in CML stem cells and BCR-ABL-induced leukemias in mice. Blood 115(3):626–635
69. Li J, Zhang J, Tang MH, Xin JP, Xu Y, Volk A et al (2016) Hematopoietic stem cell activity is regulated by Pten phosphorylation through a niche-dependent mechanism. Stem Cells 34(8):2130–2144
70. Helgason CD, Damen JE, Rosten P, Grewal R, Sorensen P, Chappel SM et al (1998) Targeted disruption of SHIP leads to hemopoietic perturbations lung pathology, and a shortened life span. Genes Dev 12(11):1610–1620

71. Sattler M, Verma S, Byrne CH, Shrikhande G, Winkler T, Algate PA et al (1999) BCR/ABL directly inhibits expression of SHIP, an SH2-containing polyinositol-5-phosphatase involved in the regulation of hematopoiesis. Mol Cell Biol 19(11):7473–7480
72. Luo J, Manning BD, Cantley LC (2003) Targeting the PI3K-Akt pathway in human cancer: rationale and promise. Cancer Cell 4(4):257–262
73. Magee JA, Ikenoue T, Nakada D, Lee JY, Guan KL, Morrison SJ (2012) Temporal changes in PTEN and mTORC2 regulation of hematopoietic stem cell self-renewal and leukemia suppression. Cell Stem Cell 11(3):415–428
74. Kalaitzidis D, Sykes SM, Wang Z, Punt N, Tang Y, Ragu C et al (2012) mTOR complex 1 plays critical roles in hematopoiesis and Pten-loss-evoked Leukemogenesis. Cell Stem Cell 11(3):429–439
75. Wang W, Sun X, Hu T, Wang L, Dong S, Gu J et al (2018) PDK1 regulates definitive HSCs via the FOXO pathway during murine fetal liver hematopoiesis. Stem Cell Res 30:192–200
76. Hu T, Li C, Zhang Y, Wang L, Peng L, Cheng H et al (2015) Phosphoinositide-dependent kinase 1 regulates leukemia stem cell maintenance in MLL-AF9-induced murine acute myeloid leukemia. Biochem Biophys Res Commun 459(4):692–698
77. Magee JA, Ikenoue T, Nakada D, Lee JY, Guan KL, Morrison SJ (2012) Temporal changes in PTEN and mTORC2 regulation of hematopoietic stem cell self-renewal and leukemia suppression. Cell Stem Cell 11(3):415–428
78. Guo F, Zhang S, Grogg M, Cancelas JA, Varney ME, Starczynowski DT et al (2013) Mouse gene targeting reveals an essential role of mTOR in hematopoietic stem cell engraftment and hematopoiesis. Haematologica 98(9):1353–1358
79. Chen C, Liu Y, Liu R, Ikenoue T, Guan KL, Liu Y et al (2008) TSC-mTOR maintains quiescence and function of hematopoietic stem cells by repressing mitochondrial biogenesis and reactive oxygen species. J Exp Med 205(10):2397–2408
80. Chen C, Liu Y, Liu Y, Zheng P (2009) mTOR regulation and therapeutic rejuvenation of aging hematopoietic stem cells. Sci Signal 2(98):ra75
81. Hoshii T, Tadokoro Y, Naka K, Ooshio T, Muraguchi T, Sugiyama N et al (2012) mTORC1 is essential for leukemia propagation but not stem cell self-renewal. J Clin Invest 122(6):2114–2129
82. Carracedo A, Pandolfi PP (2008) The PTEN-PI3K pathway: of feedbacks and cross-talks. Oncogene 27(41):5527–5541
83. Peng H, Kasada A, Ueno M, Hoshii T, Tadokoro Y, Nomura N et al (2018) Distinct roles of Rheb and raptor in activating mTOR complex 1 for the self-renewal of hematopoietic stem cells. Biochem Biophys Res Commun 495(1):1129–1135
84. Altman JK, Sassano A, Kaur S, Glaser H, Kroczynska B, Redig AJ et al (2011) Dual mTORC2/mTORC1 targeting results in potent suppressive effects on acute myeloid leukemia (AML) progenitors. Clin Cancer Res 17(13):4378–4388
85. Ghosh J, Kapur R (2017) Role of mTORC1-S6K1 signaling pathway in regulation of hematopoietic stem cell and acute myeloid leukemia. Exp Hematol 50:13–21
86. Ghosh J, Kobayashi M, Ramdas B, Chatterjee A, Ma P, Mali RS et al (2016) S6K1 regulates hematopoietic stem cell self-renewal and leukemia maintenance. J Clin Invest 126(7):2621–2625
87. Miyamoto K, Araki KY, Naka K, Arai F, Takubo K, Yamazaki S et al (2007) Foxo3a is essential for maintenance of the hematopoietic stem cell pool. Cell Stem Cell 1(1):101–112
88. Tothova Z, Kollipara R, Huntly BJ, Lee BH, Castrillon DH, Cullen DE et al (2007) FoxOs are critical mediators of hematopoietic stem cell resistance to physiologic oxidative stress. Cell 128(2):325–339
89. Yalcin S, Marinkovic D, Mungamuri SK, Zhang X, Tong W, Sellers R et al (2010) ROS-mediated amplification of AKT/mTOR signalling pathway leads to myeloproliferative syndrome in Foxo3(−/−) mice. EMBO J 29(24):4118–4131
90. Rimmele P, Liang R, Bigarella CL, Kocabas F, Xie J, Serasinghe MN et al (2015) Mitochondrial metabolism in hematopoietic stem cells requires functional FOXO3. EMBO Rep 16(9):1164–1176

91. Liang R, Rimmele P, Bigarella CL, Yalcin S, Ghaffari S (2016) Evidence for AKT-independent regulation of FOXO1 and FOXO3 in haematopoietic stem and progenitor cells. Cell Cycle 15(6):861–867
92. Miyamoto K, Miyamoto T, Kato R, Yoshimura A, Motoyama N, Suda T (2008) FoxO3a regulates hematopoietic homeostasis through a negative feedback pathway in conditions of stress or aging. Blood 112(12):4485–4493
93. Santamaria CM, Chillon MC, Garcia-Sanz R, Perez C, Caballero MD, Ramos F et al (2009) High FOXO3a expression is associated with a poorer prognosis in AML with normal cytogenetics. Leuk Res 33(12):1706–1709
94. Kornblau SM, Singh N, Qiu YH, Chen WJ, Zhang NX, Coombes KR (2010) Highly phosphorylated FOXO3A is an adverse prognostic factor in acute myeloid leukemia. Clin Cancer Res 16(6):1865–1874
95. Chapuis N, Park S, Leotoing L, Tamburini J, Verdier F, Bardet V et al (2010) I kappa B kinase overcomes PI3K/Akt and ERK/MAPK to control FOXO3a activity in acute myeloid leukemia. Blood 116(20):4240–4250
96. Naka K, Hoshii T, Muraguchi T, Tadokoro Y, Ooshio T, Kondo Y et al (2010) TGF-beta-FOXO signalling maintains leukaemia-initiating cells in chronic myeloid leukaemia. Nature 463(7281):676–U111
97. Sykes SM, Lane SW, Bullinger L, Kalaitzidis D, Yusuf R, Saez B et al (2011) AKT/FOXO signaling enforces reversible differentiation blockade in myeloid leukemias. Cell 146(5):697–708
98. Baumgartner C, Toifl S, Farlik M, Halbritter F, Scheicher R, Fischer I et al (2018) An ERK-dependent feedback mechanism prevents hematopoietic stem cell exhaustion. Cell Stem Cell 22(6):879–92 e6
99. Xue L, Pulikkan JA, Valk PJM, Castilla LH (2014) NrasG12D oncoprotein inhibits apoptosis of preleukemic cells expressing Cbfbeta-SMMHC via activation of MEK/ERK axis. Blood 124(3):426–436
100. Wang J, Kong G, Liu Y, Du J, Chang Y-I, Tey SR et al (2013) Nras(G12D/+) promotes leukemogenesis by aberrantly regulating hematopoietic stem cell functions. Blood 121(26):5203–5207
101. Nowacka JD, Baumgartner C, Pelorosso C, Roth M, Zuber J, Baccarini M (2016) MEK1 is required for the development of NRAS-driven leukemia. Oncotarget 7(49):80113–80130
102. Li Q, Bohin N, Wen T, Ng V, Magee J, Chen SC et al (2013) Oncogenic Nras has bimodal effects on stem cells that sustainably increase competitiveness. Nature 504(7478):143–147
103. Reya T, Duncan AW, Ailles L, Domen J, Scherer DC, Willert K et al (2003) A role for Wnt signalling in self-renewal of haematopoietic stem cells. Nature 423(6938):409–414
104. Fleming HE, Janzen V, Lo Celso C, Guo J, Leahy KM, Kronenberg HM et al (2008) Wnt signaling in the niche enforces hematopoietic stem cell quiescence and is necessary to preserve self-renewal in vivo. Cell Stem Cell 2(3):274–283
105. Staal FJ, Famili F, Garcia Perez L, Pike-Overzet K (2016) Aberrant Wnt signaling in leukemia. Cancers (Basel) 8(9):78
106. Cobas M, Wilson A, Ernst B, Mancini SJ, MacDonald HR, Kemler R et al (2004) Beta-catenin is dispensable for hematopoiesis and lymphopoiesis. J Exp Med 199(2):221–229
107. Kabiri Z, Numata A, Kawasaki A, Edison TDG, Virshup DM (2015) Wnts are dispensable for differentiation and self-renewal of adult murine hematopoietic stem cells. Blood 126(9):1086–1094
108. Luis TC, Ichii M, Brugman MH, Kincade P, Staal FJ (2012) Wnt signaling strength regulates normal hematopoiesis and its deregulation is involved in leukemia development. Leukemia 26(3):414–421
109. Luis TC, Naber BA, Roozen PP, Brugman MH, de Haas EF, Ghazvini M et al (2011) Canonical wnt signaling regulates hematopoiesis in a dosage-dependent fashion. Cell Stem Cell 9(4):345–356
110. Kirstetter P, Anderson K, Porse BT, Jacobsen SE, Nerlov C (2006) Activation of the canonical Wnt pathway leads to loss of hematopoietic stem cell repopulation and multilineage differentiation block. Nat Immunol 7(10):1048–1056

111. Scheller M, Huelsken J, Rosenbauer F, Taketo MM, Birchmeier W, Tenen DG et al (2006) Hematopoietic stem cell and multilineage defects generated by constitutive beta-catenin activation. Nat Immunol 7(10):1037–1047
112. Perry JM, He XC, Sugimura R, Grindley JC, Haug JS, Ding S et al (2011) Cooperation between both Wnt/{beta}-catenin and PTEN/PI3K/Akt signaling promotes primitive hematopoietic stem cell self-renewal and expansion. Genes Dev 25(18):1928–1942
113. Jamieson CH, Ailles LE, Dylla SJ, Muijtjens M, Jones C, Zehnder JL et al (2004) Granulocyte-macrophage progenitors as candidate leukemic stem cells in blast-crisis CML. N Engl J Med 351(7):657–667
114. Zhao C, Blum J, Chen A, Kwon HY, Jung SH, Cook JM et al (2007) Loss of beta-catenin impairs the renewal of normal and CML stem cells in vivo. Cancer Cell 12(6):528–541
115. Hu Y, Chen Y, Douglas L, Li S (2009) Beta-catenin is essential for survival of leukemic stem cells insensitive to kinase inhibition in mice with BCR-ABL-induced chronic myeloid leukemia. Leukemia 23(1):109–116
116. Hu J, Feng M, Liu ZL, Liu Y, Huang ZL, Li H et al (2016) Potential role of Wnt/beta-catenin signaling in blastic transformation of chronic myeloid leukemia: cross talk between beta-catenin and BCR-ABL. Tumour Biol 37:15859–15872
117. Eiring AM, Khorashad JS, Anderson DJ, Yu F, Redwine HM, Mason CC et al (2015) Beta-catenin is required for intrinsic but not extrinsic BCR-ABL1 kinase-independent resistance to tyrosine kinase inhibitors in chronic myeloid leukemia. Leukemia 29(12):2328–2337
118. Yu S, Li F, Xing S, Zhao T, Peng W, Xue HH (2016) Hematopoietic and leukemic stem cells have distinct dependence on Tcf1 and Lef1 transcription factors. J Biol Chem 291(21):11148–11160
119. Wang Y, Krivtsov AV, Sinha AU, North TE, Goessling W, Feng Z et al (2010) The Wnt/beta-catenin pathway is required for the development of leukemia stem cells in AML. Science 327(5973):1650–1653
120. Yeung J, Esposito MT, Gandillet A, Zeisig BB, Griessinger E, Bonnet D et al (2010) Beta-catenin mediates the establishment and drug resistance of MLL leukemic stem cells. Cancer Cell 18(6):606–618
121. Dietrich PA, Yang C, Leung HH, Lynch JR, Gonzales E, Liu B et al (2014) GPR84 sustains aberrant beta-catenin signaling in leukemic stem cells for maintenance of MLL leukemogenesis. Blood 124(22):3284–3294
122. Jiang X, Mak PY, Mu H, Tao W, Mak DH, Kornblau S et al (2018) Disruption of Wnt/beta-catenin exerts Antileukemia activity and synergizes with FLT3 inhibition in FLT3-mutant acute myeloid leukemia. Clin Cancer Res 24(10):2417–2429
123. Komiya Y, Habas R (2008) Wnt signal transduction pathways. Organogenesis 4(2):68–75
124. Nemeth MJ, Topol L, Anderson SM, Yang Y, Bodine DM (2007) Wnt5a inhibits canonical Wnt signaling in hematopoietic stem cells and enhances repopulation. Proc Natl Acad Sci U S A 104(39):15436–15441
125. Liang H, Chen Q, Coles AH, Anderson SJ, Pihan G, Bradley A et al (2003) Wnt5a inhibits B cell proliferation and functions as a tumor suppressor in hematopoietic tissue. Cancer Cell 4(5):349–360
126. Martin V, Valencia A, Agirre X, Cervera J, San Jose-Eneriz E, Vilas-Zornoza A, Rodriguez-Otero P, Sanz MA, Herrera C, Torres A, Prosper F, Roman-Gomez J (2010) Epigenetic regulation of the non-canonical Wnt pathway in acute myeloid leukemia. Cancer Sci 101(2):425–32
127. Schreck C, Istvanffy R, Ziegenhain C, Sippenauer T, Ruf F, Henkel L et al (2017) Niche WNT5A regulates the actin cytoskeleton during regeneration of hematopoietic stem cells. J Exp Med 214(1):165–181
128. Chattopadhyay S, Chaklader M, Law S (2018) Aberrant Wnt signaling pathway in the hematopoietic stem/progenitor compartment in experimental leukemic animal. J Cell Commun Signal 13(1):39–52
129. Pajcini KV, Speck NA, Pear WS (2011) Notch signaling in mammalian hematopoietic stem cells. Leukemia 25(10):1525–1532

130. Varnum-Finney B, Xu L, Brashem-Stein C, Nourigat C, Flowers D, Bakkour S et al (2000) Pluripotent, cytokine-dependent, hematopoietic stem cells are immortalized by constitutive Notch1 signaling. Nat Med 6(11):1278–1281
131. Stier S, Cheng T, Dombkowski D, Carlesso N, Scadden DT (2002) Notch1 activation increases hematopoietic stem cell self-renewal in vivo and favors lymphoid over myeloid lineage outcome. Blood 99(7):2369–2378
132. Karanu FN, Murdoch B, Gallacher L, Wu DM, Koremoto M, Sakano S et al (2000) The notch ligand jagged-1 represents a novel growth factor of human hematopoietic stem cells. J Exp Med 192(9):1365–1372
133. Karanu FN, Murdoch B, Miyabayashi T, Ohno M, Koremoto M, Gallacher L et al (2001) Human homologues of Delta-1 and Delta-4 function as mitogenic regulators of primitive human hematopoietic cells. Blood 97(7):1960–1967
134. Duncan AW, Rattis FM, DiMascio LN, Congdon KL, Pazianos G, Zhao C et al (2005) Integration of Notch and Wnt signaling in hematopoietic stem cell maintenance. Nat Immunol 6(3):314–322
135. Radtke F, Wilson A, Stark G, Bauer M, van Meerwijk J, MacDonald HR et al (1999) Deficient T cell fate specification in mice with an induced inactivation of Notch1. Immunity 10(5):547–558
136. Varnum-Finney B, Halasz LM, Sun M, Gridley T, Radtke F, Bernstein ID (2011) Notch2 governs the rate of generation of mouse long- and short-term repopulating stem cells. J Clin Invest 121(3):1207–1216
137. Calvi LM, Adams GB, Weibrecht KW, Weber JM, Olson DP, Knight MC et al (2003) Osteoblastic cells regulate the haematopoietic stem cell niche. Nature 425(6960):841–846
138. Butler JM, Nolan DJ, Vertes EL, Varnum-Finney B, Kobayashi H, Hooper AT et al (2010) Endothelial cells are essential for the self-renewal and repopulation of notch-dependent hematopoietic stem cells. Cell Stem Cell 6(3):251–264
139. Lobry C, Oh P, Mansour MR, Look AT, Aifantis I (2014) Notch signaling: switching an oncogene to a tumor suppressor. Blood 123(16):2451–2459
140. Chen PM, Yen CC, Wang WS, Lin YJ, Chu CJ, Chiou TJ et al (2008) Down-regulation of Notch-1 expression decreases PU.1-mediated myeloid differentiation signaling in acute myeloid leukemia. Int J Oncol 32(6):1335–1341
141. Klinakis A, Lobry C, Abdel-Wahab O, Oh P, Haeno H, Buonamici S et al (2011) A novel tumour-suppressor function for the Notch pathway in myeloid leukaemia. Nature 473(7346):230–233
142. Lobry C, Ntziachristos P, Ndiaye-Lobry D, Oh P, Cimmino L, Zhu N et al (2013) Notch pathway activation targets AML-initiating cell homeostasis and differentiation. J Exp Med 210(2):301–319
143. Kannan S, Sutphin RM, Hall MG, Golfman LS, Fang W, Nolo RM et al (2013) Notch activation inhibits AML growth and survival: a potential therapeutic approach. J Exp Med 210(2):321–337
144. Kato T, Sakata-Yanagimoto M, Nishikii H, Ueno M, Miyake Y, Yokoyama Y et al (2015) Hes1 suppresses acute myeloid leukemia development through FLT3 repression. Leukemia 29(3):576–585
145. Kode A, Manavalan JS, Mosialou I, Bhagat G, Rathinam CV, Luo N et al (2014) Leukaemogenesis induced by an activating beta-catenin mutation in osteoblasts. Nature 506(7487):240–244
146. Nakahara F, Sakata-Yanagimoto M, Komeno Y, Kato N, Uchida T, Haraguchi K et al (2010) Hes1 immortalizes committed progenitors and plays a role in blast crisis transition in chronic myelogenous leukemia. Blood 115(14):2872–2881
147. Liu X, Ma Y, Li R, Guo D, Wang N, Zhao Y et al (2018) Niche TWIST1 is critical for maintaining normal hematopoiesis and impeding leukemia progression. Haematologica 103(12):1969–1979

148. Yin DD, Fan FY, Hu XB, Hou LH, Zhang XP, Liu L et al (2009) Notch signaling inhibits the growth of the human chronic myeloid leukemia cell line K562. Leuk Res 33(1):109–114
149. Bowers M, Zhang B, Ho Y, Agarwal P, Chen CC, Bhatia R (2015) Osteoblast ablation reduces normal long-term hematopoietic stem cell self-renewal but accelerates leukemia development. Blood 125(17):2678–2688
150. Vaidya A, Kale VP (2015) TGF-beta signaling and its role in the regulation of hematopoietic stem cells. Syst Synth Biol 9(1–2):1–10
151. Ruscetti FW, Akel S, Bartelmez SH (2005) Autocrine transforming growth factor-beta regulation of hematopoiesis: many outcomes that depend on the context. Oncogene 24(37):5751–5763
152. Zhao M, Perry JM, Marshall H, Venkatraman A, Qian P, He XC et al (2014) Megakaryocytes maintain homeostatic quiescence and promote post-injury regeneration of hematopoietic stem cells. Nat Med 20(11):1321–1326
153. Karlsson G, Blank U, Moody JL, Ehinger M, Singbrant S, Deng CX et al (2007) Smad4 is critical for self-renewal of hematopoietic stem cells. J Exp Med 204(3):467–474
154. Sing GK, Keller JR, Ellingsworth LR, Ruscetti FW (1988) Transforming growth factor beta selectively inhibits normal and leukemic human bone marrow cell growth in vitro. Blood 72(5):1504–1511
155. Keller JR, McNiece IK, Sill KT, Ellingsworth LR, Quesenberry PJ, Sing GK et al (1990) Transforming growth factor beta directly regulates primitive murine hematopoietic cell proliferation. Blood 75(3):596–602
156. Sitnicka E, Ruscetti FW, Priestley GV, Wolf NS, Bartelmez SH (1996) Transforming growth factor beta 1 directly and reversibly inhibits the initial cell divisions of long-term repopulating hematopoietic stem cells. Blood 88(1):82–88
157. Soma T, Yu JM, Dunbar CE (1996) Maintenance of murine long-term repopulating stem cells in ex vivo culture is affected by modulation of transforming growth factor-beta but not macrophage inflammatory protein-1 alpha activities. Blood 87(11):4561–4567
158. Fortunel N, Batard P, Hatzfeld A, Monier MN, Panterne B, Lebkowski J et al (1998) High proliferative potential-quiescent cells: a working model to study primitive quiescent hematopoietic cells. J Cell Sci 111(Pt 13):1867–1875
159. Letterio JJ, Geiser AG, Kulkarni AB, Dang H, Kong L, Nakabayashi T et al (1996) Autoimmunity associated with TGF-beta1-deficiency in mice is dependent on MHC class II antigen expression. J Clin Invest 98(9):2109–2119
160. Kulkarni AB, Huh CG, Becker D, Geiser A, Lyght M, Flanders KC et al (1993) Transforming growth factor beta 1 null mutation in mice causes excessive inflammatory response and early death. Proc Natl Acad Sci U S A 90(2):770–774
161. Oshima M, Oshima H, Taketo MM (1996) TGF-beta receptor type II deficiency results in defects of yolk sac hematopoiesis and vasculogenesis. Dev Biol 179(1):297–302
162. Larsson J, Blank U, Helgadottir H, Bjornsson JM, Ehinger M, Goumans MJ et al (2003) TGF-beta signaling-deficient hematopoietic stem cells have normal self-renewal and regenerative ability in vivo despite increased proliferative capacity in vitro. Blood 102(9):3129–3135
163. Yamazaki S, Iwama A, Takayanagi S, Eto K, Ema H, Nakauchi H (2009) TGF-beta as a candidate bone marrow niche signal to induce hematopoietic stem cell hibernation. Blood 113(6):1250–1256
164. Blank U, Karlsson G, Moody JL, Utsugisawa T, Magnusson M, Singbrant S et al (2006) Smad7 promotes self-renewal of hematopoietic stem cells. Blood 108(13):4246–4254
165. Scandura JM, Boccuni P, Massague J, Nimer SD (2004) Transforming growth factor beta-induced cell cycle arrest of human hematopoietic cells requires p57KIP2 up-regulation. Proc Natl Acad Sci U S A 101(42):15231–15236
166. Matsumoto A, Takeishi S, Kanie T, Susaki E, Onoyama I, Tateishi Y et al (2011) p57 is required for quiescence and maintenance of adult hematopoietic stem cells. Cell Stem Cell 9(3):262–271

167. Park SM, Deering RP, Lu Y, Tivnan P, Lianoglou S, Al-Shahrour F et al (2014) Musashi-2 controls cell fate, lineage bias, and TGF-beta signaling in HSCs. J Exp Med 211(1):71–87
168. Brenet F, Kermani P, Spektor R, Rafii S, Scandura JM (2013) TGFbeta restores hematopoietic homeostasis after myelosuppressive chemotherapy. J Exp Med 210(3):623–639
169. Jiang L, Han X, Wang J, Wang C, Sun X, Xie J et al (2018) SHP-1 regulates hematopoietic stem cell quiescence by coordinating TGF-beta signaling. J Exp Med 215(5):1337–1347
170. Jonuleit T, van der Kuip H, Miething C, Michels H, Hallek M, Duyster J et al (2000) Bcr-Abl kinase down-regulates cyclin-dependent kinase inhibitor p27 in human and murine cell lines. Blood 96(5):1933–1939
171. Passegue E, Jochum W, Schorpp-Kistner M, Mohle-Steinlein U, Wagner EF (2001) Chronic myeloid leukemia with increased granulocyte progenitors in mice lacking junB expression in the myeloid lineage. Cell 104(1):21–32
172. Passegue E, Wagner EF, Weissman IL (2004) JunB deficiency leads to a myeloproliferative disorder arising from hematopoietic stem cells. Cell 119(3):431–443
173. Santaguida M, Schepers K, King B, Sabnis AJ, Forsberg EC, Attema JL et al (2009) JunB protects against myeloid malignancies by limiting hematopoietic stem cell proliferation and differentiation without affecting self-renewal. Cancer Cell 15(4):341–352
174. Quere R, Karlsson G, Hertwig F, Rissler M, Lindqvist B, Fioretos T et al (2011) Smad4 binds Hoxa9 in the cytoplasm and protects primitive hematopoietic cells against nuclear activation by Hoxa9 and leukemia transformation. Blood 117(22):5918–5930
175. Smith PG, Tanaka H, Chantry A (2012) A novel co-operative mechanism linking TGFbeta and Lyn kinase activation to imatinib resistance in chronic myeloid leukaemia cells. Oncotarget 3(5):518–524
176. Tabe Y, Shi YX, Zeng Z, Jin L, Shikami M, Hatanaka Y et al (2013) TGF-beta-neutralizing antibody 1D11 enhances cytarabine-induced apoptosis in AML cells in the bone marrow microenvironment. PLoS One 8(6):e62785
177. Wu Y, Su M, Zhang S, Cheng Y, Liao XY, Lin BY et al (2017) Abnormal expression of TGF-beta type II receptor isoforms contributes to acute myeloid leukemia. Oncotarget 8(6):10037–10049
178. Gilliland DG (2002) Molecular genetics of human leukemias: new insights into therapy. Semin Hematol 39(4 Suppl 3):6–11
179. Huntly BJ, Gilliland DG (2004) Blasts from the past: new lessons in stem cell biology from chronic myelogenous leukemia. Cancer Cell 6(3):199–201
180. Cabezas-Wallscheid N, Eichwald V, de Graaf J, Lower M, Lehr HA, Kreft A et al (2013) Instruction of haematopoietic lineage choices, evolution of transcriptional landscapes and cancer stem cell hierarchies derived from an AML1-ETO mouse model. EMBO Mol Med 5(12):1804–1820
181. Castilla LH, Garrett L, Adya N, Orlic D, Dutra A, Anderson S et al (1999) The fusion gene Cbfb-MYH11 blocks myeloid differentiation and predisposes mice to acute myelomonocytic leukaemia. Nat Genet 23(2):144–146
182. Chen SJ, Shen Y, Chen Z (2013) A panoramic view of acute myeloid leukemia. Nat Genet 45(6):586–587
183. Spangle JM, Roberts TM, Zhao JJ (2017) The emerging role of PI3K/AKT-mediated epigenetic regulation in cancer. Biochim Biophys Acta Rev Cancer 1868(1):123–131
184. Cornejo MG, Mabialah V, Sykes SM, Khandan T, Lo Celso C, Lopez CK et al (2011) Crosstalk between NOTCH and AKT signaling during murine megakaryocyte lineage specification. Blood 118(5):1264–1273
185. Fang D, Hawke D, Zheng Y, Xia Y, Meisenhelder J, Nika H et al (2007) Phosphorylation of beta-catenin by AKT promotes beta-catenin transcriptional activity. J Biol Chem 282(15):11221–11229

Chapter 2
Cellular and Molecular State of Myeloid Leukemia Stem Cells

Xueqin Xie, Mengdie Feng, Qifan Wang, Jiazhen Wang, Rong Yin, Yicun Li, and Haojian Zhang

Contents

Abstract Leukemia stem cells (LSCs) are leukemia-initiating population with the capacity to self-renew, differentiate, and stay quiescent. Human hematopoietic malignancies such as chronic myeloid leukemia (CML) and acute myeloid leukemia (AML) are derived from this cell population. LSCs are also responsible for disease relapse due to its resistance to drug treatment. This rare cell population is phenotypically and functionally heterogeneous. Increasing evidence indicates that this heterogeneous cellular state of LSCs might determine the different drug sensitivity and is the major reason for disease relapse. In here, focusing on myeloid leukemia stem cells, we describe the biological features including cellular and

X. Xie · M. Feng · Q. Wang · J. Wang · R. Yin · H. Zhang (✉)
The State Key Laboratory Breeding Base of Basic Science of Stomatology & Key Laboratory of Oral Biomedicine Ministry of Education, School & Hospital of Stomatology, Wuhan University, Wuhan, China

Frontier Science Center for Immunology and Metabolism, Medical Research Institute, Wuhan University, Wuhan, China
e-mail: haojian_zhang@whu.edu.cn

Y. Li
The State Key Laboratory Breeding Base of Basic Science of Stomatology & Key Laboratory of Oral Biomedicine Ministry of Education, School & Hospital of Stomatology, Wuhan University, Wuhan, China

H. Zhang, S. Li (eds.), *Leukemia Stem Cells in Hematologic Malignancies*, Advances in Experimental Medicine and Biology 1143,
https://doi.org/10.1007/978-981-13-7342-8_2

molecular state, heterogeneity of LSCs, and the dynamic cross talk between LSCs and bone marrow microenvironment. These specific features of LSCs highlight the dynamic cellular state of LSCs, and further exploring on it might provide potential therapeutic targets that are important for eliminating LSCs.

Keywords Leukemia stem cell · Cellular state · Molecular state · Heterogeneity · Bone marrow microenvironment

2.1 Introduction

Identification and characterization of the cells within a tumor that sustain the growth of the neoplastic clone is one of fundamental problems in the field of cancer research. The original stochastic model postulates that tumorigenesis occurs randomly and is governed by probabilities, and every single cell within tumor has a low but equal probability of proliferating extensively and being able to regrow the tumor. Base on this model, purification of a fraction enriched for tumor-initiating activity is not possible [1, 2]. By contrast, the cancer stem cell (CSC) model postulates that distinct classes of cells with different proliferation capacities exist in tumor, and only a rare definable population functions as CSCs that possess self-renewal and extensively proliferative potential, which are necessary to initiate a new tumor growth and regenerate the cellular hierarchy of tumors [3–5]. According to this model, it is possible to identify a cell population within a tumor that can predictably and reliably contain the capacity to regenerate the entire tumor.

Although the hypothesis of CSCs was first proposed over 50 years ago [4], the progress of CSC study has lagged until the past two decades, dependent on the development of the NOD/SCID xenotransplant assay, quantitative stem cell assays, and fluorescence-activated cell sorting technique (FACS). The first direct evidence for the existence of CSCs was derived by John Dick's laboratory in 1997 through the study of human acute myeloid leukemia (AML). Bone marrow-derived $CD34^{+}CD38^{-}$ cells were transplanted into NOD/SCID mice, and these cells could not only initiate AML in NOD/SCID mice but also differentiate in vivo into leukemic blasts [6]. More importantly, serial transplantation demonstrated that these cells have a capacity to self-renew and transfer AML disease into secondary recipients [6]. Therefore, this study showed for the first time that leukemia stem cells (LSCs) are exclusively $CD34^{+}CD38^{-}$ cells, which have the same cell-surface phenotype as normal human primitive cells. Subsequently, similar approaches were applied to solid tumors, leading to the identification of CSCs in multiple solid tumors, including the breast [7], brain [8], pancreas [9], colon [10], lung [11], and prostate [12]. Now, it is widely accepted that the existence of CSCs in cancer plays an essential role in cancer initiation, progression, and resistance to traditional chemotherapy or other treatments. Hematologic malignancies, in particular chronic myeloid leukemia (CML) and AML, have been served as important model diseases for CSC study [13, 14]. CML is a myeloproliferative disease that derives from an abnormal hematopoietic stem cell (HSC) harboring the Philadelphia (Ph$^+$) chromosome [13, 15, 16]. This chromosomal abnormality is generated by a reciprocal translocation between chromosome 9 and 22 [t (9; 22)(q34; q11)], forming chimeric *BCR-ABL* fused oncogene. BCR-ABL oncoprotein encoding by this fused gene has a constitutive tyrosine kinase activity, consequently resulting in leukemogenesis [15, 17]. AML is a heterogeneous clonal disorder characterized by clonal expansion of myeloid blasts with blocked differentiation into mature cells [18]. It is widely accepted that additional genetic alterations successively occur during the development of AML. These genetic changes aberrantly regulate signaling pathways and influence the function of transcriptional factors, subsequently determining the pathological features of AML and affecting the responsiveness of patients to chemotherapy. An increasing number of recurring genetic lesions have been identified [18].

LSCs are leukemia-initiating population with the capacity to self-renew, differentiate, and stay a quiescent state [14, 19] and have been extensively investigated in both CML and AML fields (Fig. 2.1). Similar to normal hematopoietic stem cells (HSCs), upon dividing, LSCs can decide to self-renew, differentiate, or enter dormancy. Over the last few years, emphasis in the LSC field has shifted more toward

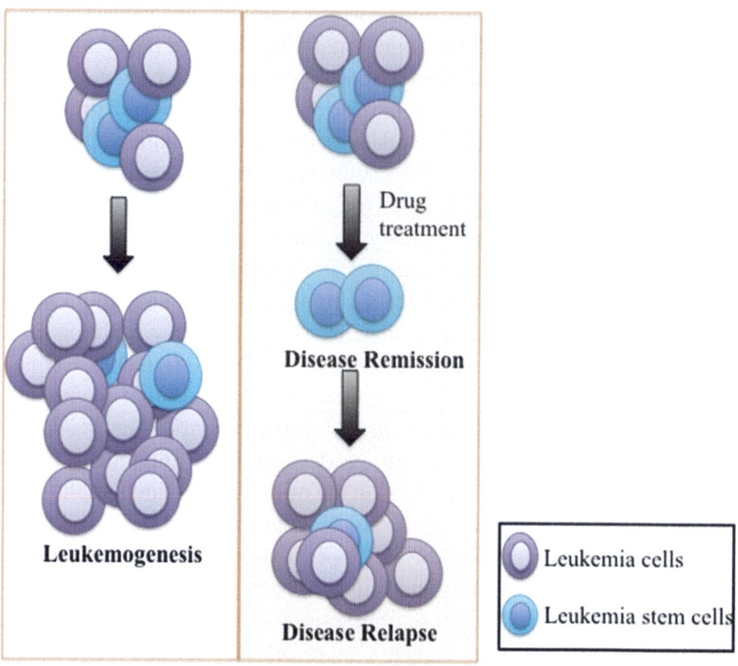

Fig. 2.1 LSCs are responsible for disease relapse. Normally, without treatment, genetic alterations such as BCR-ABL transform normal HSC and subsequently drive the expansion of leukemia cells. Drug treatment including imatinib or other TKIs blocks BCR-ABL activity and eliminates the proliferating leukemia cells. However, LSCs are resistant to TKI and survive and persistent under treatment. These LSCs lead to disease recurrence

understanding the cellular state and molecular mechanism of LSCs, including heterogeneity and drug sensitivity. Furthermore, there is increasing awareness about the critical role of microenvironment in regulating LSC survival. Here, through focusing on LSCs, we will review recent advancements in these topics, including the dynamic phenotypes of LSCs, the heterogeneity of LSCs, and the relationship between bone marrow microenvironment and LSCs.

2.2 Biological Features of LSCs

In the past decade, with self-renewal and multipotency at the hub of what defines a LSC, much research has been invested into understanding the underlying molecular and cellular processes.

2.2.1 Immunophenotype of LSCs

LSCs are generally characterized with surface immunological markers and validated by functional assays. In CML, functional LSCs in mice reside in a cell population that does not express cell lineage markers but express both c-Kit and Sca-1 ($Lin^{-}c\text{-}Kit^{+}Sca\text{-}1^{+}$, LSK) [20], recapitulating cell-surface markers expressed on normal HSCs. LSCs in human CML also reside in the HSC population [21], displaying phenotypically $Lin^{-}CD34^{+}CD38^{-}CD90^{+}$ with some specific surface markers such as IL-1RAP and CD26 [22, 23]. LSCs of human AML are also considered as $CD34^{+}CD38^{-}$ cells, which phenotypically resemble normal human primitive hematopoietic cells [6]. Recent studies identified T cell immunoglobulin mucin-3 (TIM-3) as an AML LSC marker, as TIM-3 is more highly expressed on multiple specimens of AML LSCs than on normal bone marrow HSCs [24–26]. Additional markers have been proposed, including CD123, CD47 [27], CD33 [28], CD96, CD99, CD32 [29], IL1RAP, and CD45RA [22]. It is worth to be noted that identification of patient-derived LSCs is mainly based on the establishment of xenotransplantation assay that facilitates engraftment of patient-derived cells. Thus identification and frequency of LSCs are highly dependent on the model used, which might dictate the observed biological properties of LSCs [30]. Recent study demonstrated the coexistence of at least two distinct $CD34^{+}$ LSC populations in most AML patients: a $CD34^{+}CD38^{-}$ fraction resembling normal lymphoid-primed multipotent progenitors and a $CD34^{+}CD38^{+}$ fraction resembling normal granulocyte-macrophage progenitors. Both populations have leukemic stem cell activity and are hierarchically ordered [31]. Therefore, it is reasonable to think that, with improved xenotransplant models, more immunophenotypic and functional LSCs will be identified.

2.2.2 Cellular State of LSCs

While transform normal hematopoietic stem/progenitor cells into LSCs, genetic and epigenetic alterations also change the steady cellular state of normal HSCs (Fig. 2.2) [32, 33]. It is commonly accepted that normal HSCs are largely in a state of quiescence with glycolytic, autophagy-dependent, and tightly controlled levels of protein synthesis [34–37]. While LSCs inherit common stem cell characteristics of normal HSCs, they also have some unique functional changes.

LSCs undergo unique reprogrammed cellular metabolism, a hallmark of cancers [38]. Cancer cells rely on glycolysis for energy production instead of oxidative phosphorylation (OXPHOS), a more efficient ATP-producing process. Similarly, normal HSCs also prefer to utilize glycolysis to meet their energy demands [36, 39], LSCs are uniquely reliant on OXPHOS for their maintenance and survival [40–42]. By interrogating the metabolome of human AML stem cells, recent studies found that the majority of functionally defined LSCs isolated from de novo AML patients are characterized by relatively low levels of reactive oxygen species (ROS) [41]. These ROS^{low} LSCs have higher levels of amino acids, which are metabolized in the tricarboxylic acid (TCA) cycle [40]. LSCs of AML are selectively dependent on amino acid metabolism for OXPHOS. In addition, metabolism of branched-chain

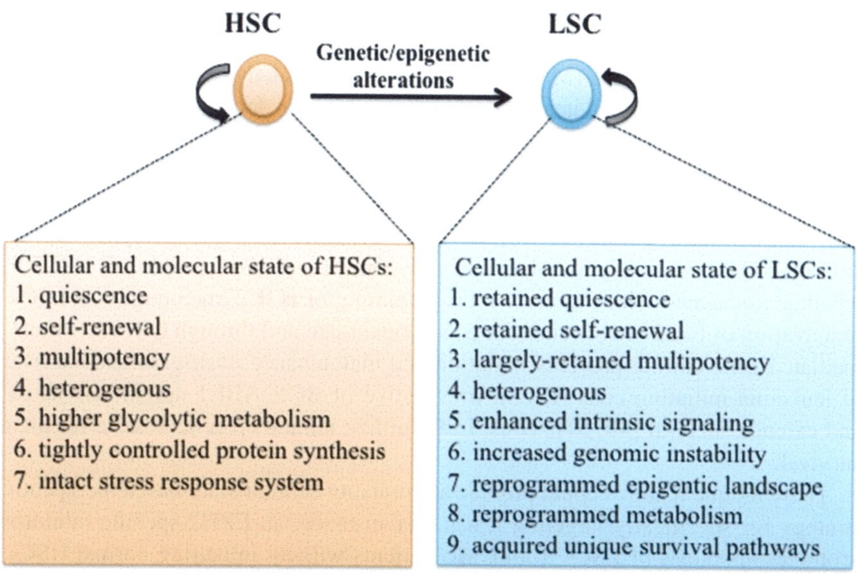

Fig. 2.2 Biological properties of normal HSCs and LSCs. Normal HSCs have three major biological features or stem cell properties: self-renewal, multipotency and quiescence. After acquiring genetic lesions, HSCs undergo cellular transformation to become LSCs that retain the major stem cell properties of HSCs with enhanced signaling activities and also acquire some unique biological features. These biological features define the cellular states of HSCs and LSCs and provide opportunities to develop strategies for specifically targeting LSCs while sparing normal HSCs

amino acids (BCAAs) is also important for LSC survival. BCAA transaminase 1 (BCAT1), a cytosolic aminotransferase for BCAAs, is activated in AML LSCs and regulates intracellular α-ketoglutarate (αKG) homeostasis by transferring α-amino groups from BCAAs to αKG. αKG is an essential cofactor for αKG-dependent dioxygenase such as Egl-9 family hypoxia-inducible factor 1 (EGLN1) and the ten-eleven translocation (TET) family of DNA demethylases [43]. Elevated BCAT1 restricts intracellular αKG and stabilizes HIF1α, which are required for LSC maintenance [44]. Similarly LSCs of CML also display higher oxidative metabolism. A metabolic analysis on both stem cell-enriched ($CD34^+$ and $CD34^+CD38^-$) and differentiated cells ($CD34^-$) derived from CML patients reveals that CML LSCs rely on upregulated oxidative metabolism for their survival [42]. LSCs show an increase in glycerol-3-phosphate, carnitine, and acylcarnitine derivatives, as well as a decrease in free fatty acid such as oleic and stearic acids. Recently, our finding also demonstrated that fatty acid metabolism is essential for the survival of CML LSCs. We found that fatty acid metabolism enzyme stearoyl-CoA desaturase-1 (Scd1) is downregulated in LSCs and plays a tumor-suppressive role in LSCs with no notable inhibitory effect on normal HSCs [45]. In addition, BCAT1 is aberrantly activated and functionally required for chronic myeloid leukemia (CML) in humans and in mouse models of CML. Blocking BCAT1 gene expression or enzymatic activity induces cellular differentiation and impairs the propagation of blast crisis CML both in vitro and in vivo. Direct supplementation with BCAAs ameliorates the defects caused by BCAT1 knockdown, indicating that BCAT1 exerts its oncogenic function through BCAA production in blast crisis CML cells. The oncogenic RNA-binding protein Musashi2 (MSI2) is physically associated with the BCAT1 transcript and positively regulates its protein expression in leukemia [46]. Together, myeloid leukemia stem cells displayed unique cellular metabolic state.

LSCs also display unique epigenetic state. Cancer cells are characterized by aberrant epigenetic landscapes and often exploit chromatin machinery to regulate gene expression programs [47, 48]. For instance, CML LSCs show higher expression of EZH2, the catalytic subunit of polycomb repressive complex 2 [49, 50], which is associated with extensive reprogramming of H3K27me3 targets in LSCs. Inactivation of Ezh2 in conventional conditional mice and through CRISPR-/Cas9-mediated gene editing prevents initiation and maintenance of disease and survival of leukemia-initiating cells (LICs), irrespective of BCR-ABL1 mutational status, and extends survival [49, 51]. Thus, LSCs utilize unique epigenetic state for their survival.

These unique features could provide a promising cellular state-based therapeutic strategy for specifically targeting LSCs. For instance, an EZH2-specific inhibitor promotes apoptosis of LSCs from CML patients without impairing normal HSCs, and combined treatment with tyrosine kinase inhibitors further potentiates these effects and results in significant loss of LSCs [50]. Additionally, targeting the unique metabolic machinery of LSCs also showed exciting effects. As ROS^{low} LSCs aberrantly overexpress BCL-2 [41], inhibition of BCL-2 with venetoclax in combination with azacitidine disrupts the tricarboxylic acid (TCA) cycle manifested by decreased α-ketoglutarate and increased succinate levels. These metabolic perturbations suppress oxidative phosphorylation (OXPHOS), which efficiently and selectively targets LSCs [40, 52]. Therefore, exploring LSC unique cellular features may result in promising therapeutic intervention that can eradicate LSCs.

2.2.3 Molecular State of LSCs

LSCs display unique state at the molecular level. Using CML as an example, the molecular evolution of CML LSCs initiates from the formation of BCR-ABL oncogene in a HSC [16]. As a result, constitutively active BCR-ABL causes uncontrolled activation of some development-related signaling pathways such as Wnt/β-catenin [53], hedgehog [54], TGFβ-FOXO [55], etc. These intrinsic genetic and signaling alterations increase the abilities of CML LSCs in self-renewal, resistance to apoptosis, and genomic instability [56]. The intrinsic molecular alterations are critical in determining the sensitivity of LSCs to drug treatment in clinic. Currently, it is accepted that the tyrosine kinase inhibitor (TKI)-refractory malignant cells in CML are LSCs [57]. Interestingly, recent studies indicate that survival of CML LSCs is independent of BCR-ABL kinase activity [58, 59], suggesting that CML stem cells utilize signaling pathways that are independent of BCR-ABL kinase activity for their maintenance and survival. Our recent studies have uncovered these signaling pathways. We show that Alox5 is upregulated by BCR-ABL, but not reduced by TKI treatment [60]. Another example is that B lymphocyte kinase (Blk) is significantly downregulated by BCR-ABL in a kinase activity-independent manner, and this pathway plays a tumor-suppressive role in regulating the survival of CML LSCs [61]. Thus, the identification of these BCR-ABL kinase activity-independent pathways may provide novel therapeutic targets for specifically eradicating LSCs.

2.2.4 Heterogeneity of LSCs

Given that clonal heterogeneity of differentiation and self-renewal properties in normal HSCs has been recognized, cellular heterogeneity of LSCs is gradually recognized. The differentiation behaviors of different HSC subtypes including myeloid-biased, lymphoid-biased, or balanced have been distinguished by measuring the contributions of individual HSCs to the circulating cell lineages in serial transplantation experiments [62, 63]. Also, clonal analysis of HSC expansion post-transplant showed that both subtypes display an extensive but variable self-renewal activity with occasional interconversion [63]. Given that HSCs are derived early from embryonic precursors, the heterogeneity of HSCs might inherit from pre-HSCs. In a recent study, the nascent pre-HSCs in the E11 mouse aorta-gonad-mesonephros (AGM) region were captured at high purity using potent surface markers and rigorously validated by single-cell-initiated serial transplantation [64]. Single-cell RNA sequencing (scRNA-seq) analysis reveals remarkable heterogeneity of pre-HSCs. Pre-HSCs show unique features in transcriptional machinery, arterial signature, metabolism state, signaling pathway, and transcription factor network [64]. Therefore, it is reasonable to consider that heterogeneity could be attributed to the inherent characterization of HSCs that is determined by the intrinsically molecular mechanisms.

LSCs may inherit this feature of normal HSCs, and display functional heterogeneity, which contributes to the different behaviors of LSCs in leukemia initiation, progression, and relapse. Residual BCR-ABL$^+$ stem cells persist but not result in recurrence in some CML patients who have maintained long-term remission even after discontinuing TKI treatment [65–67]. The lack of leukemia recurrence in these patients indicates limited leukemia-initiating potential of the residual LSCs. Normally LSCs survived post treatment have higher leukemia-initiating potential and quickly lead to CML recurrence [20, 21, 68]. The discrepancies in leukemogenic potential between these two types of LSCs can be explained by LSC heterogeneity, which might reflect unique program of LSCs determined by the intrinsically molecular machinery or extrinsically microenvironment. Consistently, using SCL-tTA/BCR-ABL mouse model of CML, a recent study demonstrates that long-term repopulation and leukemia-initiating capacity after transplantation is restricted to cells with the long-term (LT)-LSC. The leukemia-initiating potential of this CML LT-LSC is remarkably heterogeneous [69]. While LT-LSCs from leukemia mice maintain leukemia-initiating ability in serial transplantation, LT-LSCs from non-leukemic mice cannot induce leukemia [69]. However, the mechanisms underlying the heterogeneity of CML LT-LSCs are complex and still poorly understood.

Recent advances in single-cell transcriptomics are ideally placed to unravel LSC heterogeneity. Through comparing the global gene expression in LT-LSCs from leukemia and non-leukemic mice by RNA-seq, this study reveals higher expression of MPL in CML LT-LSCs, and knockdown of MPL expression reduces THPO-induced JAK/STAT signaling and diminishes growth and leukemogenic capacity of CML LT-LSCs [69]. In addition, by combining high-sensitivity detection of BCR-ABL with whole-transcriptome analysis of the same single cell, a recent study analyzed more than 2000 stem cells from CML patients throughout the disease course [70]. While revealed the heterogeneity of CML LSCs, they identified a subgroup of CML LSCs with a distinct molecular signature that selectively persisted during prolonged therapy. These drug-resistant subgroups of LSCs are more close to normal HSCs at the molecular state, showing higher expression of quiescence-related genes. However, compared to normal HSCs, drug-resistant subgroup of LSCs displays hyperactive TGFβ and TNFα signaling [70]. However, the key determinants of heterogeneous leukemia-initiating capacity and drug sensitivity of LSCs need to be further investigated.

Along with the development of in-depth genome sequencing, the true complexity of AMLs is now being understood. The heterogeneity of AMLs is multilayered and involves the genetic and epigenetic dynamics [71]. Each case of AML is a complex mosaic of cells that contain distinct genetic lesions and epigenetic aberrations [72]. Somatic mutations in the epigenetic modifiers *DNMT3A, IDH1/IDH2,* and *TET2* are initiating genetic events in AML, normally occurring in preleukemic

HSCs [73–76]. However, it seems epigenetic mutations alone are not sufficient to transform HSCs, indicating sequential acquisition of mutations is required. Therefore, the multiple levels of genetic and epigenetic heterogeneity not only endow the inherent complexity of LSCs but also challenge us understanding the biology of LSCs.

Fortunately, current technological breakthroughs significantly contribute to understanding heterogeneity of stem cells. In recent years, multiple molecular labeling tools for efficient cell tracking in vivo have been developed. For instance, the cellular barcoding offers a new clonal tracking system [77, 78]. These approaches use either lentiviruses or transposon to introduce unique barcode sequences into the genome of individual HSCs, which enables to label and track HSCs and their progeny in single mice. In the virus-mediated barcode system, cells are labeled by retroviral/lentiviral vectors with random sequence tags, which allow tracing each cell and its progeny over time. In the transposon system, a doxycycline-inducible hyperactive "Sleeping Beauty" (HSB) transposase and a single-copy non-mutagenic transposon are incorporated in the mouse genome through gene targeting [78]. Therefore, upon doxycycline administration, the transposon can randomly mobilize to a different genomic location, which creates an inheritable genomic DNA insertion and serves as a unique tag for tracking individual HSC and its progeny [78]. Therefore, these cellular tracking systems provide feasible and sensitive strategies for examining the clonality and heterogeneity of LSCs. Moreover, the technological improvements at the single-cell level have also greatly contributed to the dissection of the molecular machinery of stem cell heterogeneity [79]. Molecular profiling of the genome and transcriptome at single-cell resolution becomes feasible. For instance, in a recent study, through the single-cell transcriptome profiling of pre-granulocyte-macrophage progenitor, distinct myeloid differentiation pathways were identified, which reveal an early hematopoietic-lineage bifurcation that separates the myeloid lineages before their segregation from other hematopoietic-lineage potential [80]. Thus single-cell molecular profiling would be powerful in dissecting the heterogeneity of stem cells.

2.3 Bone Marrow Microenvironment of LSCs

It is well known that bone marrow niche is a complex ecological system, composed of multiple cell types [81–83], such as osteoblast [84], endothelial cell [85, 86], perivascular cells [64, 87], mesenchymal stem cell (MSC) [88], nonmyelinating Schwann cell [89], and megakaryocyte [90]. Osteoblasts line the surface of the endosteum on the inside of the bone cavity. Nestin$^+$ MSCs are spatially associated with HSCs and adrenergic nerve fibers and highly express HSC maintenance genes, and in vivo nestin$^+$ cell depletion rapidly reduces HSC content in the bone marrow [88]. Recent study identifies that only the Hoxb5$^+$ HSCs exhibit long-term reconstitution capacity after transplantation in primary transplant recipients and, notably, in secondary recipients and in situ imaging of mouse bone marrow, finding that the perivascular space as a near-homogenous location of LT-HSC, as majority of LT-HSCs are directly attached to VE-cadherin$^+$ cells [64]. Consistently, using a different strategy, Acar et al. found that α-catulin$^+$c-kit$^+$ cells representing HSCs are more common in the perisinusoidal niches in central marrow, and nearly all HSCs contacted leptin receptor positive (Lepr$^+$) and Cxcl12^{high} niche cells, and approximately 85% of HSCs were within 10 μm of a sinusoidal blood vessel [87]. Furthermore, nonmyelinating Schwann cells, a type of glial cells ensheathed autonomic nerves are identified in contact with a substantial proportion of HSCs and responsible for maintaining HSC hibernation [89]. Also, megakaryocytes are critical for maintaining HSC quiescence during homeostasis and promoting HSC regeneration after chemotherapeutic stress via physically associated with HSCs in the bone marrow [90]. These supporting cells secrete factors, including SCF, CXCL12, angiopoietin 1, and thrombopoietin, which promote stem cell maintenance [84, 91]. Mesenchymal stromal cell secrete HSC homing and maintenance factors. Both nonmyelinating Schwann cells and megakaryocytes are major sources of transforming growth factor β1 (TGFβ1), and activated TGFβ1/SMAD signaling is essential in maintaining HSC quiescence [89, 90]. However, in response to stress, fibroblast growth factor 1 (FGF1) signaling from megakaryocytes transiently dominates over TGF-β inhibitory signaling to stimulate HSC expansion [90]. Overall, these observations demonstrate the heterogeneous components that serve as HSC-derived niche cells to dynamically regulate HSC function (Fig. 2.3); and this dynamic nature of HSC niche might also determine the heterogeneity of stem cells.

Alterations in the bone marrow microenvironment or niche that are commonly observed in hematologic malignancies contribute to the abnormal function of LSCs [92, 93]. LSCs are presumed to reside in specific niches in the bone marrow (Fig. 2.3). Through a dynamic in vivo confocal imaging, a recent study found that the adhesion molecule E-selectin and the chemoattractant stromal-cell-derived factor 1 (SDF-1) expressed in the vasculature attract leukemia cells in the specialized vascular area of bone marrow niche [94]. Disruption of the interactions between SDF-1 and its receptor CXCR4 inhibits the homing of leukemia cells to these vessels [94], suggesting that specialized vascular structures might delineate a microenvironment with unique physiology that can be exploited by circulating malignant cells. In addition, compared to normal HSCs, LSCs use different niche signals for their engraftment. CML LSCs rely to a greater extent on selectins and their ligands for homing and engraftment than do normal stem cells. Deficiency of E-/L-selectin in the recipient bone marrow endothelium significantly reduces engraftment by LSCs of CML [95]. More interestingly, signaling from the bone marrow niche might have differential impact on different myeloid leukemias. For instance, osteoblastic cell-specific activation of the parathyroid hormone (PTH) receptor inhibits CML development but promotes MLL-AF9 oncogene-induced AML in mouse transplantation models, which is mediated by the opposing effects of increased TGF-β1 signaling on the respective LSCs, suggesting that bone marrow microenvironment of CML and AML are distinct [96]. These observations clearly indicate the abnormal bone marrow niches of leukemia.

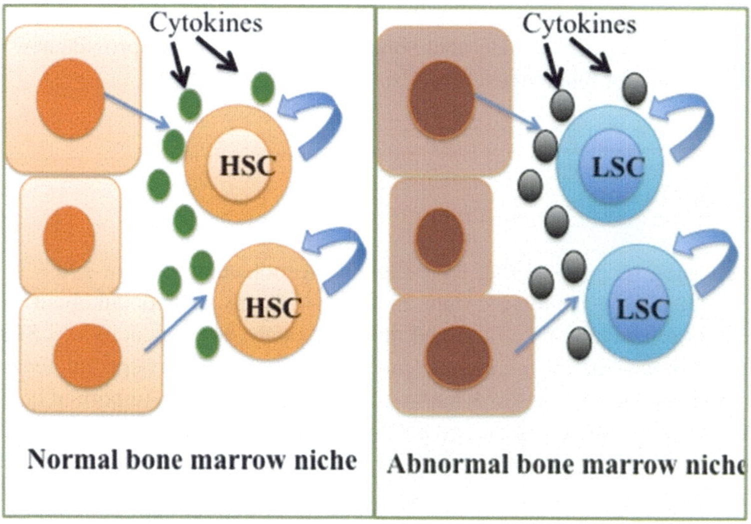

Fig. 2.3 Bone marrow microenvironment is required for LSC maintenance and survival. In both normal and abnormal condition, bone marrow niche is essential for maintaining both normal HSCs and LSCs by direct physical contact and secreting cytokines. In hematologic malignancies, genetic lesion in hematopoietic cells can remodel bone marrow niche to adopt LSC survival and maintenance. Genetic alterations occurred in the niche cells can also induce myelodysplasia and leukemia. Also compared to normal HSCs, LSCs use different niche signals for their engraftment

However, the relationship between leukemogenesis and alteration of bone marrow niche remains complex. In the one hand, bone marrow dysfunction can directly induce myelodysplasia and secondary leukemia by affecting the differentiation, proliferation, and apoptosis of heterologous cells and disrupting hematopoiesis [97]. Deletion of Dicer1 specifically in osteoprogenitors leads to myelodysplasia by reducing expression of Sbds, the gene mutated in Shwachman-Bodian-Diamond syndrome – a human bone marrow failure and leukemia predisposition condition [97]. In addition, an activating mutation of β-catenin in mouse osteoblasts alters the differentiation potential of myeloid and lymphoid progenitors and leads to development of AML by stimulating expression of the Notch ligand Jagged-1 in osteoblasts [98]. On the other hand, genetic lesions of leukemogenesis remodel the bone marrow niche. Using a xenograftment model, a recent study found that patient-derived hematopoietic cells reprogram healthy MSCs to acquire myelodysplasia syndrome (MDS) MSC-like features, which subsequently secrete cytokines to support MDS stem cells and the progeny [99]. Similarly, LSCs of CML display reduced homing and retention in the bone marrow by inhibiting CXCL12 expression of the bone marrow. This alteration also results in suppressive growth of normal HSCs but enhanced growth of LSCs [100]. Interestingly, the treatment with tyrosine kinase inhibitor, imatinib, partially corrects this abnormality of cytokine secretion, suggesting that targeting the niche might represent a new strategy for leukemia stem cells. Together, the cross talk of leukemia and bone marrow niche affects each other rather than mutually exclusive (Fig. 2.3). Thus further understanding of the differences between normal and malignant bone marrow niches might provide new therapeutic strategies for leukemia stem cells.

2.4 Conclusion

It is increasing recognized that the cellular and molecular state of LSCs is more important than their phenotypic state. In here, we discussed in detail about the biological features including cellular and molecular state, heterogeneity of LSCs, following with bone marrow microenvironment of leukemia. These specific features of LSCs provide a view about the dynamic state of LSCs, which is important in current cancer research. From the developmental perspective, understanding the molecular mechanisms of CSC maintenance and survival could uncover the process of tumorigenesis. And exploring the heterogeneity of LSCs could reveal the dynamic origin of LSCs and the mechanism of drug sensitivity of LSCs. Further investigating the relationship of LSCs with bone marrow microenvironment might enable us fully understand the ecological system of LSC and its niche. From a clinical standpoint, it is important to decipher the mechanisms of drug resistance of LSCs. Therefore, studying these different aspects of LSCs will provide novel therapeutic targets. For instance, the pathways that were identified to be essential for maintaining LSCs provide potential targets in eliminating CML LSCs [54, 101–103]. Also the recent findings of PPARγ as a novel key regulator for CML stem cell survival offer new hope for targeting CML stem cells [45, 104], although some issues such as side effects of PPARγ agonists have to be considered in clinics. Together, based on the recent scientific advances made in the LSC field, it is hopeful that we begin to understand how LSCs utilize unique cellular and molecular features to maintain their abilities of survival and self-renewal, which will lead to future clinical trials for testing new anti-LSC strategies.

Acknowledgments We are grateful to Dr. Shaoguang Li from University of Massachusetts Medical School, USA, and all the members of our laboratory for critically reading this manuscript. This work is supported by grants from the National Key Research and Development Program of China (2017YFA0505600), the National Natural Science Foundation of China (81722003, 81870124), the Wuhan Science and Technology Program for Application and Basic Research Project (2018060401011325), and the Hubei Provincial Natural Science Foundation for Creative Research Group (2018CFA018).

References

1. Till JE, McCulloch EA, Siminovitch L (1964) A stochastic model of stem cell proliferation, based on the growth of spleen Colony-forming cells. Proc Natl Acad Sci U S A 51:29–36
2. Korn AP, Henkelman RM, Ottensmeyer FP, Till JE (1973) Investigations of a stochastic model of haemopoiesis. Exp Hematol 1(6):362–375

3. Visvader JE, Lindeman GJ (2012) Cancer stem cells: current status and evolving complexities. Cell Stem Cell 10(6):717–728
4. Bruce WR, Van Der Gaag H (1963) A quantitative assay for the number of murine lymphoma cells capable of proliferation in vivo. Nature 199:79–80
5. Reya T, Morrison SJ, Clarke MF, Weissman IL (2001) Stem cells, cancer, and cancer stem cells. Nature 414(6859):105–111
6. Bonnet D, Dick JE (1997) Human acute myeloid leukemia is organized as a hierarchy that originates from a primitive hematopoietic cell. Nat Med 3(7):730–737
7. Al-Hajj M, Wicha MS, Benito-Hernandez A, Morrison SJ, Clarke MF (2003) Prospective identification of tumorigenic breast cancer cells. Proc Natl Acad Sci U S A 100(7):3983–3988
8. Singh SK, Clarke ID, Terasaki M, Bonn VE, Hawkins C, Squire J et al (2003) Identification of a cancer stem cell in human brain tumors. Cancer Res 63(18):5821–5828
9. Li C, Heidt DG, Dalerba P, Burant CF, Zhang L, Adsay V et al (2007) Identification of pancreatic cancer stem cells. Cancer Res 67(3):1030–1037
10. Ricci-Vitiani L, Lombardi DG, Pilozzi E, Biffoni M, Todaro M, Peschle C et al (2007) Identification and expansion of human colon-cancer-initiating cells. Nature 445(7123):111–115
11. Kim CF, Jackson EL, Woolfenden AE, Lawrence S, Babar I, Vogel S et al (2005) Identification of bronchioalveolar stem cells in normal lung and lung cancer. Cell 121(6):823–835
12. Goldstein AS, Huang J, Guo C, Garraway IP, Witte ON (2010) Identification of a cell of origin for human prostate cancer. Science 329(5991):568–571
13. Melo JV, Barnes DJ (2007) Chronic myeloid leukaemia as a model of disease evolution in human cancer. Nat Rev Cancer 7(6):441–453
14. Wang JC, Dick JE (2005) Cancer stem cells: lessons from leukemia. Trends Cell Biol 15(9):494–501
15. Wong S, Witte ON (2004) The BCR-ABL story: bench to bedside and back. Annu Rev Immunol 22:247–306
16. Savona M, Talpaz M (2008) Getting to the stem of chronic myeloid leukaemia. Nat Rev Cancer 8(5):341–350
17. Ren R (2005) Mechanisms of BCR-ABL in the pathogenesis of chronic myelogenous leukaemia. Nat Rev Cancer 5(3):172–183
18. Dohner H, Weisdorf DJ, Bloomfield CD (2015) Acute myeloid leukemia. N Engl J Med 373(12):1136–1152
19. Dick JE (2005) Acute myeloid leukemia stem cells. Ann N Y Acad Sci 1044:1–5
20. Hu Y, Swerdlow S, Duffy TM, Weinmann R, Lee FY, Li S (2006) Targeting multiple kinase pathways in leukemic progenitors and stem cells is essential for improved treatment of Ph+ leukemia in mice. Proc Natl Acad Sci U S A 103(45):16870–16875
21. Graham SM, Jorgensen HG, Allan E, Pearson C, Alcorn MJ, Richmond L et al (2002) Primitive, quiescent, Philadelphia-positive stem cells from patients with chronic myeloid leukemia are insensitive to STI571 in vitro. Blood 99(1):319–325
22. Wisniewski D, Affer M, Willshire J, Clarkson B (2011) Further phenotypic characterization of the primitive lineage- CD34+CD38-CD90+CD45RA- hematopoietic stem cell/progenitor cell sub-population isolated from cord blood, mobilized peripheral blood and patients with chronic myelogenous leukemia. Blood Cancer J 1(9):e36
23. Herrmann H, Sadovnik I, Cerny-Reiterer S, Rulicke T, Stefanzl G, Willmann M et al (2014) Dipeptidylpeptidase IV (CD26) defines leukemic stem cells (LSC) in chronic myeloid leukemia. Blood 123(25):3951–3962
24. Jan M, Chao MP, Cha AC, Alizadeh AA, Gentles AJ, Weissman IL et al (2011) Prospective separation of normal and leukemic stem cells based on differential expression of TIM3, a human acute myeloid leukemia stem cell marker. Proc Natl Acad Sci U S A 108(12):5009–5014
25. Kikushige Y, Miyamoto T, Yuda J, Jabbarzadeh-Tabrizi S, Shima T, Takayanagi S et al (2015) A TIM-3/Gal-9 autocrine stimulatory loop drives self-renewal of human myeloid leukemia stem cells and leukemic progression. Cell Stem Cell 17(3):341–352

26. Kikushige Y, Shima T, Takayanagi S, Urata S, Miyamoto T, Iwasaki H et al (2010) TIM-3 is a promising target to selectively kill acute myeloid leukemia stem cells. Cell Stem Cell 7(6):708–717
27. Majeti R, Chao MP, Alizadeh AA, Pang WW, Jaiswal S, Gibbs KD Jr et al (2009) CD47 is an adverse prognostic factor and therapeutic antibody target on human acute myeloid leukemia stem cells. Cell 138(2):286–299
28. Taussig DC, Pearce DJ, Simpson C, Rohatiner AZ, Lister TA, Kelly G et al (2005) Hematopoietic stem cells express multiple myeloid markers: implications for the origin and targeted therapy of acute myeloid leukemia. Blood 106(13):4086–4092
29. Saito Y, Kitamura H, Hijikata A, Tomizawa-Murasawa M, Tanaka S, Takagi S et al (2010) Identification of therapeutic targets for quiescent, chemotherapy-resistant human leukemia stem cells. Sci Transl Med 2(17):17ra9
30. Thomas D, Majeti R (2017) Biology and relevance of human acute myeloid leukemia stem cells. Blood 129(12):1577–1585
31. Goardon N, Marchi E, Atzberger A, Quek L, Schuh A, Soneji S et al (2011) Coexistence of LMPP-like and GMP-like leukemia stem cells in acute myeloid leukemia. Cancer Cell 19(1):138–152
32. Bowman RL, Busque L, Levine RL (2018) Clonal hematopoiesis and evolution to hematopoietic malignancies. Cell Stem Cell 22(2):157–170
33. Li S, Garrett-Bakelman FE, Chung SS, Sanders MA, Hricik T, Rapaport F et al (2016) Distinct evolution and dynamics of epigenetic and genetic heterogeneity in acute myeloid leukemia. Nat Med 22(7):792–799
34. Cabezas-Wallscheid N, Buettner F, Sommerkamp P, Klimmeck D, Ladel L, Thalheimer FB et al (2017) Vitamin A-retinoic acid signaling regulates hematopoietic stem cell dormancy. Cell 169(5):807–23 e19
35. Ho TT, Warr MR, Adelman ER, Lansinger OM, Flach J, Verovskaya EV et al (2017) Autophagy maintains the metabolism and function of young and old stem cells. Nature 543(7644):205–210
36. Takubo K, Nagamatsu G, Kobayashi CI, Nakamura-Ishizu A, Kobayashi H, Ikeda E et al (2013) Regulation of glycolysis by Pdk functions as a metabolic checkpoint for cell cycle quiescence in hematopoietic stem cells. Cell Stem Cell 12(1):49–61
37. Signer RA, Magee JA, Salic A, Morrison SJ (2014) Haematopoietic stem cells require a highly regulated protein synthesis rate. Nature 509(7498):49–54
38. Hanahan D, Weinberg RA (2011) Hallmarks of cancer: the next generation. Cell 144(5):646–674
39. Simsek T, Kocabas F, Zheng J, Deberardinis RJ, Mahmoud AI, Olson EN et al (2010) The distinct metabolic profile of hematopoietic stem cells reflects their location in a hypoxic niche. Cell Stem Cell 7(3):380–390
40. Jones CL, Stevens BM, D'Alessandro A, Reisz JA, Culp-Hill R, Nemkov T et al (2018) Inhibition of amino acid metabolism selectively targets human leukemia stem cells. Cancer Cell 34(5):724–40 e4
41. Lagadinou ED, Sach A, Callahan K, Rossi RM, Neering SJ, Minhajuddin M et al (2013) BCL-2 inhibition targets oxidative phosphorylation and selectively eradicates quiescent human leukemia stem cells. Cell Stem Cell 12(3):329–341
42. Kuntz EM, Baquero P, Michie AM, Dunn K, Tardito S, Holyoake TL et al (2017) Targeting mitochondrial oxidative phosphorylation eradicates therapy-resistant chronic myeloid leukemia stem cells. Nat Med 23(10):1234–1240
43. Kaelin WG Jr, McKnight SL (2013) Influence of metabolism on epigenetics and disease. Cell 153(1):56–69
44. Raffel S, Falcone M, Kneisel N, Hansson J, Wang W, Lutz C et al (2017) BCAT1 restricts alphaKG levels in AML stem cells leading to IDHmut-like DNA hypermethylation. Nature 551(7680):384–388
45. Zhang H, Li H, Ho N, Li D, Li S (2012) Scd1 plays a tumor-suppressive role in survival of leukemia stem cells and the development of chronic myeloid leukemia. Mol Cell Biol 32(10):1776–1787

46. Hattori A, Tsunoda M, Konuma T, Kobayashi M, Nagy T, Glushka J et al (2017) Cancer progression by reprogrammed BCAA metabolism in myeloid leukaemia. Nature 545(7655):500–504
47. Garraway LA, Lander ES (2013) Lessons from the cancer genome. Cell 153(1):17–37
48. Chi P, Allis CD, Wang GG (2010) Covalent histone modifications – miswritten, misinterpreted and mis-erased in human cancers. Nat Rev Cancer 10(7):457–469
49. Xie H, Peng C, Huang J, Li BE, Kim W, Smith EC et al (2016) Chronic myelogenous leukemia- initiating cells require Polycomb group protein EZH2. Cancer Discov 6(11):1237–1247
50. Scott MT, Korfi K, Saffrey P, Hopcroft LE, Kinstrie R, Pellicano F et al (2016) Epigenetic reprogramming sensitizes CML stem cells to combined EZH2 and tyrosine kinase inhibition. Cancer Discov 6(11):1248–1257
51. Zhou J, Nie D, Li J, Du X, Lu Y, Li Y et al (2018) PTEN is fundamental for elimination of leukemia stem cells mediated by GSK126 targeting EZH2 in chronic myelogenous leukemia. Clin Cancer Res 24(1):145–157
52. Pollyea DA, Stevens BM, Jones CL, Winters A, Pei S, Minhajuddin M et al (2018) Venetoclax with azacitidine disrupts energy metabolism and targets leukemia stem cells in patients with acute myeloid leukemia. Nat Med 24(12):1859–1866
53. Zhao C, Blum J, Chen A, Kwon HY, Jung SH, Cook JM et al (2007) Loss of beta-catenin impairs the renewal of normal and CML stem cells in vivo. Cancer Cell 12(6):528–541
54. Zhao C, Chen A, Jamieson CH, Fereshteh M, Abrahamsson A, Blum J et al (2009) Hedgehog signalling is essential for maintenance of cancer stem cells in myeloid leukaemia. Nature 458(7239):776–779
55. Naka K, Hoshii T, Muraguchi T, Tadokoro Y, Ooshio T, Kondo Y et al (2010) TGF-beta-FOXO signalling maintains leukaemia-initiating cells in chronic myeloid leukaemia. Nature 463(7281):676–680
56. Holyoake TL, Vetrie D (2017) The chronic myeloid leukemia stem cell: stemming the tide of persistence. Blood 129(12):1595–1606
57. Cortes J, O'Brien S, Kantarjian H (2004) Discontinuation of imatinib therapy after achieving a molecular response. Blood 104(7):2204–2205
58. Hamilton AHG, Schemionek M, Zhang B, Myssina S, Allan EK, Nicolini FE, Müller-Tidow C, Bhatia R, Brunton VG, Koschmieder S, Holyoake TL (2012) Chronic myeloid leukemia stem cells are not dependent on Bcr-Abl kinase activity for their survival. Blood 119:1501
59. Corbin AS, Agarwal A, Loriaux M, Cortes J, Deininger MW, Druker BJ (2011) Human chronic myeloid leukemia stem cells are insensitive to imatinib despite inhibition of BCR-ABL activity. J Clin Invest 121(1):396–409
60. Chen Y, Hu Y, Zhang H, Peng C, Li S (2009) Loss of the Alox5 gene impairs leukemia stem cells and prevents chronic myeloid leukemia. Nat Genet 41(7):783–792
61. Zhang H, Peng C, Hu Y, Li H, Sheng Z, Chen Y et al (2012) The Blk pathway functions as a tumor suppressor in chronic myeloid leukemia stem cells. Nat Genet 44(8):861–871
62. Dykstra B, Kent D, Bowie M, McCaffrey L, Hamilton M, Lyons K et al (2007) Long-term propagation of distinct hematopoietic differentiation programs in vivo. Cell Stem Cell 1(2):218–229
63. Benz C, Copley MR, Kent DG, Wohrer S, Cortes A, Aghaeepour N et al (2012) Hematopoietic stem cell subtypes expand differentially during development and display distinct lymphopoietic programs. Cell Stem Cell 10(3):273–283
64. Zhou F, Li X, Wang W, Zhu P, Zhou J, He W et al (2016) Tracing haematopoietic stem cell formation at single-cell resolution. Nature 533(7604):487–492
65. Mahon FX, Rea D, Guilhot J, Guilhot F, Huguet F, Nicolini F et al (2010) Discontinuation of imatinib in patients with chronic myeloid leukaemia who have maintained complete molecular remission for at least 2 years: the prospective, multicentre Stop Imatinib (STIM) trial. Lancet Oncol 11(11):1029–1035
66. Chomel JC, Bonnet ML, Sorel N, Bertrand A, Meunier MC, Fichelson S et al (2011) Leukemic stem cell persistence in chronic myeloid leukemia patients with sustained undetectable molecular residual disease. Blood 118(13):3657–3660

67. Ross DM, Branford S, Seymour JF, Schwarer AP, Arthur C, Bartley PA et al (2010) Patients with chronic myeloid leukemia who maintain a complete molecular response after stopping imatinib treatment have evidence of persistent leukemia by DNA PCR. Leukemia 24(10):1719–1724
68. Hamilton A, Helgason GV, Schemionek M, Zhang B, Myssina S, Allan EK et al (2012) Chronic myeloid leukemia stem cells are not dependent on Bcr-Abl kinase activity for their survival. Blood 119(6):1501–1510
69. Zhang B, Li L, Ho Y, Li M, Marcucci G, Tong W et al (2016) Heterogeneity of leukemia-initiating capacity of chronic myelogenous leukemia stem cells. J Clin Invest 126(3):975–991
70. Giustacchini A, Thongjuea S, Barkas N, Woll PS, Povinelli BJ, Booth CAG et al (2017) Single-cell transcriptomics uncovers distinct molecular signatures of stem cells in chronic myeloid leukemia. Nat Med 23(6):692–702
71. Li S, Mason CE, Melnick A (2016) Genetic and epigenetic heterogeneity in acute myeloid leukemia. Curr Opin Genet Dev 36:100–106
72. Paguirigan AL, Smith J, Meshinchi S, Carroll M, Maley C, Radich JP (2015) Single-cell genotyping demonstrates complex clonal diversity in acute myeloid leukemia. Sci Transl Med 7(281):281re2
73. Welch JS, Ley TJ, Link DC, Miller CA, Larson DE, Koboldt DC et al (2012) The origin and evolution of mutations in acute myeloid leukemia. Cell 150(2):264–278
74. Ding L, Ley TJ, Larson DE, Miller CA, Koboldt DC, Welch JS et al (2012) Clonal evolution in relapsed acute myeloid leukaemia revealed by whole-genome sequencing. Nature 481(7382):506–510
75. Shlush LI, Zandi S, Mitchell A, Chen WC, Brandwein JM, Gupta V et al (2014) Identification of pre-leukaemic haematopoietic stem cells in acute leukaemia. Nature 506(7488):328–333
76. Xie M, Lu C, Wang J, McLellan MD, Johnson KJ, Wendl MC et al (2014) Age-related mutations associated with clonal hematopoietic expansion and malignancies. Nat Med 20(12):1472–1478
77. Gerrits A, Dykstra B, Kalmykowa OJ, Klauke K, Verovskaya E, Broekhuis MJ et al (2010) Cellular barcoding tool for clonal analysis in the hematopoietic system. Blood 115(13):2610–2618
78. Sun J, Ramos A, Chapman B, Johnnidis JB, Le L, Ho YJ et al (2014) Clonal dynamics of native haematopoiesis. Nature 514(7522):322–327
79. Etzrodt M, Endele M, Schroeder T (2014) Quantitative single-cell approaches to stem cell research. Cell Stem Cell 15(5):546–558
80. Drissen R, Buza-Vidas N, Woll P, Thongjuea S, Gambardella A, Giustacchini A et al (2016) Distinct myeloid progenitor-differentiation pathways identified through single-cell RNA sequencing. Nat Immunol 17(6):666–676
81. Zhao M, Li L (2016) Dissecting the bone marrow hematopoietic stem cell niches. Cell Res 26:975
82. Morrison SJ, Spradling AC (2008) Stem cells and niches: mechanisms that promote stem cell maintenance throughout life. Cell 132(4):598–611
83. Wilson A, Trumpp A (2006) Bone-marrow haematopoietic-stem-cell niches. Nat Rev Immunol 6(2):93–106
84. Calvi LM, Adams GB, Weibrecht KW, Weber JM, Olson DP, Knight MC et al (2003) Osteoblastic cells regulate the haematopoietic stem cell niche. Nature 425(6960):841–846
85. Kobayashi H, Butler JM, O'Donnell R, Kobayashi M, Ding BS, Bonner B et al (2010) Angiocrine factors from Akt-activated endothelial cells balance self-renewal and differentiation of haematopoietic stem cells. Nat Cell Biol 12(11):1046–1056
86. Butler JM, Nolan DJ, Vertes EL, Varnum-Finney B, Kobayashi H, Hooper AT et al (2010) Endothelial cells are essential for the self-renewal and repopulation of Notch-dependent hematopoietic stem cells. Cell Stem Cell 6(3):251–264

87. Acar M, Kocherlakota KS, Murphy MM, Peyer JG, Oguro H, Inra CN et al (2015) Deep imaging of bone marrow shows non-dividing stem cells are mainly perisinusoidal. Nature 526(7571):126–130
88. Mendez-Ferrer S, Michurina TV, Ferraro F, Mazloom AR, Macarthur BD, Lira SA et al (2010) Mesenchymal and haematopoietic stem cells form a unique bone marrow niche. Nature 466(7308):829–834
89. Yamazaki S, Ema H, Karlsson G, Yamaguchi T, Miyoshi H, Shioda S et al (2011) Nonmyelinating Schwann cells maintain hematopoietic stem cell hibernation in the bone marrow niche. Cell 147(5):1146–1158
90. Zhao M, Perry JM, Marshall H, Venkatraman A, Qian P, He XC et al (2014) Megakaryocytes maintain homeostatic quiescence and promote post-injury regeneration of hematopoietic stem cells. Nat Med 20(11):1321–1326
91. Ding L, Saunders TL, Enikolopov G, Morrison SJ (2012) Endothelial and perivascular cells maintain haematopoietic stem cells. Nature 481(7382):457–462
92. Hoggatt J, Kfoury Y, Scadden DT (2016) Hematopoietic stem cell niche in health and disease. Annu Rev Pathol 11:555–581
93. Schepers K, Campbell TB, Passegue E (2015) Normal and leukemic stem cell niches: insights and therapeutic opportunities. Cell Stem Cell 16(3):254–267
94. Sipkins DA, Wei X, Wu JW, Runnels JM, Cote D, Means TK et al (2005) In vivo imaging of specialized bone marrow endothelial microdomains for tumour engraftment. Nature 435(7044):969–973
95. Krause DS, Lazarides K, Lewis JB, von Andrian UH, Van Etten RA (2014) Selectins and their ligands are required for homing and engraftment of BCR-ABL1+ leukemic stem cells in the bone marrow niche. Blood 123(9):1361–1371
96. Krause DS, Fulzele K, Catic A, Sun CC, Dombkowski D, Hurley MP et al (2013) Differential regulation of myeloid leukemias by the bone marrow microenvironment. Nat Med 19(11):1513–1517
97. Raaijmakers MH, Mukherjee S, Guo S, Zhang S, Kobayashi T, Schoonmaker JA et al (2010) Bone progenitor dysfunction induces myelodysplasia and secondary leukaemia. Nature 464(7290):852–857
98. Kode A, Manavalan JS, Mosialou I, Bhagat G, Rathinam CV, Luo N et al (2014) Leukaemogenesis induced by an activating beta-catenin mutation in osteoblasts. Nature 506(7487):240–244
99. Medyouf H, Mossner M, Jann JC, Nolte F, Raffel S, Herrmann C et al (2014) Myelodysplastic cells in patients reprogram mesenchymal stromal cells to establish a transplantable stem cell niche disease unit. Cell Stem Cell 14(6):824–837
100. Zhang B, Ho YW, Huang Q, Maeda T, Lin A, Lee SU et al (2012) Altered microenvironmental regulation of leukemic and normal stem cells in chronic myelogenous leukemia. Cancer Cell 21(4):577–592
101. Dierks C, Beigi R, Guo GR, Zirlik K, Stegert MR, Manley P et al (2008) Expansion of Bcr-Abl-positive leukemic stem cells is dependent on Hedgehog pathway activation. Cancer Cell 14(3):238–249
102. Irvine DA, Zhang B, Kinstrie R, Tarafdar A, Morrison H, Campbell VL et al (2016) Deregulated hedgehog pathway signaling is inhibited by the smoothened antagonist LDE225 (Sonidegib) in chronic phase chronic myeloid leukaemia. Sci Rep 6:25476
103. Irvine DA, Copland M (2012) Targeting hedgehog in hematologic malignancy. Blood 119(10):2196–2204
104. Prost S, Relouzat F, Spentchian M, Ouzegdouh Y, Saliba J, Massonnet G et al (2015) Erosion of the chronic myeloid leukaemia stem cell pool by PPARgamma agonists. Nature 525(7569):380–383

Chapter 3
Metabolic Regulations in Hematopoietic Stem Cells

Dan Huang, Chiqi Chen, Xiaoxin Hao, Hao Gu, Li Xie, Zhuo Yu, and Junke Zheng

Contents

Abstract One of the bottlenecks of the treatments for malignant hematopoietic disorders is the unavailability of sufficient amount of hematopoietic stem cells (HSCs). HSCs are considered to be originated from the aorta-gonad-mesonephros and gradually migrates into fetal liver and resides in a unique microenvironment/niche of bone marrow. Although many intrinsic and extrinsic factors (niche components) are reported to be involved in the origination, maturation, migration, and localization of HSCs at different developmental stages, the detailed molecular mechanisms still remain largely unknown. Previous studies have shown that intrinsic metabolic networks may be critical for the cell fate determinations of HSCs. For example, HSCs mainly utilize glycolysis as the main energy sources; oxidative phosphorylation is required for the homeostasis of HSCs; lipid or amino acid metabolisms may also sustain HSC stemness. Mechanistically, lots of regulatory pathways, such as MEIS1/HIF1A and PI3K/AKT/mTOR signaling, are found to fine-tune the different nutrient metabolisms and cell fate commitments of HSCs. However, more efforts are required for the optimization and establishment of precise

D. Huang · C. Chen · X. Hao · H. Gu · L. Xie · Z. Yu · J. Zheng (✉)
Key Laboratory of Cell Differentiation and Apoptosis of Chinese Ministry of Education, Shanghai Jiao Tong University School of Medicine, Shanghai, China
e-mail: zhengjunke@shsmu.edu.cn

H. Zhang, S. Li (eds.), *Leukemia Stem Cells in Hematologic Malignancies*, Advances in Experimental Medicine and Biology 1143,
https://doi.org/10.1007/978-981-13-7342-8_3

metabolic techniques specific for the HSCs with relatively rare cell frequency and understanding of the basic metabolic properties and their underlying regulatory mechanisms of different nutrients (such as glucose) during the different developmental stages of HSCs.

Keywords Hematopoietic stem cells · Metabolic regulations · Glycolysis · Oxidative phosphorylation · Stemness maintenance · Bone marrow niche

3.1 Introduction

Hematopoietic stem cells (HSCs), one of the important types of adult stem cells, mainly reside in a unique hypoxic bone marrow microenvironment (niche) [1–3] and self-renew and differentiate to all blood cells. With the development of bone marrow transplantation techniques, HSCs have been widely used in the treatments of various hematopoietic diseases, such as leukemia, lymphoma, aplastic anemia, and immune disorders. Currently, the rare HSC frequency is one of the major bottlenecks that impedes the application of HSCs for different hematopoietic disorders. Moreover, increasing evidence shows that many blood diseases are originally caused by the malfunction or malignant transformation of HSCs. Hence, understanding the intrinsic and extrinsic molecular mechanisms related to HSC self-renewal and differentiation will lead to the development of novel strategies for in vitro or in vivo HSC expansion to provide enough amounts of HSCs for the bone marrow transplantation in clinic and for the treatments of hematopoietic disorders with specific molecular targets.

HSCs originate from a precursor in the endothelial layer lining the ventral side of the dorsal aorta in the aorta-gonad-mesonephros (AGM) region, migrate into fetal liver for a dramatic expansion, and eventually move to and reside in the bone marrow. These cell fate commitments of HSCs are tightly regulated by a large number of intrinsic and extrinsic/microenvironmental factors, which control the occurrence, expansion, differentiation, homeostasis, and aging of HSCs [4–7]. Although many molecules have been identified to regulate HSC activities, it remains largely unknown how cell fate determinations are fine-tuned at different developmental stages. Intrinsic metabolic regulators are considered to play essential roles in maintaining HSC activities under both physiological and pathological situations; how different nutrient metabolisms affect HSC functions awaits for further investigations. In recent years, several lines of evidence have shown that HSCs mainly utilize glycolysis as the energy source, which is well controlled by different intrinsic or extrinsic factors [8, 9]. However, there still exist a large amount of issues to be addressed, for example, which molecules are involved in maintaining HSC glycolysis or other metabolisms, how other metabolic pathways affect HSC stemness, and whether there exist common metabolic properties of HSCs during developmental stages.

3.2 Hypoxic Bone Marrow Niches and HSC Activities

Hematopoietic cells localize in the specific bone marrow niches post birth, which has been well known to play a critical role in normal hematopoiesis. The concept of "niche" was initially proposed by Schofield R in 1978 [10]. The niche to HSC is as the earth to seed. The components/qualities of the "earth" directly influence the life cycle of the "seed" in the bone marrow. Recently, Khan JA et al. report that there also exists a HSC niche in fetal liver, which is mainly composed by Nestin$^+$NG2$^+$ pericyte cells near portal vessels [11]. The components of HSC niches may contain vascular cells, stromal cells, nerve cells, certain types of hematopoietic cells (such as megakaryocyte), extracellular matrix, and other connective tissues. Regarding that HSCs can produce all the blood cells, the hematopoietic niche is usually referred as HSC niche. Two major types of bone marrow niches have been defined including the endosteal niche and perivascular niche (Fig. 3.1). Quiescent HSCs previously have been reported to reside in the endosteal niche formed by osteoblasts, while cycling HSCs localize close to the sinusoid or arteriole orchestrated by endothelial cells, termed as a perivascular niche. Interestingly, several studies also suggest that these two types of niches are tightly connected to each other to comprise the functional niche networks. Recently, Ding et al. provide intriguing evidence showing that HSCs mainly reside in a perivascular niche, while early lymphoid progenitors localize in an endosteal niche based on the extensive analysis with a *Cxcl12* conditional knockout mouse line [12]. These results indicate HSCs and their derived progenitors may have the distinct distributions in bone marrow niches. Recent studies also imply that both dormant HSCs and proliferating HSCs can simultaneously exist in the perivascular niche. Moreover, other niches

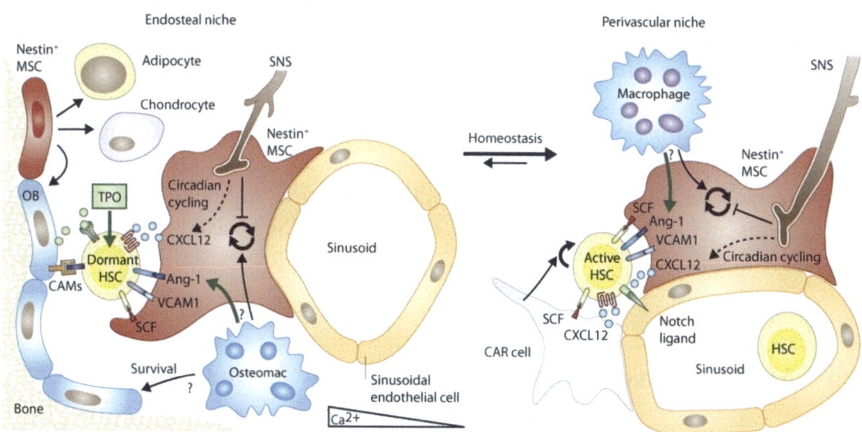

Fig. 3.1 Components of bone marrow niches (cited from Armin E et al. JEM, 2011, 421–428). Dormant HSCs mainly localize in the endosteal niche comprised by the osteoblasts (OB), while active HSC prefers residing close to the sinusoid (perivascular niche) constituted by endothelial cells. Other bone marrow niches formed by distinct stromal cells, including CAR cells, nestin$^+$ MSC, or sympathetic nerve system (SNS), are also indicated

comprised by stromal cells or hematopoietic cells, such as mesenchymal stem cells (MSC), adipocyte and megakaryocyte, are also identified to be able to functionally support HSC activities [13–17].

Bone marrow niches are considered as hypoxic niches and have relatively low oxygen tension due to the unique anatomic structure [18]. Meanwhile, the bone marrow is filled with different categories of hematopoietic cells, which leads to a much high level of total oxygen consumption. There is approximately tenfold decrease of oxygen tension in the region that is several cellular diameters away from capillaries as indicated by the oxygen diffusion model [3]. The average pO_2 in the bone marrow is around 55 mmHg, and the average oxygen saturation is around 87.5% [19]. These findings indicate that HSCs mainly localize in a relatively hypoxic niche. Previous studies showed most of dormant and primitive HSCs exist in the bone marrow niche region with a relative slow blood flow. For example, pimonidazole (a hypoxia indicator) or Hoechst can mark HSCs in the hypoxic region [20, 21]. Hypoxic condition (1–3% oxygen tension) can efficiently promote the differentiation of HSCs to megakaryocytes, erythrocytes, granulocytes, or monocyte progenitors and enhance the in vitro expansion and in vivo engraftment of HSCs [22–24]. Therefore, hypoxic conditions may contribute to the maintenance of HSC stemness.

Bone marrow niche components (such as osteoblasts, macrophages, endothelial cells, Nestin$^+$ MSCs) produce chemokines, growth factors, and adhesion molecules to support HSC activities in the bone marrow. Miharada et al. show that LT-HSCs express high level of the surface receptor, GPR78, to sustain HSCs in the endosteal niche and a relatively high level of glycolysis but low mitochondrial potential. Inhibition of GPR78 in HSCs results in a notable migration out of the endosteal niche [25]. However, it is still not clear where the physical position hypoxic area in the bone marrow is. Recent studies also indicate that the perivascular niche may be more hypoxic compared to that of the endosteal niche [26]. It is conceivable that hypoxic niche may afford a very narrow window of oxygen tension to maintain HSC activities as evidenced that certain hypoxia inducible factor 1 alpha (Hif-1α) level is essential for HSC functions [27]. In summary, these studies suggest that HSCs can tolerate the hypoxic microenvironment with the low oxygen tension in the bone marrow to remodel their metabolic properties. Although the underlying mechanisms are not fully understood, it implies that HSCs have unique metabolic characteristics.

3.3 Metabolisms of HSCs Under Hypoxic Niches

In differentiated cells, oxidative phosphorylation is the major source of ATP under aerobic conditions, which generates ~18-fold more ATP amount than that derived from the glycolysis under anaerobic situations. Oxygen is the final carrier of the electron transport/transfer chain. The proton gradient will be collapsed, followed by the dramatic decrease of ATP production if oxygen supplementation is decreased or inhibited. Under hypoxic conditions, ATP level is sustained through the cytosolic glycolysis, although only 2 mol ATP is generated from 1 mol glucose and the intermediate metabolite of pyruvate is further converted into lactate to supply NAD^+ for the driving of glycolysis. In 1861, Louis Pasteur first reported oxygen tension could regulate the rate of glucose fermentation. Thereafter, glycolysis under the anaerobic condition is as termed as "Pasteur effect". Glycolytic rate is well regulated by the expression levels and activities of various rate-limiting enzymes. Although hypoxic condition may enhance the energy production derived from glycolysis, it is less important than the fact that most types of cells can downregulate their energy demands under such stress conditions. This phenomenon indicates that it is essential to shut down unnecessary biological activities and functional processes to reserve certain levels of energy, which is also termed as "turning off the pilot light" as suggested by the comparative physiologist Kjell [28, 29]. Studies show that HSCs are metabolically quiescent to maintain their stemness under such "pilot light" condition and satisfied with the limited energy generated from glycolysis. This phenomenon is also consistent with the fact that HSCs are usually dominantly quiescent in a homeostatic microenvironment [29].

Mitochondrial respiration is considered as the major source of reactive oxygen species (ROS). ROS is mainly generated from the electron leaking, which further converts the oxygen to superoxide anion. Although superoxide anion is not a strong oxidant, it serves as the precursor of most other oxygen radicals, such as hydroxyl radical and nitrogen peroxide. ROS may lead to a marked consumption of natural antioxidants and multiple aspects of cellular damage. ROS has been reported to be tightly connected with aging and many degenerative disorders, including HSC dysfunction and aging. In fact, functional HSCs contain relatively low level of ROS, and elevated ROS level impairs HSC self-renewal capacities and quiescent status [30]. For example, loss of function of ATM and FOXOs results in the marked impairment in HSC activities [31, 32]. Therefore, we speculate that the enhanced glycolytic level of HSCs not only protects them from hypoxic stress but also avoids the ROS-mediated oxidative stress from mitochondrial respiration.

In the past decades, due to the limitation of the availabilities of HSC amount, HSC metabolic properties have not been well characterized by using conventional techniques. However, several studies have shown that HSCs may utilize glycolysis as the main energy sources. For example, HSCs with low rhodamine-123 staining (a mitochondria dye) have much higher colony-forming capacities than that of counterparts [33]; mathematics model suggests that HSCs localize in a hypoxic region of bone marrow [3]; more pimonidazole (a hypoxic marker) tends to be incorporated in HSCs [1, 8, 27]; HSCs stably express high level of Hif-1α [8, 27]; injection with the hypoxic toxic regent of tirapazamine into recipient results in the depletion of HSCs [1]. By using proteomics analysis and subsequent comparison of 145 identified differentially expressed proteins, Unwin et al. show that LSK^+ cells (enriched for HSC population) prefer to adopt glycolysis, rather than oxidative phosphorylation, for their energy source, which is opposite to that in LSK^-differentiated hematopoietic cells. Consistently, proteomics analyses from other groups further indicate that more glycolysis related pathways are involved in the HSC population than the differentiated cells [34]. We also demonstrate that Meis1 (myeloid ecotropic viral integration site 1) transactivates Hif-1α to fine-tune the glycolytic level in both murine and human HSCs, which suggests that metabolic pattern may not only be the consequence regulated by the extrinsic components of bone marrow niches but also be controlled by many intrinsic factors [8].

3.4 Mechanisms Involved in HSC Glycolysis

Hif-1α has been known as the one of the main regulators to enhance the glycolysis under hypoxic conditions, which is quite unstable and promptly undergoes degradation under normoxic situations. Hif-1α is a master regulator that controls the expressions of many downstream targets involved in oxidative stress, glycolysis, and oxidative phosphorylation. Although Hif-1α collaborates with Hif-1β to form the heterodimer and highly exists in differentiated cells under hypoxic conditions, Hif-1α seems to be stably expressed in HSCs even under normoxic conditions. The stability of Hif-1α is tightly controlled by several well-defined signaling pathways. For example, VHL can maintain Hif-1α stability through its mediated deubiquitination pathway [35]; energy sensor AMPK-mediated and redox sensor SIRT1-mediated signaling pathways are important for the certain Hif-1α level [36]; aryl hydrocarbon receptor AhR can form a nuclear localization complex of AhR/aryl to inhibit the ARNT's binding with Hif-1α to reduce its protein level, which further leads to the notion that AhR antagonist may be capable to enhance the ex vivo expansion of HSCs [37]; our previous studies also reveal that cytoskeleton protein PFN1 sustains Hif-1α level through EGR1/Gα13 pathways to regulate the glycolytic status and homeostasis of HSCs in bone marrow niches [38]. All these regulatory networks are critical for the maintenance of HSC functions under hypoxic conditions and prevent the impairment resulting from the oxidative phosphorylation.

Meis1 is one of the HOX family members and evolutionarily conserved DNA binding transcriptional factor. Meis1 is highly expressed in HSCs at different developmental stages and significantly reduced upon differentiation [39]. Meis1deletion in mice is embryonic lethal due to the defects in hematopoiesis and vasculogenesis. So far, the functions of Meis1 are still not fully understood. Our previous studies suggest that Meis1 can transcriptionally activate Hif-1α expression [8]. Moreover, Meis1 regulates HSC glycolysis and self-renewal through Hif-2α/ROS/p16 pathway (Fig. 3.2) [9]. Deletion of Meis1 results in the notable downregulation of Hif-1α and Hif-2α, enhanced ROS level, and loss of HSC quiescence. ROS scavenger NAC can rescue the loss function of Meis1-null HSCs. Interestingly, mobilized human peripheral blood HSCs also adopt glycolysis to obtain energy through Meis1/Hif-1α/Pbx1/HoxA9 pathways [40].

Many Hif-1α targeted genes have been found to play important roles in HSC function, including glycolysis related genes (GLUT1, LDHA, PKM2, PFKL, PDK2, and Cripto). Hif-1α can directly enhance the activity of pyruvate dehydrogenase kinase (PDK) to suppress the entry of pyruvate into tricarboxylic acid cycle

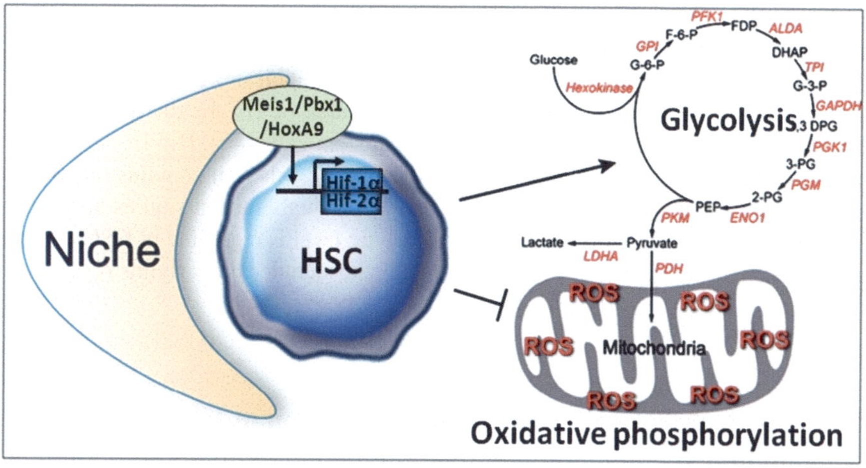

Fig. 3.2 HSCs mainly utilize glycolysis as the energy source. HSCs localize in a unique bone marrow niche and tend to utilize glycolysis, but not oxidative phosphorylation, as main energy sources, which is fine-tuned by Meis1/Pbx1/HoxA9/Hif-1α/Hif-2α at transcriptional level

(TCA cycle) [41]. Wang et al. demonstrate that the deletion of LDHA or PKM2 may lead to an impaired HSC reconstitution ability [42]. Long-term (LT) HSCs highly express Hif-1α, which stably sustains in the migrating HSCs from hypoxic bone marrow to peripheral blood [40, 43]. Takubo et al. show that Hif-1α-null adult HSCs have reduced repopulation capacities and lose quiescent cell cycle status, indicating that appropriate Hif-1α level is critical for sustaining HSC pool [27]. Mutation in the promoter binding sites of Hif-1α of vessel endothelial growth factor impairs HSC reconstitution abilities [44]. Conditional deletion of Hif-1α causes the swift from glycolysis to oxidative phosphorylation in HSCs and the upregulation of Hif-2α that may result from the compensation effect [9]. However, a recent study shows some controversial findings that loss of Hif-1α does not affect HSC stemness [45], although the authors cannot completely exclude the effect of Hif-1α from bone marrow niches.

Other downstream targets, such as VEGF, ADM, SDF1, SCF, ANG2, ANGPTLs, IGF-2, P21, and FOXOs, are also required for sustaining the quiescent status of HSCs. Interestingly, several Hif-1α targets (SDF1, SCF, ANG2, ANGPTLs) are important niche components in regulation of HSC stemness, indicating Hif-1α may regulate HSC functions through both autonomous and non-autonomous manners. Meanwhile, Hif-1α can crosstalk with Notch and Wnt signaling, two crucial pathways for HSC activities, to determine the cell fates of HSCs. Hif-1α is interacted with the intracellular domain of Notch (NICD) to sustain its stability to promote the transactivation of downstream target genes [46, 47]. Interplaying between Hif-1α and β-catenin can reduce the expressions of certain Wnt-mediated target genes while it enhances Hif-1α-targeted gene expressions [47]. This crosstalk suggests the complicated relationships between HSC metabolisms and stemness under hypoxic conditions.

In addition, Hif-1α is reported to be involved in normal hematopoiesis at different developmental stages. Hif-1α deletion in mice is embryonic lethal due to many defects in the neurogenesis, vasculogenesis, or hematopoiesis. The mRNA expression profiles of HSCs share many common features of Hif-1α distribution with human embryonic stem cells and mesenchymal stem cells [48]. Hif-1α deletion in mice causes a significant reduction in size of yolk sac, where it contains much less hematopoietic cells than the control ones [49]. Although current studies have provided intriguing evidence showing how glycolytic level is regulated in HSCs, more efforts are required for the delineation of all the detailed underlying regulatory networks.

3.5 The Roles of Mitochondrial Respiration in HSC Stemness

Although glycolysis is currently considered as the major metabolic manner of HSCs, increasing evidence shows that HSCs contain certain amount of mitochondria and oxidative phosphorylation may also contribute to HSC activities, especially in HSCs at different developmental stages or differentiated states. Bing L and his colleagues recently demonstrate that pre-HSCs originating from AGM or fetal liver HSCs seemingly have markedly higher oxidative phosphorylation level, which is different from that of adult HSCs [39]. It is possible that the metabolic status may dramatically change when dormant HSCs enter into cell cycle for the subsequent expansion, since cycling HSCs need increased demands of energy supply and macromolecular sources to generate amino acids, lipids, and nucleotides. Many of the macromolecular components are the intermediate metabolites derived from TCA cycle, indicating that mitochondrial oxidative phosphorylation lays the foundation for cell fate commitments of HSCs. Although several studies show most of mitochondria in HSCs may be inactive and mainly serve as an energy reserve under certain stress states [50], recent studies suggest that adult HSCs have high mitochondrial mass and enhanced dye-efflux abilities yet possess limited respiratory and turnover capacities [51]. Mitochondrial respiration may also be important for HSC activities [52, 53]. Although glycolysis can quickly generate ATP and to supply the energy for the biomacromolecules synthesis, the fast cell proliferation also heavily relies on many intermediate metabolites derived from TCA cycle. Therefore, slowdown of TCA cycle via the decrease of substance supply and oxygen uptake may be another efficient way to determinate HSC fates.

Mitochondrial respiration is tightly regulated by several key factors, such as LKB1 and PGC-1 [54]. Loss of LKB1 severely impairs HSC function and leads to the phenotype similar to that of Hif-1α-null HSCs, including the loss of quiescence, increase of proliferation of HSCs/progenitors, and decrease in mitochondrial mass and functions. PGC-1, including PGC-1α and PGC-1β, is the agonist of PPARγ.

PGC-1α is transcriptionally regulated by c-Myc and can significantly enhance the expression levels of several antioxidants critical for the ROS detoxication. c-Myc-deficient HSCs still can survive and expand, but not undergo subsequent differentiation, indicating that c-Myc is required for cycling progenitors instead of quiescent HSCs [55]. The E3 ligase Fbxw7 level is upregulated under hypoxic condition, which can further inhibit the expression of c-Myc to regulate cell cycle of HSCs [56]. Deletion of guanine nucleotide exchange factor binding protein 5 reduces mitochondrial potential and causes the loss of quiescence of HSCs [57]. Mitochondrial malfunction may enhance the ROS production, which further impairs HSC stemness. However, it remains unknown whether HSCs with mitochondrial malfunction have abnormal glycolysis and whether there exists crosstalk between mitochondrial regulatory pathways and Hif-1α. More efforts are required for the understanding whether or how oxidative phosphorylation plays a role in the cell fate determinations of HSCs.

Occurrence of mutations in key catalytic enzymes in TCA cycle may result in the malignant transformation of HSCs. One of the famous examples is that the mutations of IDH1 and IDH2 lead to the generation of onco-metabolite, 2-hydroxyglutarate, and initiation of the acute myeloid leukemia (AML) [58]. IDH1 and IDH2 localize in cytoplasm and mitochondria, respectively, and convert isocitric acid into α-ketoglutaric acid. Instead, isocitric acid is catalyzed into 2-hydroxyglutarate in IDH1-/IDH2-mutant cells [59]. α-Ketoglutaric acid is critical for the activities of many oxygenases. For example, TET2 catalyzes 5mC into 5hmC to initiate the DNA demethylation in α-ketoglutaric acid-dependent manner [60]. Loss of TET2 promotes the abnormal self-renewal of HSCs and myeloid-biased proliferation, which eventually leads to the development of myelodysplastic syndrome (MDS) or AML. These results indicate that IDH mutation phenocopies the loss function of TET2 and cytosine methylation acts as a common epigenetic regulation to affect HSC activities.

3.6 Reactive Oxygen Species and HSC Functions

Reactive oxygen species (ROS) is mainly derived from mitochondrial phosphorylation and has been reported to play important roles in cell fate determinations under both physiological and pathological activities. Certain level of ROS is required for sustaining physiological functions of various types of cells. Nevertheless, abnormally accumulated ROS results in the unbalance of redox and oxidative stress, which leads to the oxidation of nuclear acid, lipid, protein, and carbohydrate, aging, cell apoptosis, and malignant transformation of different cell types. Mitochondrion is the major organelle to generate ROS. ROS mainly contains superoxide, hydrogen peroxide, and oxygen radial. Superoxide is produced through the interaction between uncoupled electron released from the electron transfer chain (such as complex I and III, also termed as electron leaking) and free oxygen during mitochondrial respiration, which is subsequently catalyzed into hydrogen peroxide by superoxide dismutase. Hydrogen peroxide can be further converted into toxic hydroxyl radical or be eliminated by glutathione, peroxidase, and catalase.

As mentioned above, HSCs mainly utilize glycolysis to obtain energy and prevent the impairment derived from the overloading of ROS since HSCs are more sensitive in response to oxidative stress than the progenitors. Long-term exposure to high level of ROS causes the stemness loss or cell death of HSCs. By using DCFDA, a ROS indictor for live cells, HSCs can be divided into DCFDA^{high} and DCFDA^{low} cell populations. Subsequent in vivo transplantation analysis further shows that DCFDA^{low} cell population contributes to much higher donor repopulation than the DCFDA^{high} one, which prefers to a myeloid differentiation [30]. Because the skewness of myeloid differentiation is considered as one of the properties of aging, high level of ROS may serve as a driver in HSC aging. Interestingly, the treatments with either antioxidant or p38 MAPK inhibitor can rescue the in vitro colony-forming ability of DCFDA^{high} cells affected by the overload of ROS. These data suggest that both ROS level and exposure duration may affect the self-renewal and differentiation abilities of HSCs. However, due to the technique limitations, ROS level is usually evaluated by chemiluminescence analysis, which may not fully reflect the redox state of HSCs. It is urgent to develop novel technologies or strategies to precisely and sensitively monitor the subtle and dynamic changes of ROS level in HSCs.

Many studies have shown that ROS level is fine-tuned in different cell types through several pathways. For example, (1) polycomb complex plays an important role in histone modification to maintain ROS level. Bmi1 has been reported to be highly enriched in HSCs and maintains self-renewal abilities of HSCs by downregulating the levels of ROS, as well as p16^{Ink4a}/p19^{Arf} [61]. Bmi1 deletion results in mitochondrial dysfunction with decreased ATP production and elevated ROS level and increased DNA damage in HSCs. (2) DNA damage response is another important pathway to sustain appropriate ROS level and genome stability. ATM/p53 signaling pathways have been reported to be critical for the DNA damage repair [62], and loss of ATM impedes normal hematopoiesis, which may result from the enhanced ROS level and deficiency in ATM-mediated DNA damage response and repair. (3) FOXOs signaling pathway is also involved in the regulation of redox state of HSCs [63]. FOXO3a interplays with ATM to suppress ROS production and activates DDR. FOXO3a-deficient HSCs have markedly higher ROS level and decreased self-renewal capacities. (4) HO-1 serves as a stress-dependent antioxidase to sustain the homeostasis of redox state in HSCs. Nrf2 and Keap1 control HO-1 expression at transcriptional level to reduce the ROS production to maintain HSC activities, and HO-1 deletion leads to a dramatic decrease in repopulation abilities of HSCs [64].

3.7 Other Metabolic Signalings in Cell Fate Commitments of HSCs

Many types of stem cells can sense distinct growth signalings or different nutrient levels to activate PI3K/AKT/mTOR pathways to enhance their survival or proliferation [65]. However, constitutive activation of AKT signaling leads to the overexpansion and exhaustion of HSCs. Although AKT1 or AKT2 deletion does not affect abnormal hematopoiesis, AKT1/AKT2 double knockout HSCs cannot exit from G0 and enter cell cycle for further differentiation [66]. In contrast, abnormal activation of PI3K/AKT resulting from PTEN deletion promotes HSCs to enter into cell cycle and eventually causes the HSC exhaustion, which can be rescued by the depletion of p53 or p16 [67]. Conditional knockout of mTOR upstream suppressor TSC1 results in the overactivation of mTOR, which further leads to the notable increased mitochondrial mass and ROS level, and loss of HSC dormancy [68]. These studies indicate PTEN/TSC1/mTOR signaling pathways are crucial for the maintenance of HSC stemness.

Recently, studies also show that there exists AKT/mTOR-independent signaling in the regulation of HSC metabolisms. Tumor suppressor LKB1 can sense the changes of AMP/ATP ratio to regulate AMPK activity, and phosphorylation of AMPK can inhibit mTOR signaling to suppress cell proliferation, indicating LKB1/AMPK regulates cell metabolisms through the subtle changes of ATP level. LKB1-null mice have increased numbers of hematopoietic progenitor cells but compromised HSC self-renewal capacities. However, loss of function of LKB1 in HSCs cannot be rescued by the activation of AMPK signaling, indicating LKB1 maintains HSC pool in an AMPK-independent manner. In fact, LKB1 is required for the normal mitochondrial function to maintain HSC stemness through PGC-1α pathway [54]. Moreover, by using HPLC-MS analysis, Gurumurthy et al. also reveal that LKB1 may restrain the mitochondrial activities to strengthen the lipid metabolism of Lin^- stem/progenitor cells, indicating LKB1 can serve as a sensitive rheostat between anabolisms and catabolisms [69].

3.8 Metabolic Regulations in HSCs at Different Developmental Stages

Increasing evidence indicates that the HSC metabolisms are dynamically changed at different developmental stages. HSCs may have unique metabolic features and molecular signatures during the process of initiation, proliferation, differentiation, homeostasis, and aging [2, 70]. The metabolic dynamics may be tightly controlled by a large number of intrinsic regulators and niche components [39]. For example, the metabolic manner may be different between fetal liver HSCs and adult HSCs

[39]; Wip1 coordinates with p53 and mTORC1 signaling pathways to slowdown HSC aging [71, 72]; cytosolic glycolysis is switched to oxidative phosphorylation, while quiescent HSCs enter cell cycle and differentiate to downstream progenitors or mature hematopoietic cells. Actually, HSC differentiation heavily relies on the level of the mitochondrial respiration as evidenced by the fact that PTPMT1 deletion hinders the differentiation of HSCs and leads to the rapid failure of hematopoiesis [73, 74].

Current studies in HSC metabolisms mainly focus on the glycolysis and oxidative phosphorylation of the glucose metabolism in adult HSCs. Very rare findings related to other metabolic pathways (such as pentose phosphate pathway, lipid metabolism, and amino acid metabolism) and metabolic properties in different developmental stages of HSCs (including HSCs from AGM, fetal liver, and aging stages) have been precisely or systemically evaluated. However, different nutrient metabolisms may also be important for the cell fate commitments of HSCs. For example, Ito et al. find that PML-PPARδ-FAO pathways mediate lipid acid oxidation to preserve the capabilities of self-renewal and asymmetric divisions of HSCs [75]; Signer et al. demonstrate that the protein synthesis rate also influences HSC functions [76]; Yuki et al. and our group show that branched-chain amino acids are critical for the ex vivo expansion and homeostasis of HSCs [77, 78]; Nina et al. report vitamin A-retinoic acid signaling regulates HSC dormancy [79]; many other regulators, such as LKB1, Wnt, FOXOs, c-Myc, OCT1, AKT, and PTEN, may also participate in the metabolisms of various nutrients [54, 80–85]. To precisely dissect these complicated metabolic regulatory networks in HSCs, it is extremely urgent to develop novel and specific tools or techniques for differential analyses of each metabolic pathway.

Owing to the limitations of HSC numbers, routine assays or tools cannot be applied for the analyses of metabolic features of HSCs. In recent years, many improved techniques are being developed to facilitate the detections of different metabolic status of HSCs. For example, glycolysis and oxidative phosphorylation in HSCs can be measured and indicated as extracellular acidification rate (ECAR) and oxygen consumption rate (OCAR), respectively, by a Seahorse XF analyzer [75, 86]. Seahorse XF analyzer has also been used for the evaluation of lipid metabolisms, which may benefit for further unraveling other metabolic characteristics of HSCs with rare frequencies. LC-MS has been developed and improved to detect various intermediate metabolites with reduced cell amounts, which makes it possible to conduct metabonomic analysis in HSCs. Recent studies provide interesting evidence showing that it is feasible to perform a metabolomic analysis with less than 10^4 HSCs to explore the metabolic networks of different types of nutrients [87]. Meanwhile, anabolism or catabolism is a bioreaction of redox in nature in cells, accompanying with electron transfer between different substrates mediated by different coenzymes, such as nicotinamide adenine dinucleotide (NAD+) and nicotinamide adenine dinucleotide phosphate (NADP+), and energy formation and release. Therefore, it may be able to accurately monitor the subtle metabolic changes by the sensors that can indicate the ratio changes of NAD+/NADH or NADP+/NADPH as reported by recent studies [88].

3.9 Summary

Due to the rare frequencies and the complicated connections among various nutrients (such as glucose, lipid, and amino acid), it remains largely unknown in the metabolic properties and their underlying regulatory mechanisms of HSCs at different developmental stages. Comprehensive and specific tools/techniques are required for depicturing all the detailed metabolic profiles of HSCs in the near future, including HPLC-MS, single-cell sequencing, metabonomics, genetic murine models, and definitive functional assays. Stem cell metabolisms have become one of the leading scientific fields theses years due to many newly developed techniques and insights related to how intrinsic metabolisms regulate their cell fate determinations. The studies in metabolisms of HSCs or other types of stem cells will open a new avenue to understand the stem cell biology and its related pathogenesis, which may provide another angle to the development of novel strategies for the treatments of hematopoietic disorders or other diseases.

References

1. Parmar K et al (2007) Distribution of hematopoietic stem cells in the bone marrow according to regional hypoxia. Proc Natl Acad Sci U S A 104(13):5431–5436
2. Suda T, Takubo K, Semenza GL (2011) Metabolic regulation of hematopoietic stem cells in the hypoxic niche. Cell Stem Cell 9(4):298–310
3. Chow DC et al (2001) Modeling pO(2) distributions in the bone marrow hematopoietic compartment. II. Modified Kroghian models. Biophys J 81(2):685–696
4. Bryder D, Rossi DJ, Weissman IL (2006) Hematopoietic stem cells: the paradigmatic tissue-specific stem cell. Am J Pathol 169(2):338–346
5. Blank U, Karlsson G, Karlsson S (2008) Signaling pathways governing stem-cell fate. Blood 111(2):492–503
6. Zhang CC, Lodish HF (2008) Cytokines regulating hematopoietic stem cell function. Curr Opin Hematol 15(4):307–311
7. Zhao M, Li L (2015) Regulation of hematopoietic stem cells in the niche. Sci China Life Sci 58(12):1209–1215
8. Simsek T et al (2010) The distinct metabolic profile of hematopoietic stem cells reflects their location in a hypoxic niche. Cell Stem Cell 7(3):380–390
9. Kocabas F et al (2012) Meis1 regulates the metabolic phenotype and oxidant defense of hematopoietic stem cells. Blood 120(25):4963–4972
10. Schofield R (1978) The relationship between the spleen colony-forming cell and the haemopoietic stem cell. Blood Cells 4(1–2):7–25
11. Khan JA et al (2016) Fetal liver hematopoietic stem cell niches associate with portal vessels. Science 351(6269):176–180
12. Ding L et al (2012) Endothelial and perivascular cells maintain haematopoietic stem cells. Nature 481(7382):457–462
13. Zhou BO et al (2014) Leptin-receptor-expressing mesenchymal stromal cells represent the main source of bone formed by adult bone marrow. Cell Stem Cell 15(2):154–168
14. Mendez-Ferrer S et al (2010) Mesenchymal and haematopoietic stem cells form a unique bone marrow niche. Nature 466(7308):829–834

15. Bruns I et al (2014) Megakaryocytes regulate hematopoietic stem cell quiescence through CXCL4 secretion. Nat Med 20(11):1315–1320
16. Zhao M et al (2014) Megakaryocytes maintain homeostatic quiescence and promote post-injury regeneration of hematopoietic stem cells. Nat Med 20(11):1321–1326
17. Naveiras O et al (2009) Bone-marrow adipocytes as negative regulators of the haematopoietic microenvironment. Nature 460(7252):259–263
18. Eliasson P, Jonsson JI (2010) The hematopoietic stem cell niche: low in oxygen but a nice place to be. J Cell Physiol 222(1):17–22
19. Harrison JS et al (2002) Oxygen saturation in the bone marrow of healthy volunteers. Blood 99(1):394
20. Kubota Y, Takubo K, Suda T (2008) Bone marrow long label-retaining cells reside in the sinusoidal hypoxic niche. Biochem Biophys Res Commun 366(2):335–339
21. Lo Celso C, Wu JW, Lin CP (2009) In vivo imaging of hematopoietic stem cells and their microenvironment. J Biophoton 2(11):619–631
22. Katahira J, Mizoguchi H (1987) Improvement of culture conditions for human megakaryocytic and pluripotent progenitor cells by low oxygen tension. Int J Cell Cloning 5(5):412–420
23. Koller MR et al (1992) Reduced oxygen tension increases hematopoiesis in long-term culture of human stem and progenitor cells from cord blood and bone marrow. Exp Hematol 20(2):264–270
24. LaIuppa JA, Papoutsakis ET, Miller WM (1998) Oxygen tension alters the effects of cytokines on the megakaryocyte, erythrocyte, and granulocyte lineages. Exp Hematol 26(9):835–843
25. Miharada K et al (2011) Cripto regulates hematopoietic stem cells as a hypoxic-niche-related factor through cell surface receptor GRP78. Cell Stem Cell 9(4):330–344
26. Spencer JA et al (2014) Direct measurement of local oxygen concentration in the bone marrow of live animals. Nature 508(7495):269–273
27. Takubo K et al (2010) Regulation of the HIF-1alpha level is essential for hematopoietic stem cells. Cell Stem Cell 7(3):391–402
28. Hochachka PW et al (1996) Unifying theory of hypoxia tolerance: molecular/metabolic defense and rescue mechanisms for surviving oxygen lack. Proc Natl Acad Sci U S A 93(18):9493–9498
29. Arai F et al (2004) Tie2/angiopoietin-1 signaling regulates hematopoietic stem cell quiescence in the bone marrow niche. Cell 118(2):149–161
30. Jang YY, Sharkis SJ (2007) A low level of reactive oxygen species selects for primitive hematopoietic stem cells that may reside in the low-oxygenic niche. Blood 110(8):3056–3063
31. Miyamoto K et al (2007) Foxo3a is essential for maintenance of the hematopoietic stem cell pool. Cell Stem Cell 1(1):101–112
32. Ito K et al (2004) Regulation of oxidative stress by ATM is required for self-renewal of haematopoietic stem cells. Nature 431(7011):997–1002
33. Spangrude GJ, Johnson GR (1990) Resting and activated subsets of mouse multipotent hematopoietic stem cells. Proc Natl Acad Sci U S A 87(19):7433–7437
34. Unwin RD et al (2006) Quantitative proteomics reveals posttranslational control as a regulatory factor in primary hematopoietic stem cells. Blood 107(12):4687–4694
35. Li Z et al (2005) VHL protein-interacting deubiquitinating enzyme 2 deubiquitinates and stabilizes HIF-1alpha. EMBO Rep 6(4):373–378
36. Jeong JH et al (2011) 4-O-methylascochlorin, methylated derivative of ascochlorin, stabilizes HIF-1alpha via AMPK activation. Biochem Biophys Res Commun 406(3):353–358
37. Ichihara S et al (2009) Inhibition of ischemia-induced angiogenesis by benzo[a]pyrene in a manner dependent on the aryl hydrocarbon receptor. Biochem Biophys Res Commun 381(1):44–49
38. Zheng J et al (2014) Profilin 1 is essential for retention and metabolism of mouse hematopoietic stem cells in bone marrow. Blood 123(7):992–1001
39. Zhou F et al (2016) Tracing haematopoietic stem cell formation at single-cell resolution. Nature 533(7604):487–492

40. Kocabas F et al (2015) Hypoxic metabolism in human hematopoietic stem cells. Cell Biosci 5:39
41. Takubo K et al (2013) Regulation of glycolysis by Pdk functions as a metabolic checkpoint for cell cycle quiescence in hematopoietic stem cells. Cell Stem Cell 12(1):49–61
42. Wang YH et al (2014) Cell-state-specific metabolic dependency in hematopoiesis and leukemogenesis. Cell 158(6):1309–1323
43. Piccoli C et al (2007) The hypoxia-inducible factor is stabilized in circulating hematopoietic stem cells under normoxic conditions. FEBS Lett 581(16):3111–3119
44. Rehn M et al (2011) Hypoxic induction of vascular endothelial growth factor regulates murine hematopoietic stem cell function in the low-oxygenic niche. Blood 118(6):1534–1543
45. Vukovic M et al (2016) Adult hematopoietic stem cells lacking Hif-1alpha self-renew normally. Blood 127(23):2841–2846
46. Dunwoodie SL (2009) The role of hypoxia in development of the Mammalian embryo. Dev Cell 17(6):755–773
47. Kaidi A, Williams AC, Paraskeva C (2007) Interaction between beta-catenin and HIF-1 promotes cellular adaptation to hypoxia. Nat Cell Biol 9(2):210–217
48. Kim CG et al (2006) Profiling of differentially expressed genes in human stem cells by cDNA microarray. Mol Cell 21(3):343–355
49. Iyer NV et al (1998) Cellular and developmental control of O2 homeostasis by hypoxia-inducible factor 1 alpha. Genes Dev 12(2):149–162
50. Mantel C, Messina-Graham S, Broxmeyer HE (2010) Upregulation of nascent mitochondrial biogenesis in mouse hematopoietic stem cells parallels upregulation of CD34 and loss of pluripotency: a potential strategy for reducing oxidative risk in stem cells. Cell Cycle 9(10):2008–2017
51. de Almeida MJ et al (2017) Dye-independent methods reveal elevated mitochondrial mass in hematopoietic stem cells. Cell Stem Cell 21(6):725–729 e4
52. Anso E et al (2017) The mitochondrial respiratory chain is essential for haematopoietic stem cell function. Nat Cell Biol 19(6):614–625
53. Rimmele P et al (2015) Mitochondrial metabolism in hematopoietic stem cells requires functional FOXO3. EMBO Rep 16(9):1164–1176
54. Gan B et al (2010) Lkb1 regulates quiescence and metabolic homeostasis of haematopoietic stem cells. Nature 468(7324):701–704
55. Laurenti E, Wilson A, Trumpp A (2009) Myc's other life: stem cells and beyond. Curr Opin Cell Biol 21(6):844–854
56. Matsuoka S et al (2008) Fbxw7 acts as a critical fail-safe against premature loss of hematopoietic stem cells and development of T-ALL. Genes Dev 22(8):986–991
57. Chen Y et al (2011) Critical role for Gimap5 in the survival of mouse hematopoietic stem and progenitor cells. J Exp Med 208(5):923–935
58. Mardis ER et al (2009) Recurring mutations found by sequencing an acute myeloid leukemia genome. N Engl J Med 361(11):1058–1066
59. Ward PS et al (2010) The common feature of leukemia-associated IDH1 and IDH2 mutations is a neomorphic enzyme activity converting alpha-ketoglutarate to 2-hydroxyglutarate. Cancer Cell 17(3):225–234
60. Ko M et al (2010) Impaired hydroxylation of 5-methylcytosine in myeloid cancers with mutant TET2. Nature 468(7325):839–843
61. Park IK et al (2003) Bmi-1 is required for maintenance of adult self-renewing haematopoietic stem cells. Nature 423(6937):302–305
62. Guo Z et al (2010) ATM activation by oxidative stress. Science 330(6003):517–521
63. Storz P (2011) Forkhead homeobox type O transcription factors in the responses to oxidative stress. Antioxid Redox Signal 14(4):593–605
64. Cao YA et al (2008) Heme oxygenase-1 deficiency leads to disrupted response to acute stress in stem cells and progenitors. Blood 112(12):4494–4502

65. Yuan TL, Cantley LC (2008) PI3K pathway alterations in cancer: variations on a theme. Oncogene 27(41):5497–5510
66. Juntilla MM et al (2010) AKT1 and AKT2 maintain hematopoietic stem cell function by regulating reactive oxygen species. Blood 115(20):4030–4038
67. Lee JY et al (2010) mTOR activation induces tumor suppressors that inhibit leukemogenesis and deplete hematopoietic stem cells after Pten deletion. Cell Stem Cell 7(5):593–605
68. Chen C et al (2008) TSC-mTOR maintains quiescence and function of hematopoietic stem cells by repressing mitochondrial biogenesis and reactive oxygen species. J Exp Med 205(10):2397–2408
69. Gurumurthy S et al (2010) The Lkb1 metabolic sensor maintains haematopoietic stem cell survival. Nature 468(7324):659–663
70. Tothova Z, Gilliland DG (2007) FoxO transcription factors and stem cell homeostasis: insights from the hematopoietic system. Cell Stem Cell 1(2):140–152
71. Chen Z et al (2015) Wip1 deficiency impairs haematopoietic stem cell function via p53 and mTORC1 pathways. Nat Commun 6:6808
72. Zhang L et al (2015) Inhibition of wild-type p53-induced phosphatase 1 promotes liver regeneration in mice by direct activation of mammalian target of rapamycin. Hepatology 61(6):2030–2041
73. Yu WM et al (2013) Metabolic regulation by the mitochondrial phosphatase PTPMT1 is required for hematopoietic stem cell differentiation. Cell Stem Cell 12(1):62–74
74. Liu X et al (2015) Maintenance of mouse hematopoietic stem cells ex vivo by reprogramming cellular metabolism. Blood 125(10):1562–1565
75. Ito K et al (2012) A PML-PPAR-delta pathway for fatty acid oxidation regulates hematopoietic stem cell maintenance. Nat Med 18(9):1350–1358
76. Signer RA et al (2014) Haematopoietic stem cells require a highly regulated protein synthesis rate. Nature 509(7498):49–54
77. Taya Y et al (2016) Depleting dietary valine permits nonmyeloablative mouse hematopoietic stem cell transplantation. Science 354(6316):1152–1155
78. Liu X et al (2018) PPM1K regulates hematopoiesis and leukemogenesis through CDC20-mediated ubiquitination of MEIS1 and p21. Cell Rep 23(5):1461–1475
79. Cabezas-Wallscheid N et al (2017) Vitamin A-retinoic acid signaling regulates hematopoietic stem cell dormancy. Cell 169(5):807–823 e19
80. Shackelford DB, Shaw RJ (2009) The LKB1-AMPK pathway: metabolism and growth control in tumour suppression. Nat Rev Cancer 9(8):563–575
81. Fan Y, Dickman KG, Zong WX (2010) Akt and c-Myc differentially activate cellular metabolic programs and prime cells to bioenergetic inhibition. J Biol Chem 285(10):7324–7333
82. Shakya A et al (2009) Oct1 loss of function induces a coordinate metabolic shift that opposes tumorigenicity. Nat Cell Biol 11(3):320–327
83. Tothova Z et al (2007) FoxOs are critical mediators of hematopoietic stem cell resistance to physiologic oxidative stress. Cell 128(2):325–339
84. Yilmaz OH et al (2006) Pten dependence distinguishes haematopoietic stem cells from leukaemia-initiating cells. Nature 441(7092):475–482
85. Reya T et al (2003) A role for Wnt signalling in self-renewal of haematopoietic stem cells. Nature 423(6938):409–414
86. Pike LS et al (2011) Inhibition of fatty acid oxidation by etomoxir impairs NADPH production and increases reactive oxygen species resulting in ATP depletion and cell death in human glioblastoma cells. Biochim Biophys Acta 1807(6):726–734
87. Agathocleous M et al (2017) Ascorbate regulates haematopoietic stem cell function and leukaemogenesis. Nature 549(7673):476–481
88. Zhao Y et al (2011) Genetically encoded fluorescent sensors for intracellular NADH detection. Cell Metab 14(4):555–566

Chapter 4
RNA N^6-Methyladenosine Modification in Normal and Malignant Hematopoiesis

Hengyou Weng, Huilin Huang, and Jianjun Chen

Contents

Abstract As the most abundant internal modification in eukaryotic messenger RNAs (mRNAs), N^6-methyladenosine (m^{6}A) modification has been shown recently to posttranscriptionally regulate expression of thousands of messenger RNA (mRNA) transcripts in each mammalian cell type in a dynamic and reversible manner. This epigenetic mark is deposited by the m^{6}A methyltransferase complex (i.e., the METTL3/METTL14/WTAP complex and other cofactor proteins) and erased by m^{6}A demethylases such as FTO and ALKBH5. Specific recognition of these m^{6}A-modified mRNAs by m^{6}A-binding proteins (i.e., m^{6}A readers) determines the fate of target mRNAs through affecting splicing, nuclear export, RNA stability, and/or translation. During the past few years, m^{6}A modification has been demonstrated to play a critical role in many major normal bioprocesses including self-renewal and differentiation of embryonic stem cells and hematopoietic stem cells, tissue development, circadian rhythm, heat shock or DNA damage response, and sex determination. Thus, it is not surprising that dysregulation of the m^{6}A machinery is also

H. Weng · H. Huang · J. Chen (✉)
Department of Systems Biology, Beckman Research Institute of City of Hope, Monrovia, CA, USA

The Gehr Family Center for Leukemia Research, Beckman Research Institute of City of Hope, Monrovia, CA, USA
e-mail: jianchen@coh.org

H. Zhang, S. Li (eds.), *Leukemia Stem Cells in Hematologic Malignancies*, Advances in Experimental Medicine and Biology 1143,
https://doi.org/10.1007/978-981-13-7342-8_4

closely associated with pathogenesis and drug response of both solid tumors and hematologic malignancies. In this chapter, we summarize and discuss recent findings regarding the biological functions and underlying mechanisms of m^{6}A modification and the associated machinery in normal hematopoiesis and the initiation, progression, and drug response of acute myeloid leukemia (AML), a major subtype of leukemia usually associated with unfavorable prognosis.

Keywords m^{6}A demethylases · m^{6}A modification · Normal hematopoiesis · AML

Abbreviations

AMKL	Acute megakaryoblastic leukemia
AML	Acute myeloid leukemia
BMT	Bone marrow transplantation
CLIP	Cross-linking immunoprecipitation
CMP	Common myeloid progenitor
EHT	Endothelial-to-hematopoietic transition
GMP	Granulocyte/macrophage progenitor
HSCs	Hematopoietic stem cells
HSPCs	Hematopoietic stem and progenitor cells
lncRNA	Long noncoding RNA
LSCs	Leukemia stem cells
LSCs/LICs	Leukemia stem/initiating cells
m^{6}A	*N*^{6}-methyladenosine
miCLIP	m^{6}A individual-nucleotide-resolution cross-linking and immunoprecipitation
mRNA	Messenger RNA
MTC	Methyltransferase complex
pri-miRNA	Primary microRNA
rRNA	Ribosomal RNA
snoRNA	Small nucleolar RNA
SNPs	Single-nucleotide polymorphisms
snRNA	Small nuclear RNA
TF	Transcription factor
tRNA	Transfer RNA
TSS	Transcriptional start site

4.1 Introduction

Since the 1960s, over 150 modified RNA nucleotide variants have been identified in both protein-coding and noncoding RNAs, such as N^6-methyladenosine (m^6A) in messenger RNA (mRNA) and primary microRNA (pri-miRNA) [1–4], N^1-methyladenosine (m^1A) in mRNA and transfer RNA (tRNA) [5–7], 5-methylcytosine (m^5C) and 5-hydroxymethylcytosine (hm^5C) in mRNA and long noncoding RNA (lncRNA) [8–11], and pseudouridine (ψ) in tRNA, ribosomal RNA (rRNA), small nuclear RNA (snRNA), and small nucleolar RNA (snoRNA) [12–16] (Figs. 4.1 and 4.2). Of them, m^6A is the most prevalent and abundant internal

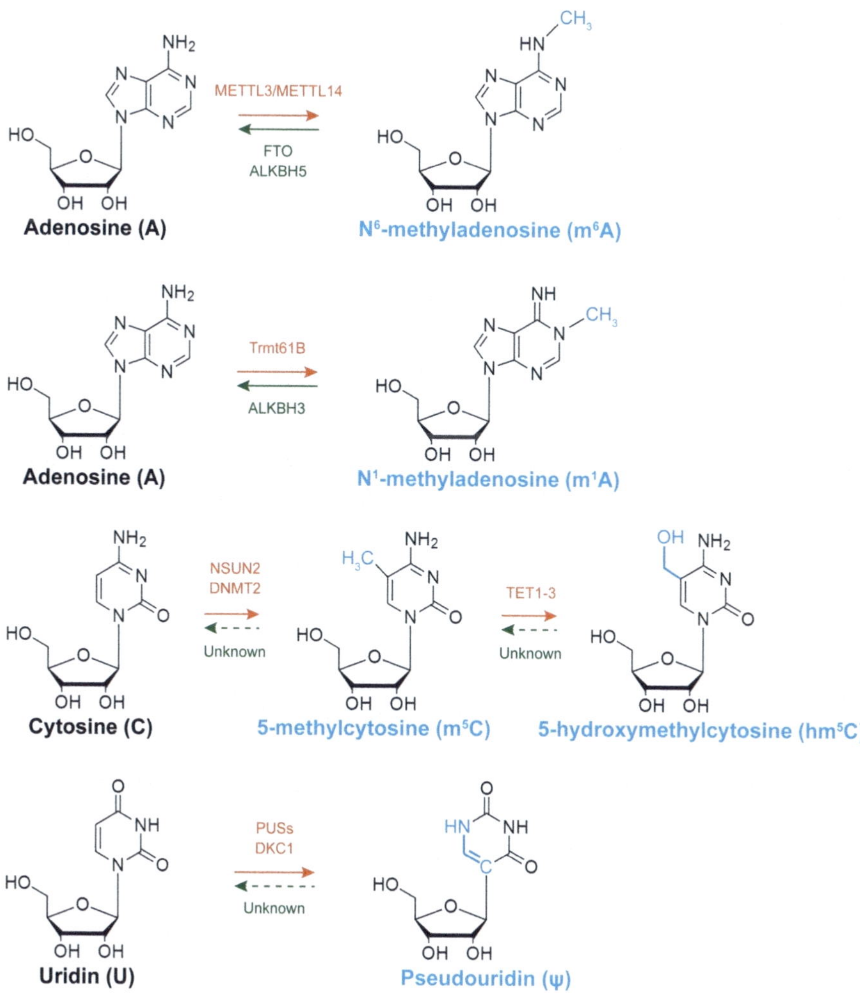

Fig. 4.1 Chemical structures of representative modified RNA nucleotide variants. Modifications are shown in blue. The known writer (in orange) and eraser (in green) proteins are also indicated

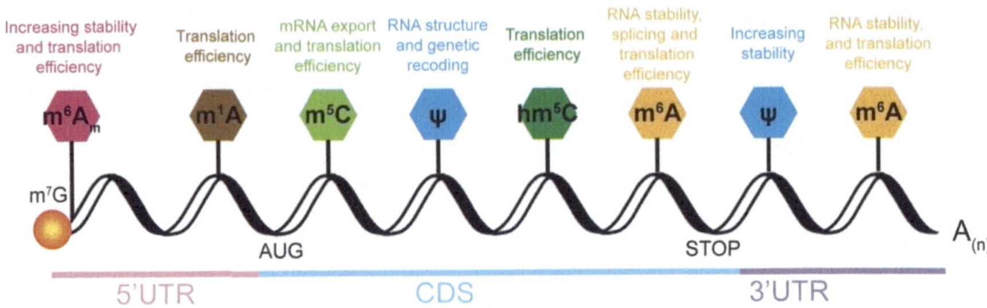

Fig. 4.2 Chemical modifications in eukaryotic mRNA. A schematic representation of common chemical modifications across eukaryotic mRNA transcript including 5′ untranslated region (5′UTR), coding region (CDS), and 3′UTR. Reported roles of these modifications are summarized on top of the corresponding modifications. Note that the same modification in different mRNA regions may have different functions in regulating mRNA fate

modification on eukaryotic mRNAs. m^6A RNA modification was first identified in the 1970s [1–3]. However, due to the lack of knowledge on its dynamic regulation and no high-throughput technology available to map m^6A modification to the RNA transcriptome, little attention had been paid to this RNA mark until 2011, when the fat mass and obesity-associated protein (FTO) was identified as a genuine demethylase of m^6A modification [17], which implies that m^6A modification is a reversible and dynamic process analogous to the well-studied modifications on DNA and histone [18]. Subsequent development of high-throughput m^6A sequencing technologies further facilitates the understanding of m^6A modification in a transcriptome-wide view, revealing that m^6A modification may affect more than 7000 mRNAs in individual transcriptomes of mammalian cells, with a special enrichment in the 3′ untranslated regions (UTRs) near the stop codons of mRNAs and with a consensus sequence of RRACH (R = G or A; H = A, C, or U) [19, 20]. Such findings strongly suggest that m^6A modification may have important biological functions. Indeed, emerging data demonstrate that m^6A modifications in mRNAs or noncoding RNAs influence RNA fate and functions and are critical for many normal and pathological bioprocesses including self-renewal and differentiation of embryonic stem cells, tissue development, circadian rhythm, heat shock or DNA damage response, sex determination, and tumorigenesis, as reviewed elsewhere [4, 16, 21–37]. Evidence is emerging that m^6A modification and the associated machinery also play essential roles in tumorigenesis and drug response (see reviews [16, 38, 39]). Here we summarize recent advances on our understanding of the functions and underlying molecular mechanisms of the m^6A machinery in normal hematopoiesis and AML pathogenesis and drug response.

4.2 Regulators of m^{6}A Modifications

Similar to other epigenetic modifications, m^{6}A modification is regulated by its formation and removal catalyzed by the methyltransferases and demethylases, known as "writers" and "erasers," respectively [16, 40, 41]. m^{6}A marks are installed by a multicomponent methyltransferase complex (MTC) consisting of the core methyltransferase-like 3 and 14 (METTL3 and METTL14) heterodimer and their cofactors including Wilms' tumor 1-associating protein (WTAP), vir like m^{6}A methyltransferase associated (VIRMA, also known as KIAA1429), RNA-binding motif protein 15 (RBM15), and zinc finger CCCH domain-containing protein 13 (ZC3H13) [22, 42–49] (Fig. 4.3). Structural studies demonstrated that METTL3 is the sole catalytic subunit, while METTL14 offers an RNA-binding scaffold to allosterically activate and enhance the catalytic activity of METTL3 [50–52]. WTAP, VIRMA, RBM15, and ZC3H13 are the regulatory subunits of the MTC to facilitate m^{6}A installation *in cellulo*. In addition, the METTL3 homolog METTL16 (methyltransferase-like 16) was recently shown to control cellular S-adenosyl methionine (SAM) level and install m^{6}A marks onto the U6 small nuclear RNA [53, 54]. FTO, previously known to function as a demethylase for N^3-methylthymidine in single-stranded DNA and N^3-methyluridine in single-stranded RNA in vitro [55, 56], was identified as the first m^{6}A demethylase that could demethylate m^{6}A in both DNA and RNA in vivo [17]. A recent study reported that FTO also demethylates m^{6}A$_m$, a modification exclusively found at the first encoded nucleotide after the 7-methylguanosine cap structure of mRNAs [57]. AlkB homolog 5 (ALKBH5) is the second identified m^{6}A demethylase that was found to be highly expressed in the testes [23]. Both FTO and ALKBH5 belong to the AlkB subfamily of the Fe(II)/2-oxoglutarate (2OG) dioxygenase superfamily that requires 2OG and molecular oxygen as co-substrates and ferrous iron Fe(II) as a cofactor to catalyze the oxidation of a substrate [56, 58].

While the prevalence and distribution of m^{6}A are determined by writers and erasers, the m^{6}A-dependent functions are mediated by m^{6}A-binding proteins, the so-called readers, which through specific recognition and binding to m^{6}A-modified mRNAs determine the fate of these transcripts [16, 21, 59] (Fig. 4.3). The YT521-B homology (YTH) domain family of proteins, including YTHDF1, YTHDF2, YTHDF3, YTHDC1, and YTHDC2, are among the first identified m^{6}A readers that possess a conserved m^{6}A-binding pocket [19, 60–66]. Of them, YTHDC1 was found to be located in the nucleus, playing a role in splicing regulation, XIST-mediated X-chromosome silencing, and nuclear export of m^{6}A-modified mRNAs [46, 63, 67]. The other YTH family proteins are all cytoplasmic m^{6}A readers regulating mRNA fate through different mechanisms: YTHDF2 promotes degradation of target mRNAs [60, 68], and YTHDF1 promotes translation of target mRNAs [61], while YTHDF3 and YTHDC2 can both mediate mRNA decay and enhance translation [66, 69–73]. We recently identified insulin-like growth factor 2 mRNA-binding proteins (IGF2BPs; including IGF2BP1/2/3) as a new family of m^{6}A readers that could promote stability and translation of their target mRNAs [32], distinct

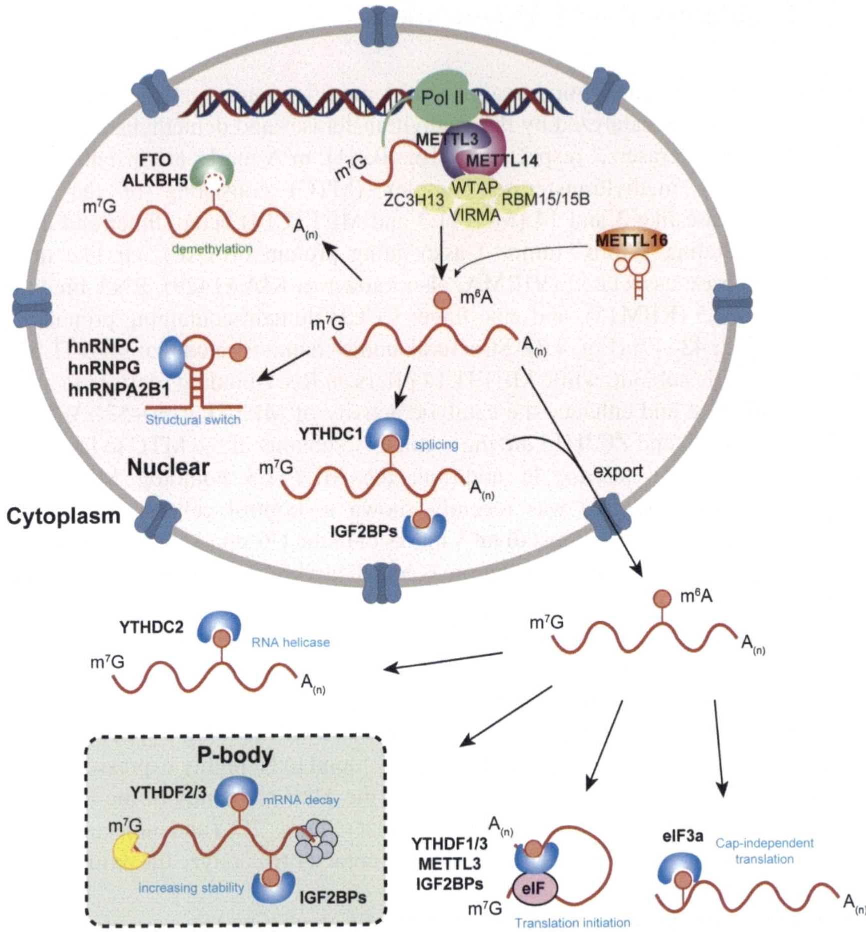

Fig. 4.3 Roles of m⁶A RNA modification in determining mRNA fate. In nuclear, the methyltransferase complex composed of METTL3, METTL14, WTAP, VIRMA, RBM15, and ZC3H13 deposits m⁶A marks co-transcriptionally onto newly transcribed RNAs, while METTL16 is responsible for m⁶A deposition on the U6 snRNA. FTO and ALKBH5 function as m⁶A demethylases to remove m⁶A marks in selected sites of RNA. M⁶A-mediated structural switch of mRNA recruits hnRNP family proteins including hnRNPC, hnRNPG, and hnRNPA2B1. Nuclear reader protein YTHDC1 recognizes m⁶A to mediate alternative splicing. IGF2BP reader proteins bind to m⁶A mRNAs to stabilize these nascent transcripts. After exporting to cytoplasm, m⁶A-modified mRNAs could be subjected to degradation by YTHDF2 or YTHDC2 or protected by IGF2BP proteins and loaded to translation machinery. YTHDF1, YTHDF3, as well as METTL3 and eIF3a could also promote translation of m⁶A mRNAs

from the functional manners of the YTH family proteins. The K homology (KH) domains of IGF2BPs are required for their recognition of m⁶A and are critical for their oncogenic functions. Interestingly, ELAV-like RNA-binding protein 1 (ELAVL1, also known as HuR), an mRNA stabilizer that was previously reported as an indirect m^{6}A-binding protein [22], was found to be a cofactor of IGF2BPs and may mediate the mRNA stabilizing effect of IGF2BPs [32]. The heterogeneous nuclear ribonucleoprotein (HNRNP) HNRNPA2B1 was previously reported to regulate alternative splicing and primary microRNA processing as an m^{6}A reader [74]; however, recent structural study suggested an "m^{6}A switch" mechanism rather than direct m^{6}A binding for the protein [75]. Two other members of the HNRNP family proteins, namely HNRNPC and HNRNPG, were shown to recognize m^{6}A-induced changes in mRNA secondary structures [76]. Other proteins were also recently reported to be m^{6}A interactors, including FMR1 and LRPPRC [77, 78], although the exact mode of binding still needs to be clarified.

4.3 Roles of m^{6}A in Normal Hematopoiesis and HSC Self-Renewal

Hematopoiesis is a tightly regulated dynamic process where mature blood cells are generated from a small pool of multipotent hematopoietic stem cells (HSCs) [79, 80]. During the past few decades, it has been well acknowledged that transcriptional regulation by a variety of hematopoietic transcription factors (TFs) plays a big role in regulating the multistep normal hematopoiesis [81–83]. In particular, during myelopoiesis where HSCs are differentiated into myeloid progenitors and eventually mature myeloid cells, the sequential actions of master TFs are required to specify and re-enforce each cell fate decision. For instance, PU.1 (also known as SPI1, the product of the oncogene SPI1) is a transcriptional master regulator of myeloid cells which plays an essential role in generating early myeloid progenitors (i.e., common myeloid progenitors, CMPs), while the basic region leucine zipper transcription factor C/EBPα is required for the production of granulocyte/macrophage progenitors (GMPs) from CMPs [80, 84].

In recent years, emerging studies reveal m^{6}A modification at the RNA level as an additional layer of the posttranscriptional regulation in governing HSC activity and normal hematopoiesis (see Fig. 4.4). During the endothelial-to-hematopoietic transition (EHT) of zebrafish embryogenesis, a key developmental event leading to the formation of the earliest hematopoietic stem and progenitor cells (HSPCs), m^{6}A modification was reported to play a role [30]. Deficiency of mettl3 in zebrafish embryos leads to decreased levels of m^{6}A and blockage of HSPC emergence, likely due to the reduced m^{6}A modification on the arterial endothelial genes *notch1a* and *rhoca* that delayed YTHDF2-mediated mRNA decay of these transcripts [30]. Vu and colleagues used short hairpin RNAs (shRNAs) to knock down *METTL3* expression in human HSPCs and observed cell growth inhibition and increase of myeloid differentiation [34]. Conversely, overexpression of wild-type, but not catalytically dead mutant of METTL3, promotes proliferation and colony formation and inhibits myeloid differentiation. We recently showed that METTL14 is highly expressed in murine HSCs and Lin$^-$ Sca-1$^+$c-kit$^+$ (LSK) cells and was downregulated during

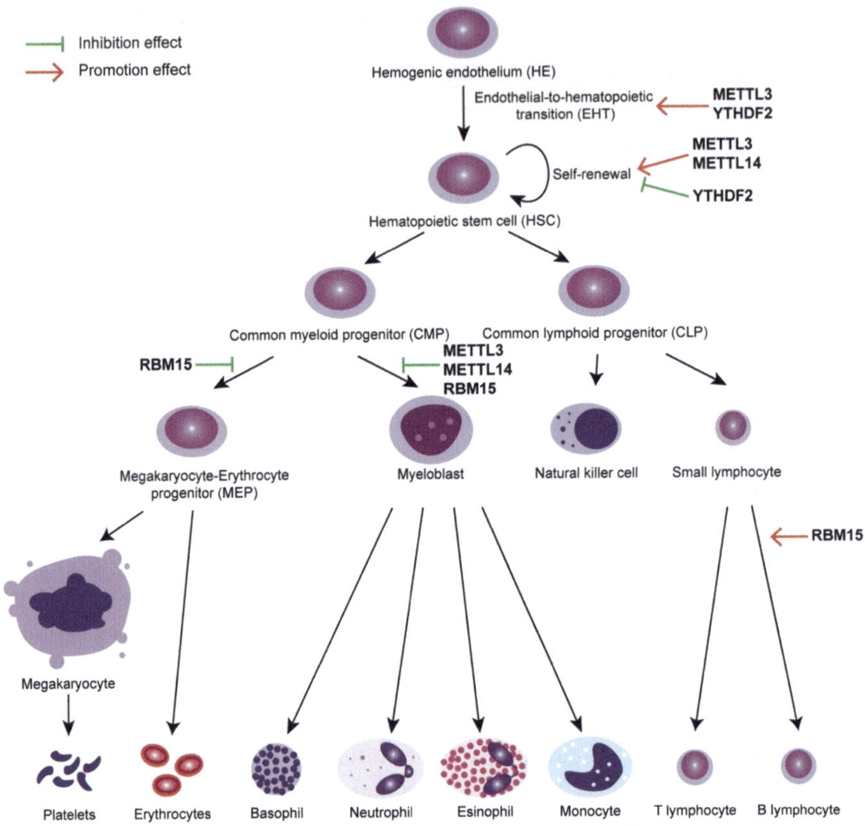

Fig. 4.4 Modifiers of m^{6}A RNA modification in normal hematopoiesis. METTL3 and METTL14 promote self-renewal of HSC and inhibit myeloid differentiation. In contrast, YTHDF2 inhibits HSC self-renewal. RBM15 inhibits myeloid and megakaryocytic differentiation while promotes B cell expansion, although it is unclear whether these effects are m^{6}A-related

myelopoiesis, showing reduced expression in CMP and GMP progenitors and especially in mature myeloid cells [33]. Consistent with the expression pattern, depletion of *METTL14* expression in human HSPCs by shRNAs promotes myeloid differentiation in vitro. Moreover, by utilizing a *Mettl14* conditional knockout mouse model, we demonstrated that induced deletion of *Mettl14* impairs HSC self-renewal ability in vivo. Such effects were mediated by decreased *MYB* and *MYC* expression owing to the reduction of m^{6}A modification on these transcripts upon *METTL14* knockdown/knockout. Considering the role of MYB [85, 86] and MYC [87, 88] transcription factors in regulating HSC self-renewal and differentiation, METTL14-mediated m^{6}A regulation on the mRNA transcripts of these TFs adds a new layer of complexity to the regulatory networks in normal hematopoiesis.

A very recent study took advantage of *Mettl3* and *Mettl14* conditional knockout mice to investigate the roles of these m^{6}A writer proteins on regulation of HSC self-renewal in adult mouse bone marrow [89]. They found that deletion of *Mettl3* alone or together with *Mettl14* in the hematopoietic system substantially increases HSC frequency in the bone marrow; in contrast, deletion of *Mettl14* alone has little effect [89]. Conditional deletion of *Mettl14* and especially that of *Mettl3* suppresses HSC self-renewal activity in recipient mice. Notably, although deletion of either *Mettl3* or *Mettl14* deletion leads to significant reduction of donor-derived myeloid cells in the peripheral blood, only deletion of *Mettl3* results in a significant reduction of B- and T-cell lineage [89].

RBM15, recently identified as a component of the m^{6}A methyltransferase complex, was also reported to play a role in normal hematopoiesis [90–92]. Conditional knockout of *Rbm15* in adult mice blocks B cell differentiation and results in an increase of the Lin$^-$Sca-1$^+$c-Kit$^+$ (LSK) HSPCs and an expansion of myeloid and megakaryocytic cells in spleen and bone marrow, demonstrating a role of RBM15 in hematopoietic development [90]. Rbm15 is expressed at highest levels in HSCs and inhibits myeloid differentiation and megakaryocytic expansion through stimulation of the Notch signaling and regulation of *MYC* expression, respectively [91, 92].

YTHDF2 is the first well-characterized m^{6}A reader that promotes mRNA decay of target transcript with m^{6}A modification [60]. Li and colleagues studied the role of YTHDF2 in adult stem cell maintenance and reported an increase of HSCs in *Ythdf2* conditional knockout mice as well as in human umbilical cord blood upon *YTHDF2* knockdown [93]. Such effects are partially mediated by the stabilization of mRNA transcripts encoding TFs critical for stem cell self-renewal. Overall, the studies of m^{6}A in normal hematopoiesis are just in the beginning. It is of great interest to explore the functions and underlying mechanisms of other m^{6}A modulators, including other writers, erasers, and readers, in stem cell biology and normal hematopoiesis.

4.4 Dysregulation of m^{6}A Regulators in Malignant Hematopoiesis

Dysregulation of the regulatory networks of normal hematopoiesis, such as incorrect activity of the hematopoietic TFs attributed to either aberrant expression or mutation, breaks the balance between HSC self-renewal and differentiation and places the progenitor cells at a higher risk of developing leukemia [80, 94]. Acute myeloid leukemia (AML) is a clonal hematopoietic disorder where a stem cell-like self-renewal capacity is gained and the differentiation capacity is blocked [95, 96]. According to the Cancer Statistics 2017, the 5-year survival of AML between 2007 and 2013 is 26.9%, much lower than other types of leukemia and many other common cancer types [97]. For instance, during the same period, the 5-year survival rate is 89.7% for breast cancer and 66.9% for chronic myeloid leukemia (CML). Therefore, there is an urgent need to better understand the mechanisms underlying AML leukemogenesis and, based on the gained knowledge, to develop effective targeted therapies.

The first evidence of a role of m^{6}A modification on leukemia, specifically on AML, came from the study of FTO, a major m^{6}A demethylase [31]. *FTO* was previously known as a gene associated with fat mass, adipogenesis, and body weight [98–100], and single-nucleotide polymorphisms (SNPs) in *FTO* were linked to higher risk of developing cancers including leukemia and lymphoma by large-scale epidemiology studies [101–103]. We found that *FTO* is highly expressed in certain subtypes of AML, including those carrying t(11q23)/*MLL* rearrangements, t(15;17)/*PML-RARA*, *FLT3*-ITD, and/or *NPM1* mutations. Modulation of *FTO* expression by depletion of *FTO* or forced expression of wild-type FTO (but not catalytically inactive mutant) could significantly influence AML cell survival and leukemogenesis and affect the response of AML cells to all-trans retinoic acid (ATRA) [31]. Importantly, the oncogenic function of FTO in AML relies on its m^{6}A demethylase activity. By reducing m^{6}A abundance on the transcripts of *ASB2* and *RARA*, two genes with reported roles in cell proliferation and drug response of leukemia cells, FTO posttranscriptionally regulates expression of *ASB2* and *RARA* through reducing m^{6}A abundance, thereby decreasing mRNA stability of these transcripts. These data provide compelling evidence on the role of FTO in leukemogenesis and establish a first link between m^{6}A modification and leukemia pathogenesis (Fig. 4.5) [31].

More recently, we showed that, by targeting FTO directly, R-2-hydroxyglutarate (R-2HG), a previously reported oncometabolite produced by mutant isocitrate dehydrogenase 1/2 (IDH1/2) enzymes [104–106], exhibits a broad and intrinsic antitumor activity in leukemia. By inhibiting the m^{6}A demethylase activity of FTO, R-2HG results in an increase of m^{6}A abundance on FTO target genes, such as *MYC* and *CEBPA*, leading to decay of these transcripts in R-2HG sensitive cells (see Fig. 4.5). Our data indicate that FTO/MYC homeostasis controls the sensitivity of leukemic cells to 2HG: high abundance of FTO confers R-2HG sensitivity, whereas hyper-activation of MYC signaling renders leukemic cells resistant to R-2HG [37]. Consistent with this notion, pharmaceutical or genetic inhibition of MYC signaling by JQ1 or *MYC* shRNAs can resensitize R-2HG-resistant leukemic cells to R-2HG. In addition, we showed that R-2HG treatment or FTO inhibition also sensitized AML cells to first-line chemotherapy drugs such as ATRA, Azacitidine (AZA), Decitabine, and Daunorubicin ([37] and unpublished data). Collectively, our studies demonstrate the critical role of FTO in leukemia pathogenesis and drug response and also highlight the therapeutic potential of targeting FTO signaling for AML treatment.

Subsequently, the dysregulation of the m^{6}A installing machinery was also reported to be involved in AML pathogenesis [107] (see Fig. 4.6). METTL3 and METTL14 were both expressed at a higher level in AML than in the vast majority of other cancer types according to The Cancer Genome Atlas (TCGA) genome-wide gene expression datasets [33–35]. We found that METTL14 is aberrantly overexpressed in certain subtypes of AMLs, such as those carrying t(11q23)/*MLL* rearrangements, t(15;17)/*PML-RARA*, and t(8;21)/*AML1-ETO* [33]. We next conducted in vitro and in vivo gain- and loss-of-function studies and demonstrated that METTL14 plays a critical oncogenic role in AML pathogenesis [33]. Depletion of

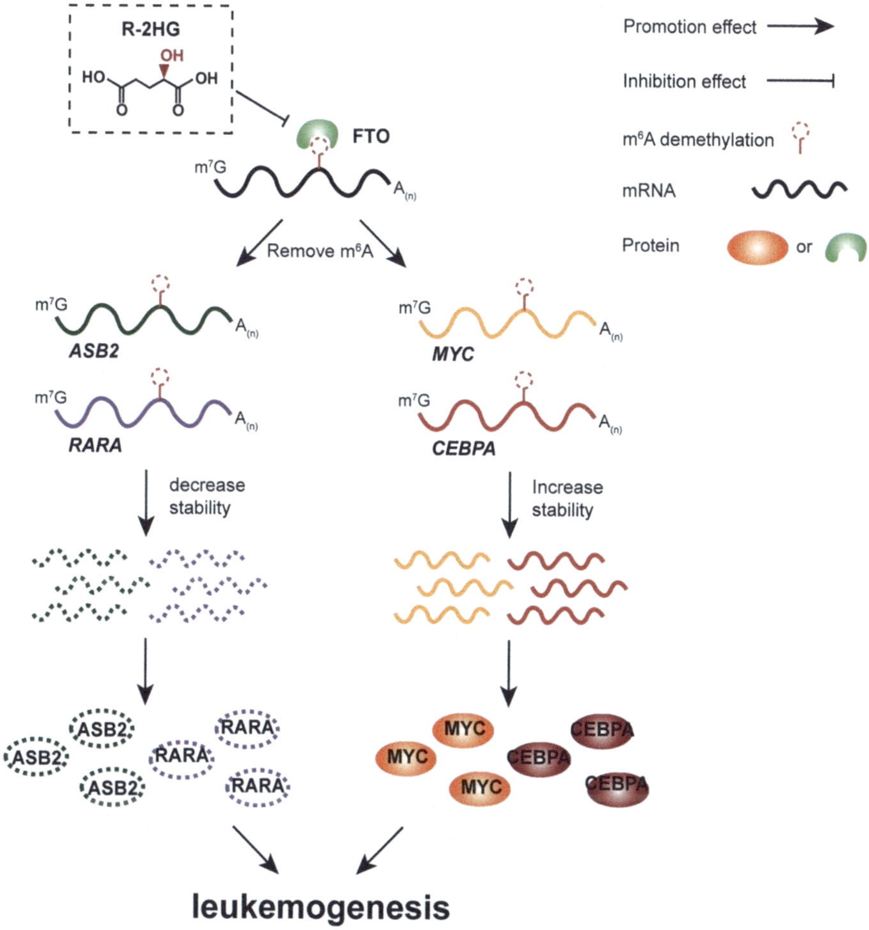

Fig. 4.5 Role of FTO in leukemogenesis. FTO removes m^{6}A modification on its target mRNAs (e.g., *ASB2*, *RARA*, *MYC*, and *CEBPA*), resulting in the decreased or increased stability of these transcripts, and promotes leukemogenesis. R-2HG inhibits FTO and exhibits antileukemia effects

METTL14 inhibited survival and proliferation of AML cells, promoted myeloid differentiation, and suppressed leukemic oncofusion protein (e.g., MLL-AF9, MLL-AF10, and AML1-ETO9a)-mediated immortalization of normal HSPCs. The opposite is true when wild type, but not mutant METTL14 (i.e., R298P), was forced expressed. Moreover, knockdown of *METTL14* significantly inhibited progression of human AML cells in xenotransplantation recipient mice, while inducible knockout of *Mettl14* greatly inhibited AML development and maintenance in bone marrow transplantation (BMT) recipient mice [33]. High-throughput RNA-seq and transcriptome-wide m^{6}A-seq, coupled with gene-specific m^{6}A-qPCR assays, crosslinking and immunoprecipitation (CLIP) assays, luciferase reporter and mutagenesis assays, mRNA stability assays, and polysome profiling assays, demonstrated

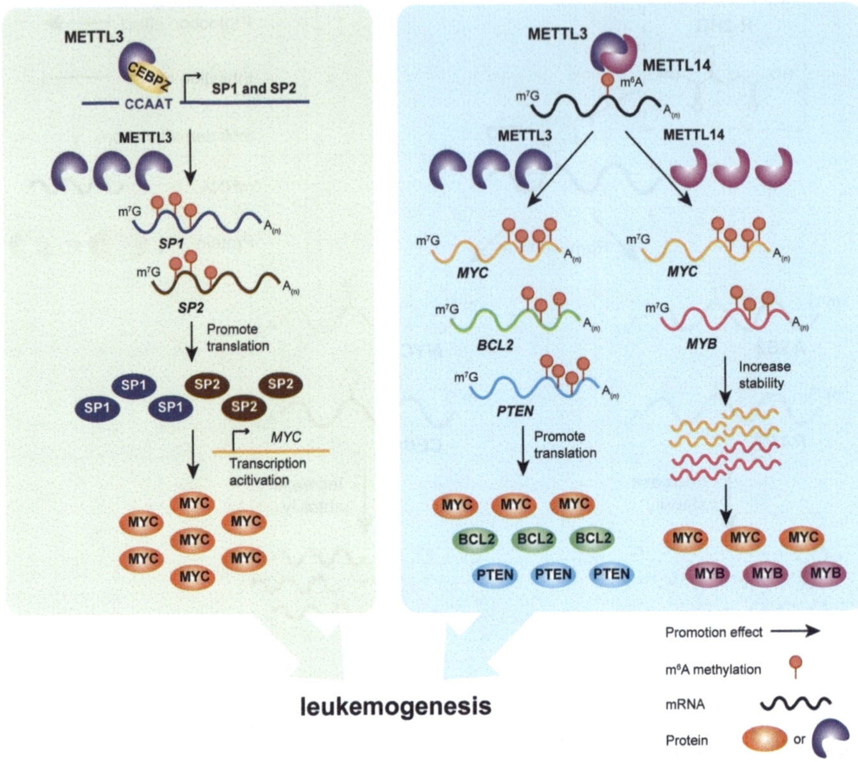

Fig. 4.6 Roles of m^{6}A writer genes in leukemogenesis. METTL3 and METTL14 function as oncogenes in AML through depositing m^{6}A modification on their target transcripts (e.g., *MYC*, *BCL2*, and *PTEN* for METTL3 and *MYB* and *MYC* for METTL14) to enhance translation and/or increase mRNA stability of these transcripts. METTL3 can also be recruited to promoter of its target gene by CEBPZ and deposit m^{6}A modification on the coding region of the target transcripts (e.g., *SP1* and *SP2*) to promote their translation and eventually leads to *MYC* activation

that METTL14 posttranscriptionally regulates the expression of its critical target mRNA transcripts, such as *MYB* and *MYC*, two well-known TF genes involved in leukemogenesis, in an m^{6}A-dependent manner [33]. Silencing of *METTL14* reduces m^{6}A abundance of *MYB* and *MYC* transcripts, especially near the 3′ end of the mRNAs, resulting in decreased mRNA stability and translation of these transcripts. Such effects were not mediated by the YTH family proteins, and therefore other readers (such as IGF2BPs [32]) may mediate the effect of METTL14 on *MYB* and *MYC*. Furthermore, we also identified SPI1 (also called PU.1), a transcriptional master regulator of myelopoiesis, as a negative regulator of *METTL14* expression in AML [33]. Taken together, our work reveals a previously unappreciated SPI1-METTL14-MYB/MYC signaling axis in leukemogenesis and highlights the critical roles of METTL14 and m^{6}A modification in malignant hematopoiesis [33].

Meanwhile, two other groups have demonstrated that METTL3 also plays an essential oncogenic role in AML, by showing that depletion of *METTL3* expression results in cell growth inhibition, cell cycle arrest, and induction of differentiation and apoptosis, whereas overexpression of wild-type *METTL3*, but not a catalytically inactive mutant, promotes proliferation in AML cells and primary blasts [34, 35]. Similar to METTL14, METTL3 is also required for AML development as demonstrated by the xenotransplantation assay data that shRNA-mediated knockdown or CRISPR/Cas9-mediated knockout of *METTL3* substantially inhibited AML progression and prolonged survival in recipient mice [34, 35]. Such findings support the oncogenic role of METTL3 as an m^{6}A-catalyzing enzyme. Despite of the similar functions of METTL3 reported by the two groups, distinct underlying mechanisms of METTL3 in AML were reported [34, 35]. Vu et al. performed m^{6}A individual-nucleotide-resolution cross-linking and immunoprecipitation (miCLIP), RNA-seq, and Ribo-Seq, and identified *MYC*, *BCL2* and *PTEN* as direct RNA targets of METTL3, whose m^{6}A abundances were substantially reduced when *METTL3* was knocked down [34]. However, upon *METTL3* depletion, while expression of *MYC*, *BCL2*, and *PTEN* transcripts showed a great increase, the corresponding protein levels decreased on day 3 after shRNA transduction and recovered 1 day later [34]. Although the authors proposed that an alternative internal ribosome entry site (IRES)-mediated translational mechanism may be involved to reactivate translation of *MYC* and *BCL2* under METTL3 depletion-induced cell apoptosis, it remains unclear why the expression changes of these target genes at the mRNA level are inconsistent with that at the protein levels. A systematic understanding of the changes and functions of m^{6}A readers under this circumstance may help to address this question. Different from Vu's study [34], Barbieri et al. found that METTL3 and METTL14 could bind chromatin, mainly localizing to the transcriptional start sites (TSSs) of coding genes characterized by bimodal H3K4me3 peaks [35]. However, METTL3 and METTL14 did not bind the same TSSs, suggesting independent regulation on target genes. CEBPZ, a transcription factor critical for hematopoietic differentiation, recruits METTL3, but not METTL14, to chromatin [35]. Promoter-bound METTL3 induces m^{6}A modification within the coding region of the associated mRNA transcript, such as *SP1* and *SP2*, and enhances their translation by relieving ribosome stalling [35].

In addition to METTL3 and METTL14, other m^{6}A methyltransferases or cofactors of the m^{6}A methyltransferase complex have also been implicated to function in AML. A genome-wide CRISPR-Cas9 screening identified METTL16, as well as METTL3 and METTL14, as critical genes for AML survival [35], while detailed study is yet to be done. Before it was identified as a component of the m^{6}A methyltransferase complex [43], WTAP was shown to be upregulated in AML and is a novel client protein of HSP90 [108]. Knockdown of WTAP in AML cells led to reduced proliferation and promotion of phorbol 12-myristate 13-acetate (PMA)-induced myeloid differentiation and inhibited growth of human leukemia cells in xenograft mice [108]. RBM15 was found to be a fusion partner of the MKL1 gene in t(1,22)(p13;q13) acute megakaryoblastic leukemia (AMKL), a subtype of pediatric AML [109]. It is therefore reasonable to speculate that RBM15 may play a role in leukemogenesis, possibly related to m^{6}A modification, for which further studies are warranted.

4.5 m^{6}A and Leukemia Stem Cells

Malignant stem cells are considered as the potential origin of and a key therapeutic target for AML, similar to what is believed for other cancer types [110–115]. Leukemia stem cells (LSCs) are defined as cells with two important properties: (1) capable of engrafting and initiating the disease when transplanted into immunodeficient animals and can self-renew by giving rise to leukemia in serial transplantations and (2) produce non-LSC bulk blasts that resemble the original disease but are unable to self-renew [116]. During the last few decades, investigators have been dedicated to the characterization of LSCs using different combinations of cell surface markers. It was found that LSCs immunophenotypically resemble certain normal hematopoietic progenitor populations and usually reside in the CD34$^+$CD38$^-$ fraction, although approximately 25% of AML cases lack CD34 expression [116]. It is believed that therapeutic targeting and eliminating of LSC is the key to eradicate leukemia and achieve long-term remissions.

Retroviral transduction of the *MLL-AF9* oncofusion gene into mouse HSPCs followed by transplantation into recipient mice represents one of the best mouse models for AML LSC studies [116]. By using this model, we have shown that depletion of either *Fto* or *Mettl14* in mouse HSPCs could inhibit leukemia initiation in primary BMT recipients and maintenance in secondary transplantations [31, 33], suggesting a role of these m^{6}A modifying proteins on AML LSC function. Indeed, we conducted limiting dilution assays using bone marrow cells harvested from MLL-AF9 primary leukemia mice to directly evaluate the effect of *Mettl14* depletion on the frequency of leukemia stem/initiating cells (LSCs/LICs) and found that the estimated LSC/LIC frequency was significantly reduced when *Mettl14* was knocked out [33]. While our data provide the first link between m^{6}A modification and LSC self-renewal [33], further systematic studies are warranted to better understand the role of m^{6}A modification on LSC biology.

4.6 Conclusions and Perspectives

Evidence is emerging that m^{6}A modification and the associated machinery play essential roles in both normal and malignant hematopoiesis, including the self-renewal and differentiation of normal and malignant HSCs. Systematical studies of the functions and underlying molecule mechanisms of individual m^{6}A writer, eraser, and reader genes will further advance our understanding of the complex networks and molecular mechanisms underlying normal and malignant hematopoiesis. In addition, it is also important to understand how expression of individual m^{6}A regulators is regulated during normal and malignant hematopoiesis.

The impact of m^{6}A modification on the fate of its embedded RNA mediates the functions of such modification. Therefore, it is of great importance to identify key functionally important RNA targets (including mRNA, lncRNA, etc.) that when manipulated could phenocopy or reverse the effects of knockdown or overexpression of individual m^{6}A regulators. For instance, recent studies show the oncogenic *MYC* transcript as an important target of many m^{6}A regulators in cancer including leukemia [32–34, 37], highlighting the importance of precise regulation of *MYC* expression at the posttranscription level in normal development and the big impact of its dysregulation in tumorigenesis. Further systematic identification and functional studies of all critical target genes of m^{6}A modification by combined use of high-throughput sequencing such as CLIP-seq, m^{6}A-seq, and RNA-seq, followed by the validation assays and in vitro and in vivo functional studies, are warranted. Such knowledge will be not only important for the understanding of the effects of manipulation of individual m^{6}A regulator but also may lead to the discovery of novel therapeutic targets for AML and other types of leukemia.

As demonstrated by the published data thus far, m^{6}A regulators represent promising targets for treatment of cancer, including leukemia. Development of effective and selective inhibitors targeting oncogenic m^{6}A regulators (e.g., FTO and METTL14) that play more essential roles in leukemogenesis, especially in self-renewal of LSCs/LICs, than in normal hematopoiesis may hold great therapeutic potential in treating leukemia, especially when in combination with other therapeutic agents.

References

1. Wei CM, Moss B (1977) Nucleotide sequences at the N6-methyladenosine sites of HeLa cell messenger ribonucleic acid. Biochemistry 16:1672–1676
2. Adams JM, Cory S (1975) Modified nucleosides and bizarre 5′-termini in mouse myeloma mRNA. Nature 255:28–33
3. Krug RM, Morgan MA, Shatkin AJ (1976) Influenza viral mRNA contains internal N6-methyladenosine and 5′-terminal 7-methylguanosine in cap structures. J Virol 20:45–53
4. Alarcon CR, Lee H, Goodarzi H, Halberg N, Tavazoie SF (2015) N6-methyladenosine marks primary microRNAs for processing. Nature 519:482–485
5. Dunn DB (1961) The occurrence of 1-methyladenine in ribonucleic acid. Biochim Biophys Acta 46:198–200
6. Hall RH (1963) Method for isolation of 2 -O-methylribonucleosides and N1-methyladenosine from ribonucleic acid. Biochim Biophys Acta 68:278–283
7. El Yacoubi B, Bailly M, de Crecy-Lagard V (2012) Biosynthesis and function of posttranscriptional modifications of transfer RNAs. Annu Rev Genet 46:69–95
8. Squires JE et al (2012) Widespread occurrence of 5-methylcytosine in human coding and non-coding RNA. Nucleic Acids Res 40:5023–5033
9. Amort T et al (2013) Long non-coding RNAs as targets for cytosine methylation. RNA Biol 10:1003–1008
10. Fu L et al (2014) Tet-mediated formation of 5-hydroxymethylcytosine in RNA. J Am Chem Soc 136:11582–11585
11. Huber SM et al (2015) Formation and abundance of 5-hydroxymethylcytosine in RNA. Chembiochem 16:752–755
12. Carlile TM et al (2014) Pseudouridine profiling reveals regulated mRNA pseudouridylation in yeast and human cells. Nature 515:143–146

13. Cohn WE, Volkin E (1951) Nucleoside-5′-phosphates from ribonucleic acid. Nature 167:483–484
14. Charette M, Gray MW (2000) Pseudouridine in RNA: what, where, how, and why. IUBMB Life 49:341–351
15. Roundtree IA, Evans ME, Pan T, He C (2017) Dynamic RNA modifications in gene expression regulation. Cell 169:1187–1200
16. Deng X et al (2018) RNA N(6)-methyladenosine modification in cancers: current status and perspectives. Cell Res 28:507–517
17. Jia G et al (2011) N6-methyladenosine in nuclear RNA is a major substrate of the obesity-associated FTO. Nat Chem Biol 7:885–887
18. Zhao BS, Roundtree IA, He C (2016) Post-transcriptional gene regulation by mRNA modifications. Nat Rev Mol Cell Biol 18:31–42
19. Dominissini D et al (2012) Topology of the human and mouse m6A RNA methylomes revealed by m6A-seq. Nature 485:201–206
20. Meyer KD et al (2012) Comprehensive analysis of mRNA methylation reveals enrichment in 3′ UTRs and near stop codons. Cell 149:1635–1646
21. Zhao BS, Roundtree IA, He C (2017) Post-transcriptional gene regulation by mRNA modifications. Nat Rev Mol Cell Biol 18:31–42
22. Wang Y et al (2014) N6-methyladenosine modification destabilizes developmental regulators in embryonic stem cells. Nat Cell Biol 16:191–198
23. Zheng G et al (2013) ALKBH5 is a mammalian RNA demethylase that impacts RNA metabolism and mouse fertility. Mol Cell 49:18–29
24. Zhao X et al (2014) FTO-dependent demethylation of N6-methyladenosine regulates mRNA splicing and is required for adipogenesis. Cell Res 24:1403–1419
25. Geula S et al (2015) Stem cells. m6A mRNA methylation facilitates resolution of naive pluripotency toward differentiation. Science 347:1002–1006
26. Chen T et al (2015) m(6)A RNA methylation is regulated by microRNAs and promotes reprogramming to pluripotency. Cell Stem Cell 16:289–301
27. Zhou J et al (2015) Dynamic m(6)A mRNA methylation directs translational control of heat shock response. Nature 526:591–594
28. Xiang Y et al (2017) RNA m(6)A methylation regulates the ultraviolet-induced DNA damage response. Nature 543:573–576
29. Zhao BS et al (2017) m(6)A-dependent maternal mRNA clearance facilitates zebrafish maternal-to-zygotic transition. Nature 542:475–478
30. Zhang C et al (2017) m(6)A modulates haematopoietic stem and progenitor cell specification. Nature 549:273–276
31. Li Z et al (2017) FTO plays an oncogenic role in acute myeloid leukemia as a N(6)-methyladenosine RNA demethylase. Cancer Cell 31:127–141
32. Huang H et al (2018) Recognition of RNA N6-methyladenosine by IGF2BP proteins enhances mRNA stability and translation. Nat Cell Biol 20:285–295. In Press
33. Weng H et al (2018) METTL14 inhibits hematopoietic stem/progenitor differentiation and promotes leukemogenesis via mRNA m(6)A modification. Cell Stem Cell 22:191–205 e199
34. Vu LP et al (2017) The N(6)-methyladenosine (m(6)A)-forming enzyme METTL3 controls myeloid differentiation of normal hematopoietic and leukemia cells. Nat Med 23:1369–1376
35. Barbieri I et al (2017) Promoter-bound METTL3 maintains myeloid leukaemia by m(6)A-dependent translation control. Nature 552:126–131
36. Zhang S et al (2017) m(6)A demethylase ALKBH5 maintains tumorigenicity of glioblastoma stem-like cells by sustaining FOXM1 expression and cell proliferation program. Cancer Cell 31:591–606 e596
37. Su R et al (2018) R-2HG exhibits anti-tumor activity by targeting FTO/m(6)A/MYC/CEBPA signaling. Cell 172:90–105 e123
38. Deng X, Su R, Stanford S, Chen J (2018) Critical enzymatic functions of FTO in obesity and cancer. Front Endocrinol (Lausanne) 9:396

39. Deng X, Su R, Feng X, Wei M, Chen J (2018) Role of N(6)-methyladenosine modification in cancer. Curr Opin Genet Dev 48:1–7
40. Bird A (2002) DNA methylation patterns and epigenetic memory. Genes Dev 16:6–21
41. Bannister AJ, Kouzarides T (2011) Regulation of chromatin by histone modifications. Cell Res 21:381–395
42. Bokar JA, Shambaugh ME, Polayes D, Matera AG, Rottman FM (1997) Purification and cDNA cloning of the AdoMet-binding subunit of the human mRNA (N6-adenosine)-methyltransferase. RNA 3:1233–1247
43. Ping XL et al (2014) Mammalian WTAP is a regulatory subunit of the RNA N6-methyladenosine methyltransferase. Cell Res 24:177–189
44. Schwartz S et al (2014) Perturbation of m6A writers reveals two distinct classes of mRNA methylation at internal and 5′ sites. Cell Rep 8:284–296
45. Liu J et al (2014) A METTL3-METTL14 complex mediates mammalian nuclear RNA N6-adenosine methylation. Nat Chem Biol 10:93–95
46. Patil DP et al (2016) m(6)A RNA methylation promotes XIST-mediated transcriptional repression. Nature 537:369–373
47. Wen J et al (2018) Zc3h13 regulates nuclear RNA m(6)A methylation and mouse embryonic stem cell self-renewal. Mol Cell 69:1028–1038 e1026
48. Knuckles P et al (2018) Zc3h13/Flacc is required for adenosine methylation by bridging the mRNA-binding factor Rbm15/Spenito to the m(6)A machinery component Wtap/Fl(2)d. Genes Dev 32:415–429
49. Guo J, Tang HW, Li J, Perrimon N, Yan D (2018) Xio is a component of the Drosophila sex determination pathway and RNA N(6)-methyladenosine methyltransferase complex. Proc Natl Acad Sci U S A 115:3674–3679
50. Wang P, Doxtader KA, Nam Y (2016) Structural basis for cooperative function of Mettl3 and Mettl14 methyltransferases. Mol Cell 63:306–317
51. Wang X et al (2016) Structural basis of N(6)-adenosine methylation by the METTL3-METTL14 complex. Nature 534:575–578
52. Sledz P, Jinek M (2016) Structural insights into the molecular mechanism of the m(6)A writer complex. Elife 5:e18434
53. Warda AS et al (2017) Human METTL16 is a N(6)-methyladenosine (m(6)A) methyltransferase that targets pre-mRNAs and various non-coding RNAs. EMBO Rep 18:2004–2014
54. Pendleton KE et al (2017) The U6 snRNA m(6)A methyltransferase METTL16 regulates SAM synthetase intron retention. Cell 169:824–835 e814
55. Jia G et al (2008) Oxidative demethylation of 3-methylthymine and 3-methyluracil in single-stranded DNA and RNA by mouse and human FTO. FEBS Lett 582:3313–3319
56. Gerken T et al (2007) The obesity-associated FTO gene encodes a 2-oxoglutarate-dependent nucleic acid demethylase. Science 318:1469–1472
57. Mauer J et al (2017) Reversible methylation of m(6)Am in the 5′ cap controls mRNA stability. Nature 541:371–375
58. Kurowski MA, Bhagwat AS, Papaj G, Bujnicki JM (2003) Phylogenomic identification of five new human homologs of the DNA repair enzyme AlkB. BMC Genomics 4:48
59. Yang Y, Hsu PJ, Chen YS, Yang YG (2018) Dynamic transcriptomic m(6)A decoration: writers, erasers, readers and functions in RNA metabolism. Cell Res 28:616–624
60. Wang X et al (2014) N6-methyladenosine-dependent regulation of messenger RNA stability. Nature 505:117–120
61. Wang X et al (2015) N(6)-methyladenosine modulates messenger RNA translation efficiency. Cell 161:1388–1399
62. Xu C et al (2014) Structural basis for selective binding of m6A RNA by the YTHDC1 YTH domain. Nat Chem Biol 10:927–929
63. Xiao W et al (2016) Nuclear m(6)A reader YTHDC1 regulates mRNA splicing. Mol Cell 61:507–519
64. Luo S, Tong L (2014) Molecular basis for the recognition of methylated adenines in RNA by the eukaryotic YTH domain. Proc Natl Acad Sci U S A 111:13834–13839

65. Zhu T et al (2014) Crystal structure of the YTH domain of YTHDF2 reveals mechanism for recognition of N6-methyladenosine. Cell Res 24:1493–1496
66. Hsu PJ et al (2017) Ythdc2 is an N(6)-methyladenosine binding protein that regulates mammalian spermatogenesis. Cell Res 27:1115–1127
67. Roundtree IA et al (2017) YTHDC1 mediates nuclear export of N(6)-methyladenosine methylated mRNAs. eLife 6:e31311
68. Du H et al (2016) YTHDF2 destabilizes m(6)A-containing RNA through direct recruitment of the CCR4-NOT deadenylase complex. Nat Commun 7:12626
69. Shi H et al (2017) YTHDF3 facilitates translation and decay of N(6)-methyladenosine-modified RNA. Cell Res 27:315–328
70. Li A et al (2017) Cytoplasmic m(6)A reader YTHDF3 promotes mRNA translation. Cell Res 27:444–447
71. Bailey AS et al (2017) The conserved RNA helicase YTHDC2 regulates the transition from proliferation to differentiation in the germline. Elife 6:e26116
72. Wojtas MN et al (2017) Regulation of m(6)A transcripts by the 3′-->5′ RNA helicase YTHDC2 is essential for a successful meiotic program in the mammalian germline. Mol Cell 68:374–387 e312
73. Jain D et al (2018) ketu mutant mice uncover an essential meiotic function for the ancient RNA helicase YTHDC2. Elife 7:e30919
74. Alarcon CR et al (2015) HNRNPA2B1 is a mediator of m(6)A-dependent nuclear RNA processing events. Cell 162:1299–1308
75. Wu B et al (2018) Molecular basis for the specific and multivariant recognitions of RNA substrates by human hnRNP A2/B1. Nat Commun 9:420
76. Liu N et al (2015) N(6)-methyladenosine-dependent RNA structural switches regulate RNA-protein interactions. Nature 518:560–564
77. Edupuganti RR et al (2017) N(6)-methyladenosine (m(6)A) recruits and repels proteins to regulate mRNA homeostasis. Nat Struct Mol Biol 24:870–878
78. Arguello AE, DeLiberto AN, Kleiner RE (2017) RNA chemical proteomics reveals the N(6)-methyladenosine (m(6)A)-regulated protein-RNA interactome. J Am Chem Soc 139:17249–17252
79. Doulatov S, Notta F, Laurenti E, Dick JE (2012) Hematopoiesis: a human perspective. Cell Stem Cell 10:120–136
80. Rosenbauer F, Tenen DG (2007) Transcription factors in myeloid development: balancing differentiation with transformation. Nat Rev Immunol 7:105–117
81. Koschmieder S, Rosenbauer F, Steidl U, Owens BM, Tenen DG (2005) Role of transcription factors C/EBPalpha and PU.1 in normal hematopoiesis and leukemia. Int J Hematol 81:368–377
82. Goode DK et al (2016) Dynamic gene regulatory networks drive hematopoietic specification and differentiation. Dev Cell 36:572–587
83. Rosmarin AG, Yang Z, Resendes KK (2005) Transcriptional regulation in myelopoiesis: hematopoietic fate choice, myeloid differentiation, and leukemogenesis. Exp Hematol 33:131–143
84. Dakic A et al (2005) PU.1 regulates the commitment of adult hematopoietic progenitors and restricts granulopoiesis. J Exp Med 201:1487–1502
85. Sandberg ML et al (2005) c-Myb and p300 regulate hematopoietic stem cell proliferation and differentiation. Dev Cell 8:153–166
86. Mucenski ML et al (1991) A functional c-myb gene is required for normal murine fetal hepatic hematopoiesis. Cell 65:677–689
87. Wilson A et al (2004) c-Myc controls the balance between hematopoietic stem cell self-renewal and differentiation. Genes Dev 18:2747–2763
88. Satoh Y et al (2004) Roles for c-Myc in self-renewal of hematopoietic stem cells. J Biol Chem 279:24986–24993
89. Yao QJ et al (2018) Mettl3-Mettl14 methyltransferase complex regulates the quiescence of adult hematopoietic stem cells. Cell Res 28:952–954

90. Raffel GD et al (2007) Ott1(Rbm15) has pleiotropic roles in hematopoietic development. Proc Natl Acad Sci U S A 104:6001–6006
91. Ma X et al (2007) Rbm15 modulates Notch-induced transcriptional activation and affects myeloid differentiation. Mol Cell Biol 27:3056–3064
92. Niu C et al (2009) c-Myc is a target of RNA-binding motif protein 15 in the regulation of adult hematopoietic stem cell and megakaryocyte development. Blood 114:2087–2096
93. Li Z et al (2018) Suppression of m(6)A reader Ythdf2 promotes hematopoietic stem cell expansion. Cell Res 28:904–917
94. Ye M et al (2015) Hematopoietic differentiation is required for initiation of acute myeloid leukemia. Cell Stem Cell 17:611–623
95. Testa U (2011) Leukemia stem cells. Ann Hematol 90:245–271
96. Dohner H, Weisdorf DJ, Bloomfield CD (2015) Acute myeloid leukemia. N Engl J Med 373:1136–1152
97. Siegel RL, Miller KD, Jemal A (2017) Cancer statistics, 2017. CA Cancer J Clin 67:7–30
98. Fischer J et al (2009) Inactivation of the Fto gene protects from obesity. Nature 458:894–898
99. Church C et al (2010) Overexpression of Fto leads to increased food intake and results in obesity. Nat Genet 42:1086–1092
100. Merkestein M et al (2015) FTO influences adipogenesis by regulating mitotic clonal expansion. Nat Commun 6:6792
101. Soderberg KC et al (2009) Overweight, obesity and risk of haematological malignancies: a cohort study of Swedish and Finnish twins. Eur J Cancer 45:1232–1238
102. Castillo JJ, Mull N, Reagan JL, Nemr S, Mitri J (2012) Increased incidence of non-Hodgkin lymphoma, leukemia, and myeloma in patients with diabetes mellitus type 2: a meta-analysis of observational studies. Blood 119:4845–4850
103. Hernandez-Caballero ME, Sierra-Ramirez JA (2015) Single nucleotide polymorphisms of the FTO gene and cancer risk: an overview. Mol Biol Rep 42:699–704
104. Wang F et al (2013) Targeted inhibition of mutant IDH2 in leukemia cells induces cellular differentiation. Science 340:622–626
105. Sasaki M et al (2012) IDH1(R132H) mutation increases murine haematopoietic progenitors and alters epigenetics. Nature 488:656–659
106. Losman JA et al (2013) (R)-2-hydroxyglutarate is sufficient to promote leukemogenesis and its effects are reversible. Science 339:1621–1625
107. Martin GH, Park CY (2018) Meddling with METTLs in normal and leukemia stem cells. Cell Stem Cell 22:139–141
108. Bansal H et al (2014) WTAP is a novel oncogenic protein in acute myeloid leukemia. Leukemia 28:1171–1174
109. Ma Z et al (2001) Fusion of two novel genes, RBM15 and MKL1, in the t(1;22)(p13;q13) of acute megakaryoblastic leukemia. Nat Genet 28:220–221
110. Pollyea DA, Jordan CT (2017) Therapeutic targeting of acute myeloid leukemia stem cells. Blood 129:1627–1635
111. Misaghian N et al (2009) Targeting the leukemic stem cell: the Holy Grail of leukemia therapy. Leukemia 23:25–42
112. Yang ZJ, Wechsler-Reya RJ (2007) Hit ‘em where they live: targeting the cancer stem cell niche. Cancer Cell 11:3–5
113. Borovski T, De Sousa EMF, Vermeulen L, Medema JP (2011) Cancer stem cell niche: the place to be. Cancer Res 71:634–639
114. Plaks V, Kong N, Werb Z (2015) The cancer stem cell niche: how essential is the niche in regulating stemness of tumor cells? Cell Stem Cell 16:225–238
115. Beck B, Blanpain C (2013) Unravelling cancer stem cell potential. Nat Rev Cancer 13:727–738
116. Thomas D, Majeti R (2017) Biology and relevance of human acute myeloid leukemia stem cells. Blood 129:1577–1585

Chapter 5
Leukemia Stem Cells in the Pathogenesis, Progression, and Treatment of Acute Myeloid Leukemia

Kanak Joshi, Lei Zhang, Peter Breslin S.J., and Jiwang Zhang

Contents

Abstract Despite the significant progress that has been made in understanding the biology of leukemia stem cells (LSCs), some key questions regarding the concept of LSCs have not as yet been satisfactorily addressed experimentally. As a result, the clinical relevance of LSCs remains less than clear due to controversies caused largely by technical limitations in efficiently identifying LSCs. This has impeded

---

K. Joshi · L. Zhang · J. Zhang (✉)
Oncology Institute, Cardinal Bernardin Cancer Center, Loyola University Chicago, Chicago, IL, USA

Department of Cancer Biology, Cardinal Bernardin Cancer Center, Loyola University Chicago, Chicago, IL, USA
e-mail: jzhang@luc.edu

P. Breslin S.J.
Oncology Institute, Cardinal Bernardin Cancer Center, Loyola University Chicago, Chicago, IL, USA

Department of Cancer Biology, Cardinal Bernardin Cancer Center, Loyola University Chicago, Chicago, IL, USA

Department of Biology, Loyola University Chicago, Chicago, IL, USA

H. Zhang, S. Li (eds.), *Leukemia Stem Cells in Hematologic Malignancies*, Advances in Experimental Medicine and Biology 1143,
https://doi.org/10.1007/978-981-13-7342-8_5

our ability to fully address the features of genetic heterogeneity and metabolic/epigenetic plasticity of pre-LSCs and LSCs. With the development and use of humanized immunocompromised mice, we are able to more precisely analyze LSCs for their functions and interaction with the bone marrow niche. In addition, some promising targets in LSCs have recently been identified, including Sonic Hedgehog (SHH) and BCL-2, which are highly expressed in AML cells. It is hopeful that new anti-LSC compounds will be tested fully in clinical trials for their efficacy in treating human leukemias.

Keywords Leukemia stem cell · Humanized immunocompromised mice · AML

5.1 Introduction

Acute myeloid leukemia (AML) is a common type of hematopoietic malignancy which primarily develops in the elderly population. The average age at diagnosis for AML is 67 years. The 7 + 3 regimen of intensive chemotherapy consists of 3 days of an anthracycline antibiotic or an anthracenedione + 7 days of standard-dose cytarabine (ARA-C). This has been considered the first line of treatment for AML for over 45 years. 7 + 3 treatment can induce an overall response (OR) including complete remission (CR) or CR with incomplete count recovery (CRi) in 60–80% of patients. However, most elderly patients cannot tolerate such an intense regimen of antineoplastic agents. Currently, the 5-year overall rate of survival (OS) for AML is 35–40% for patients under 60 years of age and <10% for patients over the age of 60 years at the time of diagnosis [1–6].

For patients who fail to achieve CR/CRi and may die of either treatment-related complications or primary drug resistance-related disease progression, novel medications and medication combinations are critically needed, especially for elderly patients. Over the past several years, many new regimens such as CPX-351 (a liposomal formulation of cytarabine and daunorubicin encapsulated at a synergistic 5:1 molar ratio) and hypomethylating agents (HMAs), including azacitidine (AZA) and decitabine (DEC), have been approved by the FDA for the treatment of elderly patients who cannot tolerate 7+3 chemotherapy [7]. CPX-351 treatment significantly increases the CR/CRi rate (47.7% for CPX-351 vs. 33.3% for 7+3) and OS (median 9.56 months for CPX-351 vs. 5.95 for standard 7+3) in elderly AML patients [7]. Although HMAs can only induce CR/CRi in 15–25% of elderly AML patients when used as a monotherapy, such treatment can significantly extend OS (median 10.4 months for HMAs vs. 6.5 months for 7+3) [8–12]. Most recently, the successful use of the BCL-2 inhibitor Venetoclax (VEN) and AZA in combination to treat elderly patients represents a potentially revolutionary change in the treatment of AML [13–16].

For those AML patients who have achieved CR/CRi, disease relapse is one of the major causes of death, owing to the inability to achieve complete elimination of leukemia-regenerating cells (LRCs), which form the basis of minimal residual disease (MRD) and are responsible for the relapse of leukemia [17–22]. The prevention of disease relapse in such patients by maximally eliminating LRCs in order to reduce the potential for MRD is the most urgent issue that needs to be addressed clinically in order to arrive at the goal of a cure for this disease. Most previous studies suggested that LRCs are derived from leukemia stem cells (LSCs), which are resistant to chemotherapy and are responsible for disease relapse. Thus, a combination strategy involving targeting LSCs plus chemo-debulking therapy has been proposed as an effective treatment for AML [23–31]. However, such a concept is challenged by two recent research reports which demonstrate that these chemoresistant LRCs in MRD are not derived from naïve LSCs [32, 33]. Thus a more complete understanding of the nature of the LRCs in MRD and an elucidation of the mechanism by which MRD occurs will help us to develop therapies to better target LRCs in MRD by (1) preventing the possibility of MRD occurring, (2) eliminating MRD when it does develop, and (3) inhibiting the propagation of LRCs to significantly decrease the likelihood of disease relapse.

Although the classical LSC model provides a plausible explanation for clinical observations in AML and also includes relevant implications for therapy, yet due to technical limitations as well as the dynamic and heterogeneous nature of LSCs, this model still lacks direct evidence to fully support the notion that LRCs in MRD are derived from LSCs. As a consequence, inconsistencies and even controversies have been associated with previous studies [34–36]. In this review, we will update the most recent advances in LSC research and discuss the significance and limitations of the LSC concept, as well as those related to the potential targeting of LSCs to treat AML. Thus, we hope to be able to provide reasonable explanations for the controversial aspects of LSC research, as well as cancer stem cells (CSCs) in solid tumors, and propose useful concepts that might guide future LSC and CSC studies.

5.2 The Concept of LSCs

It is now well-established that almost any cancer, including AML, can be conceptualized as a heterogeneous mixture of malignant types of cells which share some properties but which also allow them to be distinguished morphologically, genetically, and functionally [37]. The LSC concept was introduced in the 1990s, first for AML and subsequently for other hematopoietic malignancies, to explain the heterogeneity of malignant cells by reference to the hierarchical model of cellular organization of tissues described for normal hematopoiesis [23, 38, 39]. According to this model, there is a subpopulation of leukemic cells (LCs) in AML tissues which is located at the apex of the differentiation hierarchy which gives rise to highly proliferative leukemic progenitors (LPs) and subsequently to partially differentiated leukemic blasts (LBs) which maintain disease progression and allow for metastasis [40]. It is still generally accepted that most current chemotherapeutic agents target rapidly dividing LPs and LBs, which constitute the bulk of malignant cells in leukemia. But these medications cannot kill LSCs because the latter are relatively slow cycling (i.e., quiescent) cells [40]. The LSC concept is primarily intended to focus on and explain several critical questions in leukemia research and attempts to identify factors which lie at the root of attempts to cure leukemia. Such questions include the following: (1) From which cell(s) do(es) the malignancy originate? (2) During the progression of leukemia, which cells propagate the leukemic tissue in both primary tumors and distant metastases? (3) During treatment, which cells are resistant to therapy? (4) After therapy-induced remission of leukemia, which cells remain and form the basis of MRD and ultimately lead to disease relapse? [23, 41]

Using fluorescence-conjugated antibody staining and fluorescence-activated cell sorting (FACS) techniques, molecular oncologists are able to separate hematopoietic cells into multiple subpopulations based on their unique combinations of cell surface antigens. Then, using transplantation assays, they can identify a small subset of hematopoietic cells with a unique phenotype, the hematopoietic stem cells (HSCs), which are capable of reconstituting multiple lineages of hematopoiesis in recipient animals over the long term [42, 43]. Using exactly this technical strategy, John Dick and colleagues demonstrated in 1994 and again in 1997 that there is a small subset of malignant cells in the bone marrow (BM) or peripheral blood (PB) of most AML patients which can regenerate the same type of AML in recipient animals upon transplantation [38, 39]. Because severe combined immune deficiency mice (SCID mice) were used as recipients in their first study, Dick and colleagues named such cells SCID leukemia-initiating cells or SL-ICs [38]. These cells are $Lin^-CD34^{++}CD38^-$ and share a similar phenotype with normal HSCs [39]. Due to their unique ability to regenerate AML in recipient mice, it was speculated that SL-SCs are the LSCs which lie at the root of AML development. In addition, many follow-up studies suggested that LSCs are relatively resistant to most types of chemotherapeutic agents [44–47]; thus it was conjectured that LSCs are the cause of relapse of leukemia following treatment-initiated remission. Based on these early studies, the combination of FACS purification and in vivo transplantation has been used as the standard method to define the phenotype of LSCs, the limited dilution and competitive transplantation assay has been used as a standard method to evaluate the frequency of LSCs in a given population of AML cells, while the serial transplantation assay has been used to detect the self-renewal capacity of LSCs [48]. Exactly the same series of assays has been extended to studies of CSCs in solid tumors [41]. This LSC concept has been most clearly documented in chronic myeloid leukemia (CML) in the chronic stage and provides the best explanation for the clinical observations seen in malignancies such as CML, which shows a relatively simple genetic abnormality and a linear clonal evolution program [49–55]. LSCs in CML are genetically homogeneous with only one driver mutation, *BCR-ABL*, in the chronic stage [56].

It was known that CML is initiated from an HSC that harbors a chromosomal translocation (t9:21) that creates the BCR-ABL fusion protein [57]. The LSCs in CML have a phenotype similar to normal HSCs as demonstrated by similar cell surface markers, slow cell cycling kinetics, and low levels of reactive oxygen species (ROS). Tyrosine kinase inhibitors (TKIs) such as imatinib [58], the first type of drug to directly target BCR-ABL, effectively induce a major or complete molecular remission (CMR) and prolong survival in most patients with CML. However, only

10–20% of patients achieve successful treatment-free remission (TFR). Approximately one-fourth of patients fail TKI therapy due to unique *BCR-ABL* kinase mutations, alternative oncogene activation, or because of progression to an accelerated phase or blast crisis [49], whereas the remaining 50–60% of patients experience disease relapse after discontinuation of TKI therapy due to the insensitivity of their LSCs to TKIs [53, 59, 60]. LSCs in CML are also phenotypically and epigenetically heterogeneous, just as normal HSCs are. The most TKI-resistant LSCs are $Lin^-CD34^+CD38^{-/low}CD45RA^-cKIT^-CD26^+$ [61]. In most patients, TKI treatment fails to completely eliminate LSCs, thus rendering TKI therapy a lifelong requirement for them [62–65]. Studies have suggested that in contradistinction to LPs and LBs, LSCs in CML are not reliant upon BCR-ABL activity for their survival [53, 59]. CML LSCs express high levels of the thrombopoietin (THPO) receptor MPL and high levels of BCR-ABL protein, and also show high levels of autophagic activity [66–71]. It was proposed that LSCs in CML might depend upon THPO-MPL-stimulated activation of JAK-STAT-BCL-xL and MAPK signaling, BCR-ABL kinase activity-independent activation of the JAK2/PP2A/β-catenin network, or digested products of autophagy for their survival [66–71]. Thus to achieve a sustained, deep CMR and TFR, a combination of a TKI together with inhibitors of these signaling pathways is necessary in order to completely eliminate LSCs in CML [66–71].

However, in AML, the LSC concept is complicated by the existence of a founder clone and the subsequent development of subclones which can be found in patient samples and is explainable by the complicated clonal evolutionary history and complex genetic abnormalities that occur during the pathogenesis of AML. In addition, to complicate matters further, compared to LSCs in CML, LSCs in AML possess relatively greater plasticity [23–25, 27, 29, 31]. The controversies which arose from previous studies might be traceable to limitations imposed by the techniques used, which resulted in a failure to integrate these phenotypic and genetic heterogeneities with the functions of LSCs.

5.3 Genetic Heterogeneity of LSCs in AML

AML can be classified into several distinct subtypes based on cellular morphology and stage of differentiation. The phenotypes and xenograft capacities of LSCs from different subtypes of AML are not the same [72–74]. More importantly, genetic mutations observable in different patients' blood and BM are also not identical even when they are found within the same classification subtype.

Recurrent somatic myeloid malignancy-associated mutations (including chromosomal translocations, cytogenetic abnormalities, and point mutations) are detected in BM and peripheral blood (PB) samples from almost all AML patients. Studies have suggested that AML develops from an original mutant HSC by clonal evolution and dynamic disease developmental processes [75–78]. Mutations of DNA methylation regulator proteins, such as DNMT3A, TET2, and IDH1/IDH2, histone modifiers such as ASXL1, splicing factors (including SF3B1, SRSF2, and U2AF1), and p53 are detected in ~10% of all individuals by the age of 70 and increase further as aging proceeds further [79–81]. This is referred to as age-related clonal hematopoiesis (ARCH) or clonal hematopoiesis of indeterminate potential (CHIP), a pre-myelodysplastic syndrome (MDS) condition without a clear diagnosis of any specific hematologic abnormality. HSCs with such founder mutations are predisposed to lead eventually to MDS, a preleukemic condition which can subsequently transform to actual AML by accumulating additional mutations. The average number of mutations observed in MDS patients' cells is 3, while AML patients' cells harbor 5–12 mutations. In cases of ARCH, if mutations of *SF3B1*, *SRSF2*, *U2AF1,* and *TP53* are observed, it is predictive of progression to MDS and subsequent transformation to AML. However, if mutations of *EZH2*, *IDH2*, *ASXL1*, *DNMT3A*, *TP53*, and *SRSF2* are detected in MDS patients, the likely predictable outcome will be AML transformation and shorter survival time [82–88]. The most common additional mutations during the MDS-to-AML transformation process are inactivating mutations in core hematopoietic transcription factor genes (such as *RUNX1*, *CEBPA*, *EVI1*, *GATA2*, and *ETV6*) and activating mutations in proliferative signaling pathways (such as *FLT3*, *N-RAS/K-RAS*, *PTPN11*, *KIT*, *JAK2*, *MPL*, *CBL*, and *NF1* family members) in a subclone of diseased HSCs or hematopoietic progenitor cells (HPCs) [89–91]. Studies suggested that it might take >10 years for HSCs with founder mutations to accumulate sufficient additional mutations to lead to a full transformation to AML [92]. This suggests that these hematologic malignancies proceed by a dynamic developmental process of clonal evolution from mutant HSCs in ARCH to MDS and finally to AML (Fig. 5.1). Importantly, such clonal evolution is not a strictly linear process. Multiple subclones are generated along the way. Thus, in addition to the founder clone (with 1–2 mutations) which is generated in ARCH and the intermediate preleukemic clone(s) that is (are) generated during the MDS stage, more than one leukemic subclone likely exists in any AML patient. Each of these leukemic subclones shares the same founder mutation(s) but likely has distinct additional mutations. In most cases, one of the subclones might become dominant due to the growth advantage it has acquired by virtue of its unique set of mutant genes, while in other cases there exists more than one detectable codominant clone. Theoretically, each subclone of LCs might have its own distinct LSCs. The phenotype and leukemogenic activity of LSCs from different subclones might differ. In addition, there are also stem cells for the founder clone (pre-pre-LSCs) and pre-LSCs for the preleukemic clones [93–95]. The phenotype and leukemogenic capacity of pre-LSCs in low-risk MDS clones are closer to pre-pre-LSCs, whereas the phenotype and leukemogenic capacity of pre-LSCs in high-risk MDS clones are closer to LSCs [96, 97]. Thus LSCs in AML patients are heterogeneous both phenotypically and genetically. In the same patient's specimen, multiple phenotypic LSCs can be detected. LSCs from most cases have an lymphoid-primed multipotent progenitor (LMPP)-like, GMP-like, or even committed myeloid precursor-like phenotype, while pre-pre-LSCs and pre-LSCs normally present with an HSC phenotype [27, 98, 99]. To further complicate these issues, distinct genomic abnormalities can be detected in LSCs of the same phenotype. Thus, in some cases,

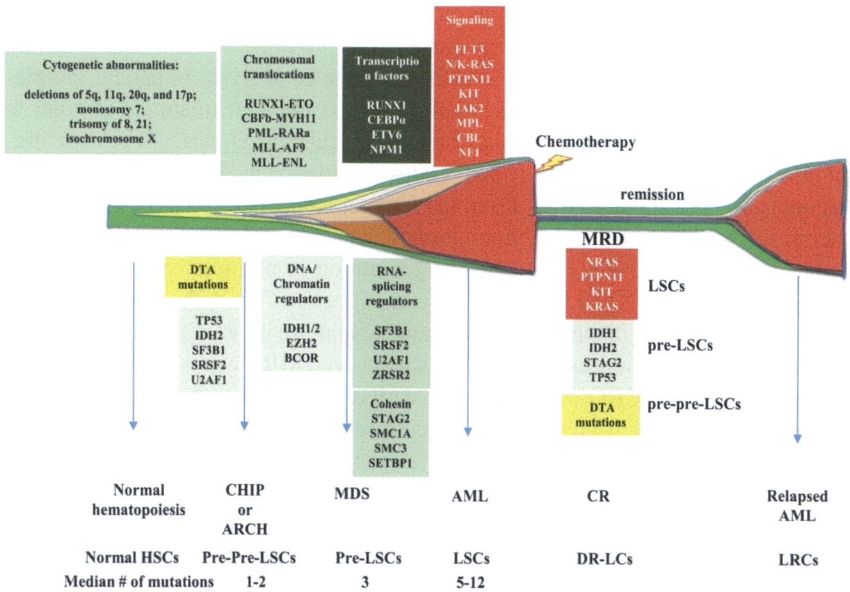

Fig. 5.1 The dynamic pathogenesis and genetic heterogeneity of AML. From the founder clone, harboring the first genetic alteration, to AML transformation, multiple genetic abnormalities must accumulate sequentially, which leads to the generation of multiple subclones of pre-LCs and LCs. Pre-LCs and LCs from different subclones share the founder genetic alteration(s) and have distinct additional genetic abnormalities. Chemotherapy kills most of the LCs and induces CR in patients. However, MRD can be detected in most CR patients. If the MRD contains LSCs derived from either the major or minor leukemic clones, it will predict early disease relapse, while if the MRD contains pre-LSCs, it will predict a later disease relapse. If the MRD contains pre-pre-LSCs, it will predict no disease relapse

only the LSCs for the dominant clone(s) might able to survive and regenerate AML in xenograft models, while in some other cases, LSCs from both the dominant and minor clone(s) can grow. In addition, in most cases, pre-pre-LSCs and pre-LSCs are not able to grow in xenograft models [100]. Thus the AML seen in xenografts based on current models (which will be discussed below) might not fully represent the true genetic heterogeneity that is a feature of all AML in human patients.

In AML patients, morphologic CR can be achieved by eliminating >95% of LCs. The persistence of LRCs from dominant and minor clones in MRD predicts early disease relapse [21, 101]. To achieve deep CR and long-term disease-free survival, all LCs in the dominant and minor subclones, including LSCs, need to be completely eliminated. Fortunately, in most cases, the LCs are relatively sensitive to standard chemotherapies due to their high proliferation status. LSCs can also be eliminated in many cases. However, pre-pre-LSCs and pre-LSCs are quiescent in terms of the cell cycle and are therefore much less susceptible to chemotherapeutic interventions [88, 99, 102]. Thus evidence of pre-pre-leukemic clone and pre-leukemic clones can be detected in MRD in most patients after chemotherapy-induced CR/CRi. The persistence of pre-LSCs in MRD predicts a later disease relapse due to the additional mutations that must accumulate in order for active AML to be regenerated. Fortunately, the persistence of pre-pre-LSCs in MRD does not predict disease relapse [103, 104]. Thus, more detailed analyses of patient samples using more sensitive single-cell techniques prior to and after treatment in order to separate founder clone, pre-leukemic clones, and leukemic clones will provide useful information to more accurately predict disease relapse and better instruct clinical treatment choices and decisions. From the founder clone, harboring the first genetic alteration, to AML transformation, multiple genetic abnormalities need to be accumulated sequentially, which leads to the generation of multiple subclones of pre-LCs and LCs. Pre-LCs and LCs from different subclones share the founder genetic alteration(s) and have distinct additional genetic abnormalities. Chemotherapy kills most of the LCs and induces CR in patients. However, MRD can be detected in most CR patients. If the MRD contains LSCs derived from either the major or minor leukemic clones, it will predict early disease relapse, while if the MRD contains pre-LSCs, it will predict a later disease relapse. In the most favorable cases, the MRD contains pre-pre-LSCs and no disease relapse will be predicted.

It should be emphasized that, although approximately 80% of AML are de novo and develop without an MDS stage (lack of hematopoietic symptoms before AML development), the dynamic and multistep clonal evolution process is also confirmed in the pathogenesis of de novo AML by demonstrating the hierarchies and patterns by which the mutant clones evolve [92, 93, 95, 102, 105–108]. The reason why only ~20% of AML patients have an MDS stage might have to do with the types, the emergence time sequence, and the variant allele frequency of the genetic abnormalities. For example, mutations of epigenetic regulators such as *DNMT3A*, *ASXL1*, and *TET2* are most commonly detected in ARCH; these are founder mutations for most hematopoietic malignancies including MDS and AML [80]. Such mutations promote HSC self-renewal and clonal expansion by resetting the epigenetic landscape. In addition, HSCs with *DNMT3A*, *ASXL1*, and *TET2* mutations are also predisposed to MDS and AML due to increased genomic instability. However, mutations in genes for splicing factors (including *SF3B1*, *SRSF2*, *U2AF1*, and *ZRSR2*) and chromatin regulators (such as *EZH2*, *IDH1/IDH2*, and *BCOR*), as well as cytogenetic abnormalities (including deletions of 5q, 11q, 20q, and 17p, monosomy 7, trisomies of 8 and 21, and isochromosome X), are the most commonly detected gene alterations in MDS. Such genetic abnormalities lead to inefficient hematopoiesis and cytopenia by inducing programmed death in hematopoietic progenitor cells such as MEPs and GMPs. In support of this notion, significantly more cell death was detected in BM hematopoietic cells in low- and intermediate-risk MDS patients. It was found that extrinsic pathway apoptosis was increased in *SF3B1*- and *SFSR2*-mutant HPCs due to the abnormal splicing of *MAP3K7* and *Caspase-8* [109, 110]. As a consequence, despite the expansion of the primitive HSC compartment observed in most MDS patients, both the long-term and short-term hematopoietic reconstitutive capacities of MDS HSCs are comprised [111, 112]. This explains why it has been so challenging to generate MDS xenografts using current animal models. Additional mutations are required for disease progression and AML transformation by preventing cell death and promoting proliferation of HPCs. In addition, when AML presents with such MDS-related genetic abnormalities, it usually suggests secondary AML which developed from MDS [85, 88, 97, 113, 114]. However, in de novo AML, chromosomal translocation-associated leukemic fusion genes and/or mutations in *NPM1*, transcription factors such as *RUNX1* and *CEBPα*, and/or proliferative signaling pathway genes (including *N-Ras*, *K-Ras*, *FLT3* or *c-Kit*) are typically observed. Such genetic abnormalities promote self-renewal in HSCs and proliferation in HPCs without inducing cell death. This explains why AML patients having such mutations normally do not go through an MDS stage [106, 107].

5.4 Plasticity of LSCs

Compared to HPCs and mature blood cells, HSCs are relatively more plastic, which explains why they are able to give rise all lineages of blood cells. During normal hematopoiesis, HSCs differentiate to produce MPPs and, subsequently, to committed progenitors, which then produce mature blood cells, a process in which they lose both their self-renewal capacity and multi-lineage plasticity. Such processes are regulated by BM microenvironmental factors which induce dynamic gene expression by initiating specific transcriptional and epigenetic programs [115].

Mutations of epigenetic modifiers are commonly detected in AML patients (described above). In addition, mutations of signaling molecules and transcription factors also induce changes in epigenetic patterns. Thus during AML development, genetic lesions induce a dynamic epigenetic reprogramming (including histone modifications and DNA methylation), unleashing cellular plasticity and leading to oncogenic cellular reprogramming. Therefore, compared to normal hematopoietic counterparts, LCs show increased epigenetic plasticity. Although distinct epigenetic alterations and gene expression profiles were identified in patient LCs with distinct genetic mutations, suggesting a genetic mutation-driven, unique epigenetic reprogramming [106, 116, 117], LSCs from almost all patients share a similar "stemness" epigenetic signature which is largely independent of genetic mutations [118, 119]. Such an epigenetic signature is correlated to a unique expression profile of "stemness" genes and xenograft ability which predicts adverse patient outcome [119, 120]. These studies suggested that LSC properties are regulated by a unique epigenetic signature which is independent of genetic mutations. Such an epigenetic signature and stemness gene expression pattern can be induced by the BM niche microenvironment in non-LSCs which may convert them to LSCs, suggesting a relatively high degree of plasticity for LCs [36]. Interestingly, such LSC signatures are associated with relative hypomethylation compared to LBs, which might explain why HMA therapy fails to induce a durable CR in AML and MDS patients since HMAs are unable to eradicate LSCs [17].

5.5 Technical Limitations of Research on LSCs

Although the immune rejection of transplanted human AML cells is significantly reduced in NOD/SCID mice due to the lack of T- and B-cell-mediated adaptive immune responses, the fact that natural killer (NK) and Mφ-/dendritic cell-mediated innate immune reactions also restrict the growth of human AML cells explains the low efficiency of engraftment and low frequency of LSCs detected when utilizing NOD/SCID mice as transplantation recipients. In addition, due to the short lifespan of NOD/SCID mice (~8 months), such mice are not useful for the study of LSCs when long-term time courses are required in order to regenerate leukemia. To solve such a problem, several additional mouse strains were generated. NSG and NRG mice are the result of crossing NOD.CB17-Prkdc^{scid}/J mice with B6.129S4-IL-2Rγ^{null} or NOD-Rag1^{null} and NOD-SCID IL-2Rγ^{null} mice. These mouse strains are deficient not only in mature lymphocytes but also in NK cell cytotoxic activity and Mφ/dendritic cell activity due to the targeted mutation of the *interleukin 2 (IL-2) receptor common gamma chain* gene (IL-2Rγ^{null}); thus these newer strains are much more efficient in supporting the growth of patients' LSCs compared to NOD/SCID mice. In addition, the lifespans of both NSG and NRG mice are significantly extended compared to NOD/SCID mice, which makes them suitable hosts for studying the long-term effects of transplanted xenoengraftments [121–123]. As a consequence, higher frequencies of LSCs in AML samples can be detected using NSG and NRG mice compared to what had been predicted by earlier studies using NOD/SCID mice [73]. Moreover, multiple phenotypes of LSCs were identified and coexisted within the same patient samples. It was documented that LSCs exist not only among LCs within the HSC phenotype but also in LCs with phenotypes of multipotent progenitors (MPPs), granulocyte-monocyte progenitors (GMP), and even more committed granulocyte or monocyte precursors. For example, Goardon et al. demonstrated that in most cases LSCs are enriched in the CD34^{+} population. At least two types of LSCs, Lin^{-}CD34^{+}CD38^{-}CD90^{-}CD45RA^{+} LMPP-like and Lin^{-}CD34^{+}CD38^{+}CD123$^{+/lo}$CD110^{-}CD45RA^{+} GMP-like, were found to coexist within the same patient sample [98]. In some cases, especially in *NPM1*-mutant and *TET2*-mutant AML, LSCs are enriched in the CD34^{-} population [124, 125]. By comparing gene expression profiles among LSCs, non-LSCs, and HSCs, several cell surface markers for LSCs have been identified, including CD123 (IL-3R), CLL-1, TIM3, CD44, CD96, CD47, CD133, IL-1RAP, CD117, CD32, CD93, GPR25, and CD25. Some of these have been used to further separate true functional LSCs from non-LSCs [108, 126–142]. Some of these markers, such as CD123, are associated with *FLT3-ITD* mutations [96, 143, 144] and together predict disease progression and relapse in both AML [45, 145] and MDS [92, 93]. These studies all suggest that there is significantly greater phenotypic diversity among LSCs from AML patients compared to those from CML patients.

Technically, NSG and NRG mice are comparable in terms of xenograft efficiency (Table 5.1). However, due to the lack of the DNA damage repair gene *Prkdc*, NSG mice are highly sensitive to treatment with DNA-damaging agents such as radiation

Table 5.1 Animal models used in LSC research

Mouse strains	Mutant or transgenic genes	T	B	NK	Mϕ	DNA repair	Lifespan
Nod/Scid	NOD/Prkdc^{scid}	–	–	+	+/–	Defective	~8.5 mos
NSG/NOG	NOD/Cg-Prkdc^{scid}/Il2Rγ^{null}/SzJ	–	–	–	+/–	Defective	~20 mos
NRG	NOD/Cg-Rag1^{null}/IL2Rγ^{null}	–	–	–	+/–		~20 mos
NSGS	NSG-Tg hIL3, SCF, GM-CSF	–	–	–	+/–	Defective	~18 mos
NRGS	NRG-Tg hIL3, SCF, GM-CSF	–	–	–	+/–		~18 mos
MISTRG	Rag2^{null}IL2Rγ^{null} -Tg hM-CSF,IL-3,GM-CSF, SIRPα,TPO	–	–	–	+/–		~7 mos
SCID-Hu BLT mouse	NSG or NRG mice with humanized BM, liver, and thymus	–	–	+	+/–		~7 mos
NSG-human BM ossicle	NSG mice with ectopic human BM ossicles	–	–	–	+/–		

and chemotherapeutic medications, more so than NRG mice are [146]. Thus NRG mice might be more appropriate experimental subjects when radiation and/or chemotherapy are used in follow-up treatment studies. However, due to the relatively low homology of many key cytokines such as IL-3, GM-SCF, and SCF, between humans and mice, murine cytokines cannot effectively stimulate the growth of human AML cells. *To wit.*, low ratios of AML chimeras were detected in the PB of NSG and NRG mice. In many cases, although successful engraftment of AML cells was detected in the BM, low numbers of human AML cells were detected in spleens and PB [147]. To solve such problems, scientists developed NSGS (NSG-SGM3) and NRGS (NRG-SGM3) mice by crossing NSG and NRG mice with mice that transgenically express human IL-3, GM-SCF, and SCF (~2000–4000 pg/ml). Both NSGS and NRGS mice express the human IL-3, GM-SCF, and SCF cytokines, which significantly enhance the growth of human AML cells (~1.5 to fivefold relative to NSG mice). The success of engraftment can be detected within 4 months, which is a shorter latency period than exists for NSG and NRG mice. In addition, a higher ratio of AML chimeras can be detected in the BM, spleens, and PB of successfully engrafted mice, therefore better recapitulating the patients' AML [147]. However, there is also a concern that higher physiological levels of these cytokines in such mice might be altering the behaviors of LCs and LSCs. Moreover, the addition of more human cytokines might further promote the growth of human LCs such as in MISTRG mice [148]. However, more human cytokines seem not to be good for the normal development of mice, which consequently reduces their lifespan. Thus, better models, in which normal physiological levels of key human cytokines are expressed, such as in NOG-EXL mice (which express ~35 pg/ml hGM-CSF and ~80 pg/ml hIL-3), would appear to be indicated for future studies.

It should be noted that only 50–75% of patient samples were able to successfully generate xenografts in the models described above [98, 147, 149–152]. More recently, in order to better recapitulate AML development as it occurs in human patients, several humanized animal models have been developed. These models were developed by prior implantation of human hematopoietic cells, such as in SCID-Hu BLT mice, or by ectopic BM environmental tissues such as the humanized BM ossicle xenotransplantation model [153–155]. These can significantly enhance the successful engraftment of human AML cells in localized regions. With such models, some AML samples which failed to grow in NSG or NSGS mice can still be successfully engrafted. It has also been speculated that a higher frequency of LSCs might be detectable in such models compared to the NSG and NSGS models. It is also expected that even pre-LSCs, which normally cannot grow in most other model animals, might be capable of growth in such humanized models. Thus humanized mouse models likely represent a better system with which to study LSCs as they not only increase the possibility of successful engraftment but may also maintain the heterogeneous features of AML cells [154, 155]. The limitation of these models is that they are technically very sophisticated and therefore require experienced personnel to conduct experiments; they are also difficult to be standardized.

Despite the significant efforts that have been made during the past two decades, LSC research has been significantly hampered by a lack of adequate models. Theoretically, all LSCs should be capable of regenerating leukemia if an optimized model is used. However, up until now, even when humanized models were used, in order to obtain successful engraftment and regeneration of leukemia, almost all studies involved the transplantation of large numbers of cells ($>10^3$), suggesting that the precise identification of LSCs remains a challenge. In addition, due to the genetic heterogeneity of leukemic clones and also due to the high degree of plasticity of LSCs, LSCs are also genetically heterogeneous even within the same phenotype, which blurs precise determination of distinct leukemic clones. Such a complicated situation might explain some of the controversies that hamper current LSC research because the leukemogenic capacities of LSCs with different genetic mutations are not the same [34]. Recently, Craig Jordan's laboratory demonstrated that low levels of intracellular ROS might be a better bioindicator of LSCs than cell surface markers by showing that functional LSCs are significantly enriched in the ROS-low subpopulation. The phenotypes of LSCs are not always correlated to their ROS levels and leukemogenic activities. In order to more precisely understand the molecular pathways and genetic/epigenetic diversities of LSCs, an optimized model system that can detect LSCs and pre-LSCs at the resolution of single cells is urgently needed. This will require the use of more advanced techniques such as mass cytometry [156] and single-cell transcriptomics. In addition, an improved system is required in order to best integrate both the phenotypic and genetic heterogeneities of LSCs with their biologic properties and leukemogenic capabilities.

5.6 LSCs in Drug Resistance, MRD, and Disease Relapse

Many early studies suggested that LSCs from AML patients are relatively quiescent with respect to the cell cycle and are also resistant to chemotherapy when compared to the bulk of leukemic blasts [157, 158]. A high frequency of LSCs at the time of diagnosis of AML (naïve LSCs, as defined by either phenotype or function, as well as high stem cell gene expression programs) predicts a high probability for MRD, disease relapse, and a poor prognosis [150, 159–165]. These studies might provide useful biomarkers to predict outcomes for AML patients. For example, Metzeler et al. identified a panel of 86 stem cell-like genes the expression signature of which predicts survival in cytogenetically normal AML [166, 167]. Ng et al. further limited such a panel to 17 stemness genes and found that the expression pattern of such genes can reliably determine risk in AML [168, 169]. Paired studies of AML samples at diagnosis and relapse using deep sequencing analysis of genetic abnormalities suggested that the LRCs in MRD could be derived from LSCs of the dominant genetic subclone, LSCs of minor genetic subclones, or pre-LSCs [102, 170–172]. These studies suggested that the origin of LRCs in AML patients is potentially heterogeneous [37, 51, 52]. Using a primary human AML NSG xenotransplantation model, Ishikawa et al. demonstrated that LSCs home back to the endosteal region of the BM niche shortly after transplantation to propagate for leukemia regeneration. After leukemia developed, the endosteal BM niche maintained self-renewal and a quiescent state in these LSCs and protected them from the cytotoxic, cell cycle-dependent effects of ARA-C treatment [3]. They also demonstrated that when G-CSF stimulation was used to induce LSCs to enter the cell cycle, it enhanced their sensitivity to ARA-C treatment in such models [138, 149]. These studies suggested that therapy-resistant LSCs are already present in the BM at diagnosis and that these cells are protected by the endosteal stem cell niche. In addition, LSCs might also have an intrinsic partial resistance to chemotherapy owing to their distinct biological properties (Fig. 5.2). LSCs are responsible for disease relapse. However, the

	LSCs	**bulk of LCs**	**healthy HSCs**
OXPHOS	High	Low	Low
Glycolysis	Low	High	High
BCL-2	High	Low	Low
MCL1	patient-dependent	patient-dependent	High
Mito mass	Reduced	Increased	Reduced
Mitophagy	High	Low	High
ROS	Low	High	Low
AA	High	Low	Low
Glutathione	High	Low	High
Proliferation	Low	High	Quiescent
CFU ability	High	Low/absent	High
Leukemogenesis	High	Low/absent	N/A
Chemo-resistance	Relative	None	High

Fig. 5.2 The biological properties of LSCs, bulk LCs, and normal HSCs

prevailing explanation for the role of LSCs in leukemia relapse was challenged by two very recent studies.

Using patient-derived xenograft models, two recent studies showed that by 2–3 days post-chemotherapy, drug-resistant leukemic cells (CR-LCs) in MRD are neither quiescent nor LSCs [32, 33]. Both studies demonstrated that at the time of maximal chemotherapy-induced AML repression, which occurred 2–3 days after the last ARA-C administration, neither quiescent cells nor LSCs were enriched in MRD in the BM as determined by phenotypic analyses, colony-forming assays, functional transplantation studies, and gene expression signatures. In fact, the frequency of LSCs is lower in BM early during chemotherapy-induced AML inhibition and rebound to pretreatment levels at the onset leukemia relapse (15–18 days after the last administration of ARA-C) [33]. The LRCs derived from CR-LCs also reacquire almost all of the phenotypes and stemness gene expression profiles found in naïve LSCs. Thus the authors claimed that CR-LCs are not derived from naïve LSCs. However, despite their significantly reduced colony-forming ability in vitro and leukemia-regenerating capacity in vivo, CR-LCs primarily use oxidative phosphorylation (OXPHOS) to meet their energy needs, in which behavior they are comparable to both naïve LSCs and LRCs but are distinct from the bulk of LCs which primarily use glycolysis for their energy needs. In addition, CR-LCs share most of the features of ROS-low LRCs described by Jordan's group, including high expression of the fatty acid (FAA) transporter CD36 and increased intracellular FAA metabolism (Fig. 5.3). Studies suggested that the stem cell properties of LSCs are better correlated to low ROS levels rather than to any specific phenotype. Thus there is a possibility that the CR-LCs identified in these two studies might be still selected by chemotherapy from among the naïve ROS-low LSCs. Chemotherapeutic drugs convert LSCs to CR-LCs by inducing DNA damage and transient oxidative stress which temporarily repress stem cell properties, including cell surface CD34 expression, stem cell gene expression, and clonogenic/leukemogenic ability. It was suggested that the increased ROS levels might mediate drug resistance in CR-LCs. During the interphase between treatments, CR-LCs reacquire stemness properties through metabolic and epigenetic reprogramming to reset to a ROS-low status. In the future, a label-tracing assay or single-cell technique will be essential to determine the exact origins of CR-LCs and LRCs. In addition, it will be necessary to carefully account for both the phenotypic and genetic heterogeneity features of LSCs and pre-LSCs.

5.7 Mitochondrial Metabolic Regulation of LSCs

Most recent studies have suggested that the activities of LSCs are most likely regulated by mitochondrial dynamics, through metabolic and/or epigenetic reprogramming mechanisms. Functionally transplantable LSCs are enriched in the ROS-low state. The maintenance of low ROS levels is required for preserving the undifferentiated/self-renewal-competent status of LSCs. ROS-low LSCs maintain a low rate

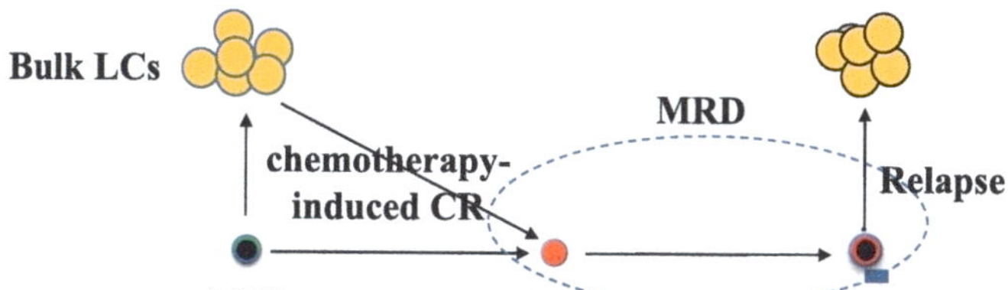

	LSCs	DR-LCs	LRCs
OXPHOS	High	High	High
Glycolysis	Low	Low	Low
BCL-2	High	High	High
Mitochond. mass	Reduced	Increased	Reduced
Mitophagy	High	Low	High
ROS	Low	High	Low
AA con.	High	Low	Low
CD36	Low	High	High
FAO	Low	High	High
Proliferation	Low	High	Low
CFU capacity	High	Low/absent	High
Leukemogenesis	High	Low/absent	High
Chemo-resistance	relative	High	High

Fig. 5.3 The biological properties of CR-LCs compared to LSCs and LRCs

of energy metabolism which renders them relatively quiescent. In addition, compared to ROS-high non-LSCs, ROS-low LSCs express high levels of BCL-2 and are resistant to chemotherapy. Interestingly, although more $CD34^{+}CD38^{-}$ cells can be detected in the ROS-low population than in the ROS-high population, yet there is no clear association of other LSC surface markers such as including CD123, CD33, CD117, CD90, and CD44 with the lower oxidative state. This suggested that, compared to phenotypic biomarkers, biological properties of LSCs such as mitochondrial dynamics, oxidative status, and energy production are more consistent among all patients irrespective of disease stage and genetic heterogeneity. Studies suggested that the ROS-low state is most likely a common feature for both normal tissue stem cells and CSCs [173, 174]. Thus, it might be more reliable to detect LSCs by examining their unique biological properties such as ROS level or relative quiescence with respect to the cell cycle rather than by indexing the molecular markers on their surfaces.

In normal hematopoietic tissue, HPCs are primarily reliant upon mitochondrial OXPHOS of glucose, an economical way to produce energy, while HSCs efficiently utilize anaerobic glycolysis for energy homeostasis, thereby producing low ROS levels [175, 176]. ROS are a by-product of OXPHOS, which is required for the proliferation and differentiation of HPCs [177]. However, maintaining a low reliance on OXPHOS, lower mitochondrial biogenesis, and a ROS-low status is very important for maintaining the quiescent state and self-renewal ability of HSCs [178–180]. Inducing OXPHOS in HSCs such as by cytokine stimulation or oxygen stress will increase ROS production and comprise the self-renewal ability of HSCs. It is believed that HSCs are localized in a hypoxic BM niche which protects HSCs from oxygen stress [175]. In addition, HSCs maintain a ROS-low status by using intrinsic mechanisms such as HIF-1α-PDK pathway-mediated OXPHOS repression [176, 181] and clearing of damaged mitochondria by mitophagy [182, 183]. Such mechanisms are regulated by the nutrient-sensitive PI3K-AKT-mTOR and LKB1-AMPK pathways, as well as by glutamine and PML-PPARγ- fatty acid metabolism [184].

Alterations in cellular metabolism and energetics are among the hallmarks of cancer [185, 186]. It was generally accepted that, in contrast to their normal counterparts, most types of tumor cells, including AML cells, are reliant upon aerobic glycolysis and not OXPHOS for energy production even in oxygen-rich conditions, a phenomenon known as the Warburg effect [187]. Studies suggested that glycolytic metabolism is required for proliferation and survival in most cancer cells because in the presence of sufficient nutrients, in addition to providing sufficient cellular ATP, anaerobic glycolysis can also provide important metabolites necessary for the biosynthetic demands of continuous cellular proliferation. However, recent studies demonstrated that CSCs from multiple tumor types including AML, glioblastoma, melanoma, and pancreatic cancer display distinct metabolic features compared to cancer cells without stem cell features (cancer blasts) [174, 188]. CSCs from most cancer types including LSCs are dependent on OXPHOS for energy production and have lower glycolytic reserves than cancer blasts [189–191]. Thus the use of OXPHOS is greater in CSCs than in cancer blasts. Intriguingly, despite the high OXPHOS feature, CSCs maintain a ROS-low status, suggesting a very efficient ROS-scavenging system is used in such cells for maintaining redox balance. For example, increased antioxidant responses have been detected in some CSCs owing to the upregulation of HIF-1α expression. Studies suggested that LSCs might maintain ROS-low status and mitochondrial dynamics by inducing AMPK-FIS-mitophagy pathway-mediated clearance of damaged mitochondria [192], which might be induced by the hypoxic BM niche [193, 194].

In most tissue cells, OXPHOS is fueled by glucose; however naïve LSCs are primarily dependent on amino acids, primarily glutamine, glutamate, and proline, to fuel OXPHOS and ATP synthesis. More intriguingly, when exposed to chemotherapy, LSCs reprogram their metabolic machinery and utilize FAA to fuel OXPHOS and energy production. It was known that most cancer cells consume FAA for survival when glucose metabolism becomes limiting [195–198]. Through β-oxidation, FAA not only sustain ATP levels but also produce cytosolic NADPH, a key to counteracting oxidative stress [199, 200]. This might explain why more LSCs can be detected in either BM or extramedullary adipose tissues after chemotherapy [44]. It might also help to explain the poorer prognosis for AML in obese patients [201]. Furthermore, because LSCs appear to be deficient in their ability to employ glycolysis and are highly reliant on OXPHOS, it is suggested that inhibition of mitochondrial OXPHOS might be an effective therapeutic strategy to treat human AML by targeting LSCs [202].

5.8 Targeting LSCs as an Important Component of Therapy for AML

Therapeutic targeting of LSCs is still difficult to accomplish [25]. To develop drugs that can eliminate LSCs, significant efforts have been made to identify their unique vulnerabilities. By comparing gene expression profiles among LSCs, bulk LBs, and normal HSCs, several LSC-specific surface molecules (including CD123, IL-1RAP, CD47, TIM3, and CD44) and self-renewal/proliferation pathways (such as NF-κB, Wnt-β-catenin, Sonic Hedgehog, Bmi1, and PI3K-AKT/mTOR) [203–205] have been identified. Antibody-drug conjugates and CAR-T cells that are specific for some of these cell surface molecules, as well as small molecule inhibitors that are specific for components of these pathways, have been developed and tested in preclinical studies for anti-LSC efficacy. There have been some promising preliminary results, especially when these types of molecules were combined with various classes of chemotherapeutic agents [128, 131, 134, 139, 206–208]. In addition, targeting the microenvironment in order to enhance anti-LSC therapy by inhibiting BM niche-emergent signals (e.g., CXCL12-CXCR4 and HIF-1α) has also been tested preclinically [209, 210]. Most of these new treatments are currently in the early stages of clinical trials. Detailed information regarding these studies has been summarized well and is discussed in several review articles [29, 206, 208, 211–214]. Here we will discuss only the inhibition of Hedgehog signaling, one of the most promising therapies to target LSCs.

5.9 Targeting the Hedgehog Pathway to Treat AML

Sonic Hedgehog (SHH) signaling is activated in LSCs isolated from most AML samples. This signaling entity is essential for the maintenance of LSCs [215, 216]. Overexpression of GLI1, the key transcriptional mediator of SHH signaling, is associated with disease relapse, drug resistance, poor prognosis, and reduced OS in AML patients [217]. Preclinical studies have suggested that the inhibition of SHH sensitizes LSCs to ARA-C or AZA and represses AML development [218]. Based on these studies, clinical trials were conducted to test the anti-AML effects of glasdegib, an SHH signaling inhibitor, in combination with low-dose cytarabine (LDAC) or standard 7 + 3 treatment. In a Phase Ib study, it was shown that glasdegib + LDAC induced CR/CRi in 31% (16/54) of patients, with an average OS of 8.3-months in patients with either AML or high-risk MDS; this CR/CRi rate is significantly better than the 11%–28% rate previously reported in studies using either an HMA or LDAC alone [219]. Based on this study, in November, 2018, the USFDA granted accelerated approval of glasdegib + LDAC for the treatment of newly-diagnosed AML in adults 75 years of age or older. However, in a Phase II randomized multicenter study, glasdegib + LDAC induced CR/CRi in only 15% of patients (17/132). Nevertheless, the 8.8-month median OS reached in the glasdegib + LDAC arm is significantly better than the 4.9-month OS in the LDAC-alone arm [220]. Although another study suggested that glasdegib plus standard 7 + 3 therapy induced CR/CRi in 46.4% of patients with a 14.9-month median OS, such treatment is limited only to those patients who can sustain the standard 7 + 3 regimen of chemotherapy [221]. Interestingly, a preclinical study suggested that glasdegib + AZA might synergistically inhibit AML and MDS [222]. Future studies will need to determine whether glasdegib can facilitate the elimination of LSCs and prevent disease relapse when combined with VEN + AZA (described below).

5.10 Inhibition of BCL-2 to Target LSCs in AML

BCL-2 is highly expressed in most types of cancer cells [223, 224]. Unlike normal HSCs which depend on MCL1 for survival [225], most cancer cells, including AML cells, are reliant on BCL-2 for their survival [223, 224, 226–228]. Thus, BCL-2 has been approved as a safe and reliable target for the treatment of BCL-2-expressing cancer [223, 224, 229]. Several BCL-2 inhibitors have been developed and tested in clinical trials as antineoplastic therapy. ABT-737 [230] and its derivative Navitoclax (NAV, ABT-263) [231] are small molecule mimetics of the BH3 domain of the BH3-only sensitizer protein BAD, which efficiently binds to BCL-2, BCL-xL, and BCL-W, thus permitting the release of mitochondrial-bound pro-apoptotic proteins to induce apoptosis in BCL-2-dependent cancer cells [231]. Venetoclax (VEN, ABT-199), another modified derivative of ABT-737 developed by AbbVie in 2010, maintains specificity for BCL-2 but lacks affinity for BCL-xL [232]. Although preclinical studies suggested that ABT-737 and NAV might have greater anticancer activity in BCL-2-expressing cancers compared to VEN, clinical use of ABT-737 and NAV was hindered by dose-limiting thrombocytopenia induced by these drugs, as platelets are dependent upon BCL-xL for their survival [233]. VEN treatment spared platelets with tolerable side effects at its effective anticancer dosage, and in 2016 it was granted accelerated approval by the FDA for the treatment of chronic lymphocytic leukemia (CLL) patients whose cells show a 17p deletion; full approval was granted in June 2018 for patients with CLL or small lymphocytic leukemia regardless of 17p deletion [232, 234–236]. Recent studies demonstrated that VEN treatment is also effective in myeloid malignancies including AML, as demonstrated in cell lines, primary patient samples, and clinical patient treatment.

In AML, LSCs express high levels of BCL-2 compared to non-LSCs [51]. In addition to its anti-apoptotic activity, BCL-2 also regulates energy production in LSCs by maintaining their relatively high reliance on OXPHOS without increasing ROS production. BCL-2 inhibition impedes OXPHOS and thus hinders ATP production [190]. Thus BCL-2 inhibition selectively kills LSCs by inhibiting the production of energy-storing molecules and also by inducing mitochondrial apoptosis, all while having minimal effects on normal HSCs [13, 51]. Preclinical studies suggested that targeting BCL-2 leads to the preferential elimination of LSCs in xenografts from most AML patient samples. However, as a single-agent treatment, VEN induced CR/CRi in only 19% of AML patients, and its treatment effects were not durable [237]. It was suggested that elevated MCL1 is the major reason for VEN resistance in AML cells and patient samples [115, 116, 238–241]. Supporting this notion, it was demonstrated that MCL1 inhibition sensitizes VEN-resistant AML cells to treatment with this drug [238, 239, 241–243]. The anti-AML effect of VEN in combination with an MCL1 inhibitor needs to be evaluated clinically, however. Recent studies suggest that HMA, PI3K/AKT inhibitors, MEK inhibitors, cytarabine, or idarubicin might indirectly repress MCL1 expression and synergize with VEN to increase the latter's overall anti-AML efficacy [242, 244–247].

As single agents, the HMAs azacitidine and decitabine only induce a response in 15–25% of elderly AML patients [248]. Interestingly, the combination of VEN + AZA or DEC induced CR/CRi in 67% (97/145) and 85% (28/33) of elderly patients with newly-diagnosed AML, respectively [13–15], and in 64% (21/33) of patients with chemotherapy-relapsed or refractory AML [16, 226]. This is significantly superior to results obtainable with traditional chemotherapy. Of importance, ~20% of patients achieved an MRD-negative deep CR, suggesting that LSCs may indeed have been maximally ablated. These studies suggested that 400 mg is an effective and tolerable/safe dose of VEN and the VEN + AZA combination is better than the VEN + DEC combination as indicated by earlier responsiveness and longer survival. Although the long-term effects of such treatment need to be further evaluated by randomized clinical trials and long-term observations, we believe that VEN + AZA ought to and will likely become the first-line of treatment for elderly AML patients and may alter the treatment strategy for younger AML patients as well. Mechanistically, AZA and DEC both sensitize AML cells to VEN by upregulating NOX1 and repressing MCL1 [246, 249–251]. Thus VEN + either AZA or DEC synergistically eliminates the bulk of LCs by inducing apoptosis, which leads to disease remission. AZA + VEN also synergistically removes LSCs in many patients by inhibiting OXPHOS and disrupting glutathione-dependent ETC complex II-related energy metabolism [13, 14]. Patients with *IDH1*/*IDH2* mutations were more likely to respond better due to the inhibition of cytochrome C oxidase by 2-hydroxyglutarate, the abnormal product of mutant IDH1/IDH2 [237, 252], whereas *K-RAS*/*N-RAS* or *PTPN11* mutations may predict a poorer response or disease progression because of the upregulation of MCL1 in such patients. It is notable that, in contrast to chemotherapy which induces p53-dependent apoptosis, the VEN + AZA treatment program kills AML cells in a p53-independent fashion [236]. Thus most patients with *TP53* mutations might still respond to VEN + AZA therapy. In addition, VEN + LDAC induces CR/CRi in 54% (44/82) of patients with a median OS of 10-months, which is superior to LDAC alone (CR/CRi 11% and OS 5 months) [253]. Whether VEN + LDAC can also eliminate LSCs and induce a deep and long-term CR in patients needs to be evaluated. In response to these studies, which were reported in November of 2018, the USFDA has granted accelerated approval to VEN + AZA and VEN + LDAC for the treatment of newly-diagnosed AML in adults age 75 years and older.

5.11 Prospective

Despite the significant progress that has been achieved in research into LSCs, several key questions regarding the concept of an LSC have not been satisfactorily addressed experimentally. Thus the clinical relevance of LSCs remains in question due to controversies extant from many previous studies. These controversies are primarily attributable to the technical limitations inherent in trying to unequivocally identify LSCs, which results in the inability to fully address the features of the genetic heterogeneity and metabolic/epigenetic plasticity of pre-LSCs and LSCs.

It is well-accepted that the founder mutations leading to AML normally occur in primitive HSCs, suggesting the HSC origin of AML. However, the immunophenotypes of LSCs are {LMPP-like} or GMP-like or even resemble more differentiated precursors in almost all AML cases, which implies that LSCs likely arise from a mutant progenitor that has acquired self-renewal capabilities rather than originating directly from a single HSC. It is also well-accepted that the pre-LSCs in most AML cases share a phenotype similar to normal HSCs and are quiescent with regard to the cell cycle, which explains the relatively chemoresistant characteristic of pre-LSCs. However, direct evidence is still lacking to demonstrate that the tendency toward cell cycle quiescence is also a consistent feature of LSCs because earlier studies failed to consider in their analyses the possibility that pre-LSCs are separate entities from LSCs [99, 138, 149, 254].

Studies have suggested that leukemia could relapse from either a preleukemic clone or leukemic subclones [102]. Future studies need to address whether the CR-LCs and LRCs detectable during MRD are derived from LSCs, pre-LSCs, or certain types of drug-resistant LBs. Fully addressing this question will allow us to design and develop still better strategies to treat AML by killing the bulk of LCs to induce disease remission and, importantly, eliminating LRCs from MRD to prevent disease relapse (Fig. 5.4).

Furthermore, due to the dynamic metabolic reprogramming of LSCs in response to exposure to chemotherapeutic pharmaceuticals, and also at the onset of disease relapse, different treatment strategies will almost certainly be necessary in order to target LSCs at different treatment stages. For example, naïve LSCs, CR-LCs, and LRCs all use OXPHOS to generate ATP; therefore the inhibition of mitochondrial OXPHOS might be a useful strategy to kill all of these cells. However, naïve LSCs rely on amino acids to fuel OXPHOS reactions for the generation of ATP, while after chemotherapy, CR-LCs and LRCs convert to the use of FAA to fuel OXPHOS for energy production. Therefore, to target LSCs, CR-LCs, and LRCs, distinct metabolic pathways such as amino acid-OXPHOS or CD36-FAO-OXPHOS must be targeted [255]. In addition, compared to LSCs and LRCs, the CR-LC subpopulation lacks leukemogenic capacity and shows some chemotherapy stress-induced transient metabolic features such as high ROS levels. Thus targeting CR-LCs might represent a very critical therapeutic component to prevent disease relapse either by

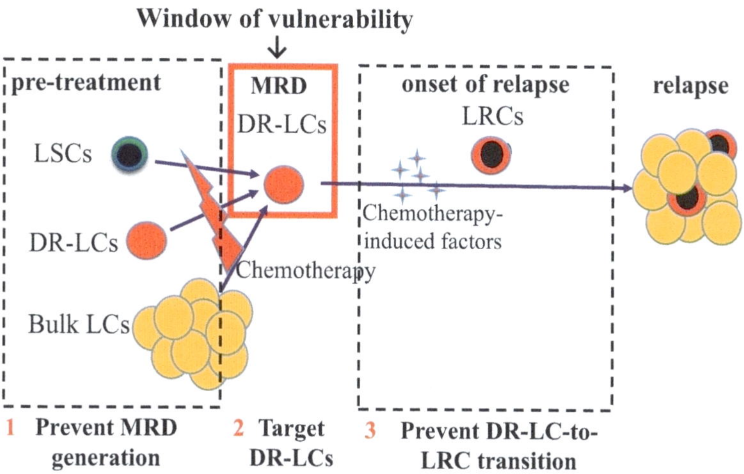

Fig. 5.4 Strategies to prevent disease relapse by targeting drug-resistant cells. (1) There are three potential origins for drug-resistant leukemic cells (DR-LCs): LSCs, pre-existing DR-LCs, or cells which may be selected from the existing bulk of LCs. Thus, determining the actual origin of DR-LCs will help in the design of specific strategies to prevent the development of MRD because LSCs, DR-LCs, and bulk LCs might have distinct biological properties (see Figs. 5.2 and 5.3). We may able to add a specific targeted therapy to standard chemotherapy to prevent the production of MRD without affecting the activity of normal HSCs. (2) Immediately after chemotherapy, the DR-LCs in MRD have distinct metabolic properties (see Fig. 5.3) which might provide a window to target DR-LCs by inhibiting these unique metabolic pathways. (3) Since DR-LCs need to regain leukemogenic potential in order to become LRCs for disease relapse, we might be able to prevent such relapse by inhibiting the DR-LC-to-LRC transition

eliminating CR-LCs or by inhibiting the CR-LC-to-LRC transition. One study suggested that patients' serum factors such as dopamine and its receptor DRD2 stimulate CR-LCs to reacquire the properties of LSCs. Thus targeted inhibition of DRD2 signaling was shown to lead to the suppression of AML growth, providing direct evidence for the potential benefit of anti-CR-LC targeted therapy.

Lastly, although BCL-2 inhibition targets LSCs by inhibiting mitochondrial OXPHOS, AZA might not target LSCs due to their hypomethylated epigenetic feature. Of interest, a combination of VEN + AZA seems to synergistically kill both bulk LCs and LSCs. In the future, more detailed molecular mechanisms need to be elucidated in order to fully explain how AZA enhances the anti-AML feature of VEN. Such information will help us to design more effective strategies to improve treatment of therapy-resistant AML and also for patients who fail to achieve a deep CMR. Taken together, it is hoped that these newer therapeutic strategies, based as they must be on a clearer, sounder technical appreciation of the nature of bulk LCs as well as the component cells of MRD, will significantly improve the outlook for AML patients and will do so sooner rather than later.

References

1. Burnett AK et al (2010) Attempts to optimize induction and consolidation treatment in acute myeloid leukemia: results of the MRC AML12 trial. J Clin Oncol 28:586–595. https://doi.org/10.1200/JCO.2009.22.9088
2. Fernandez HF et al (2009) Anthracycline dose intensification in acute myeloid leukemia. N Engl J Med 361:1249–1259. https://doi.org/10.1056/NEJMoa0904544
3. Burnett A, Wetzler M, Lowenberg B (2011) Therapeutic advances in acute myeloid leukemia. J Clin Oncol 29:487–494. https://doi.org/10.1200/JCO.2010.30.1820
4. Pulte D, Gondos A, Brenner H (2010) Expected long-term survival of patients diagnosed with acute myeloblastic leukemia during 2006–2010. Ann Oncol 21:335–341. https://doi.org/10.1093/annonc/mdp309
5. Dohner H, Weisdorf DJ, Bloomfield CD (2015) Acute myeloid leukemia. N Engl J Med 373:1136–1152. https://doi.org/10.1056/NEJMra1406184
6. Ferrara F, Schiffer CA (2013) Acute myeloid leukaemia in adults. Lancet 381:484–495. https://doi.org/10.1016/S0140-6736(12)61727-9
7. Lancet JE et al (2018) CPX-351 (cytarabine and daunorubicin) liposome for injection versus conventional cytarabine plus daunorubicin in older patients with newly diagnosed secondary acute myeloid leukemia. J Clin Oncol 36:2684–2692. https://doi.org/10.1200/JCO.2017.77.6112
8. Kantarjian HM et al (2012) Multicenter, randomized, open-label, phase III trial of decitabine versus patient choice, with physician advice, of either supportive care or low-dose cytarabine for the treatment of older patients with newly diagnosed acute myeloid leukemia. J Clin Oncol 30:2670–2677. https://doi.org/10.1200/JCO.2011.38.9429
9. Dombret H et al (2015) International phase 3 study of azacitidine vs conventional care regimens in older patients with newly diagnosed AML with >30% blasts. Blood 126:291–299. https://doi.org/10.1182/blood-2015-01-621664
10. Lubbert M et al (2012) A multicenter phase II trial of decitabine as first-line treatment for older patients with acute myeloid leukemia judged unfit for induction chemotherapy. Haematologica 97:393–401. https://doi.org/10.3324/haematol.2011.048231
11. Fenaux P et al (2010) Azacitidine prolongs overall survival compared with conventional care regimens in elderly patients with low bone marrow blast count acute myeloid leukemia. J Clin Oncol 28:562–569. https://doi.org/10.1200/JCO.2009.23.8329
12. Bohl SR, Bullinger L, Rucker FG (2018) Epigenetic therapy: azacytidine and decitabine in acute myeloid leukemia. Expert Rev Hematol 11:361–371. https://doi.org/10.1080/17474086.2018.1453802
13. Pollyea DA et al (2018) Venetoclax with azacitidine disrupts energy metabolism and targets leukemia stem cells in patients with acute myeloid leukemia. Nat Med 24:1859–1866. https://doi.org/10.1038/s41591-018-0233-1
14. DiNardo CD et al (2018) Safety and preliminary efficacy of venetoclax with decitabine or azacitidine in elderly patients with previously untreated acute myeloid leukaemia: a non-randomised, open-label, phase 1b study. Lancet Oncol 19:216–228. https://doi.org/10.1016/S1470-2045(18)30010-X
15. DiNardo CD et al (2019) Venetoclax combined with decitabine or azacitidine in treatment-naive, elderly patients with acute myeloid leukemia. Blood 133:7–17. https://doi.org/10.1182/blood-2018-08-868752
16. Aldoss I et al (2018) Efficacy of the combination of venetoclax and hypomethylating agents in relapsed/refractory acute myeloid leukemia. Haematologica 103:e404–e407. https://doi.org/10.3324/haematol.2018.188094
17. Craddock C et al (2013) Azacitidine fails to eradicate leukemic stem/progenitor cell populations in patients with acute myeloid leukemia and myelodysplasia. Leukemia 27:1028–1036. https://doi.org/10.1038/leu.2012.312

18. Sarkozy C et al (2013) Outcome of older patients with acute myeloid leukemia in first relapse. Am J Hematol 88:758–764. https://doi.org/10.1002/ajh.23498
19. Balsat M et al (2017) Postinduction minimal residual disease predicts outcome and benefit from allogeneic stem cell transplantation in acute myeloid leukemia with NPM1 mutation: a study by the Acute Leukemia French Association group. J Clin Oncol 35:185–193. https://doi.org/10.1200/JCO.2016.67.1875
20. Chen X et al (2015) Relation of clinical response and minimal residual disease and their prognostic impact on outcome in acute myeloid leukemia. J Clin Oncol 33:1258–1264. https://doi.org/10.1200/JCO.2014.58.3518
21. Buccisano F et al (2012) Prognostic and therapeutic implications of minimal residual disease detection in acute myeloid leukemia. Blood 119:332–341. https://doi.org/10.1182/blood-2011-08-363291
22. Oran B, de Lima M (2011) Prevention and treatment of acute myeloid leukemia relapse after allogeneic stem cell transplantation. Curr Opin Hematol 18:388–394
23. Thomas D, Majeti R (2017) Biology and relevance of human acute myeloid leukemia stem cells. Blood 129:1577–1586. https://doi.org/10.1182/blood-2016-10-696054
24. Siveen KS, Uddin S, Mohammad RM (2017) Targeting acute myeloid leukemia stem cell signaling by natural products. Mol Cancer 16:13. https://doi.org/10.1186/s12943-016-0571-x
25. Pollyea DA, Jordan CT (2017) Therapeutic targeting of acute myeloid leukemia stem cells. Blood 129:1627–1635. https://doi.org/10.1182/blood-2016-10-696039
26. Pollyea DA, Gutman JA, Gore L, Smith CA, Jordan CT (2014) Targeting acute myeloid leukemia stem cells: a review and principles for the development of clinical trials. Haematologica 99:1277–1284. https://doi.org/10.3324/haematol.2013.085209
27. Pandolfi A, Barreyro L, Steidl U (2013) Concise review: preleukemic stem cells: molecular biology and clinical implications of the precursors to leukemia stem cells. Stem Cells Transl Med 2:143–150. https://doi.org/10.5966/sctm.2012-0109
28. Horton SJ, Huntly BJ (2012) Recent advances in acute myeloid leukemia stem cell biology. Haematologica 97:966–974. https://doi.org/10.3324/haematol.2011.054734
29. Felipe Rico J, Hassane DC, Guzman ML (2012) Acute myelogenous leukemia stem cells: from bench to bedside. Cancer Lett 338:4–9. https://doi.org/10.1016/j.canlet.2012.05.034
30. Lane SW, Scadden DT, Gilliland DG (2009) The leukemic stem cell niche: current concepts and therapeutic opportunities. Blood 114:1150–1157
31. Dick JE (2005) Acute myeloid leukemia stem cells. Ann N Y Acad Sci 1044:1–5
32. Farge T et al (2017) Chemotherapy-resistant human acute myeloid leukemia cells are not enriched for leukemic stem cells but require oxidative metabolism. Cancer Discov 7:716–735. https://doi.org/10.1158/2159-8290.CD-16-0441
33. Boyd AL et al (2018) Identification of chemotherapy-induced leukemic-regenerating cells reveals a transient vulnerability of human AML recurrence. Cancer Cell 34:483–498 e485. https://doi.org/10.1016/j.ccell.2018.08.007
34. Klco JM et al (2014) Functional heterogeneity of genetically defined subclones in acute myeloid leukemia. Cancer Cell 25:379–392. https://doi.org/10.1016/j.ccr.2014.01.031
35. Magee JA, Piskounova E, Morrison SJ (2012) Cancer stem cells: impact, heterogeneity, and uncertainty. Cancer Cell 21:283–296. https://doi.org/10.1016/j.ccr.2012.03.003
36. Meacham CE, Morrison SJ (2013) Tumour heterogeneity and cancer cell plasticity. Nature 501:328–337. https://doi.org/10.1038/nature12624
37. Reya T, Morrison SJ, Clarke MF, Weissman IL (2001) Stem cells, cancer, and cancer stem cells. Nature 414:105–111
38. Lapidot T et al (1994) A cell initiating human acute myeloid leukaemia after transplantation into SCID mice. Nature 367:645–648
39. Bonnet D, Dick JE (1997) Human acute myeloid leukemia is organized as a hierarchy that originates from a primitive hematopoietic cell. Nat Med 3:730–737
40. Lowenberg B (2017) Introduction to the review series on leukemic stem cells. Blood 129:1567. https://doi.org/10.1182/blood-2017-01-753319

41. Kreso A, Dick JE (2014) Evolution of the cancer stem cell model. Cell Stem Cell 14:275–291. https://doi.org/10.1016/j.stem.2014.02.006
42. Weissman IL (2002) The road ended up at stem cells. Immunol Rev 185:159–174
43. Spangrude GJ, Aihara Y, Weissman IL, Klein J (1988) The stem cell antigens Sca-1 and Sca-2 subdivide thymic and peripheral T lymphocytes into unique subsets. J Immunol 141:3697–3707
44. Ye H et al (2016) Leukemic stem cells evade chemotherapy by metabolic adaptation to an adipose tissue niche. Cell Stem Cell 19:23–37. https://doi.org/10.1016/j.stem.2016.06.001
45. Ho TC et al (2016) Evolution of acute myelogenous leukemia stem cell properties after treatment and progression. Blood 128:1671–1678. https://doi.org/10.1182/blood-2016-02-695312
46. Jordan CT (2005) Targeting the most critical cells: approaching leukemia therapy as a problem in stem cell biology. Nat Clin Pract Oncol 2:224–225. https://doi.org/10.1038/ncponc0164
47. Mikkola HK, Radu CG, Witte ON (2010) Targeting leukemia stem cells. Nat Biotechnol 28:237–238. https://doi.org/10.1038/nbt0310-237
48. Dick JE, Lapidot T (2005) Biology of normal and acute myeloid leukemia stem cells. Int J Hematol 82:389–396
49. Holyoake TL, Vetrie D (2017) The chronic myeloid leukemia stem cell: stemming the tide of persistence. Blood 129:1595–1606. https://doi.org/10.1182/blood-2016-09-696013
50. Zhang B et al (2013) Microenvironmental protection of CML stem and progenitor cells from tyrosine kinase inhibitors through N-cadherin and Wnt-beta-catenin signaling. Blood 121:1824–1838. https://doi.org/10.1182/blood-2012-02-412890
51. Goff DJ et al (2013) A Pan-BCL2 inhibitor renders bone-marrow-resident human leukemia stem cells sensitive to tyrosine kinase inhibition. Cell Stem Cell 12:316–328. https://doi.org/10.1016/j.stem.2012.12.011
52. Zhang B et al (2012) Altered microenvironmental regulation of leukemic and normal stem cells in chronic myelogenous leukemia. Cancer Cell 21:577–592. https://doi.org/10.1016/j.ccr.2012.02.018
53. Hamilton A et al (2012) Chronic myeloid leukemia stem cells are not dependent on Bcr-Abl kinase activity for their survival. Blood 119:1501–1510. https://doi.org/10.1182/blood-2010-12-326843
54. Gerber JM et al (2011) Characterization of chronic myeloid leukemia stem cells. Am J Hematol 86:31–37. https://doi.org/10.1002/ajh.21915
55. Chu S et al (2011) Persistence of leukemia stem cells in chronic myelogenous leukemia patients in prolonged remission with imatinib treatment. Blood 118:5565–5572
56. Chen Y, Peng C, Li D, Li S (2010) Molecular and cellular bases of chronic myeloid leukemia. Protein Cell 1:124–132. https://doi.org/10.1007/s13238-010-0016-z
57. Arrigoni E et al (2018) Concise review: chronic myeloid leukemia: stem cell niche and response to pharmacologic treatment. Stem Cells Transl Med 7:305–314. https://doi.org/10.1002/sctm.17-0175
58. Hochhaus A, Larson RA, Guilhot F, Radich JP, Branford S, Hughes TP, Baccarani M, Deininger MW, Cervantes F, Fujihara S, Ortmann C-E, Menssen HD, Kantarjian H, O'Brien SG, Druker BJ (2017) Long-term outcomes of imatinib treatment for chronic myeloid leukemia. N Engl J Med 376(10):917–927. https://doi.org/10.1056/NEJMoa1609324
59. Corbin AS et al (2011) Human chronic myeloid leukemia stem cells are insensitive to imatinib despite inhibition of BCR-ABL activity. J Clin Invest 121:396–409. https://doi.org/10.1172/JCI35721
60. Geelen IGP et al (2017) Treatment outcome in a population-based, 'real-world' cohort of patients with chronic myeloid leukemia. Haematologica 102:1842–1849. https://doi.org/10.3324/haematol.2017.174953
61. Warfvinge R et al (2017) Single-cell molecular analysis defines therapy response and immunophenotype of stem cell subpopulations in CML. Blood 129:2384–2394. https://doi.org/10.1182/blood-2016-07-728873

62. Campiotti L et al (2017) Imatinib discontinuation in chronic myeloid leukaemia patients with undetectable BCR-ABL transcript level: a systematic review and a meta-analysis. Eur J Cancer 77:48–56. https://doi.org/10.1016/j.ejca.2017.02.028
63. Mahon FX et al (2010) Discontinuation of imatinib in patients with chronic myeloid leukaemia who have maintained complete molecular remission for at least 2 years: the prospective, multicentre Stop Imatinib (STIM) trial. Lancet Oncol 11:1029–1035. https://doi.org/10.1016/S1470-2045(10)70233-3
64. Holtz M, Forman SJ, Bhatia R (2007) Growth factor stimulation reduces residual quiescent chronic myelogenous leukemia progenitors remaining after imatinib treatment. Cancer Res 67:1113–1120. https://doi.org/10.1158/0008-5472.CAN-06-2014
65. Holtz MS, Forman SJ, Bhatia R (2005) Nonproliferating CML CD34+ progenitors are resistant to apoptosis induced by a wide range of proapoptotic stimuli. Leukemia 19:1034–1041. https://doi.org/10.1038/sj.leu.2403724
66. Zhang B et al (2016) Heterogeneity of leukemia-initiating capacity of chronic myelogenous leukemia stem cells. J Clin Invest 126:975–991. https://doi.org/10.1172/JCI79196
67. Bellodi C et al (2009) Targeting autophagy potentiates tyrosine kinase inhibitor-induced cell death in Philadelphia chromosome-positive cells, including primary CML stem cells. J Clin Invest 119:1109–1123. https://doi.org/10.1172/JCI35660
68. Karvela M et al (2016) ATG7 regulates energy metabolism, differentiation and survival of Philadelphia-chromosome-positive cells. Autophagy 12:936–948. https://doi.org/10.1080/15548627.2016.1162359
69. Neviani P et al (2005) The tumor suppressor PP2A is functionally inactivated in blast crisis CML through the inhibitory activity of the BCR/ABL-regulated SET protein. Cancer Cell 8:355–368. https://doi.org/10.1016/j.ccr.2005.10.015
70. Baquero P et al (2019) Targeting quiescent leukemic stem cells using second generation autophagy inhibitors. Leukemia 33:981–994. https://doi.org/10.1038/s41375-018-0252-4
71. Neviani P et al (2013) PP2A-activating drugs selectively eradicate TKI-resistant chronic myeloid leukemic stem cells. J Clin Invest 123:4144–4157. https://doi.org/10.1172/JCI68951
72. Miyamoto T, Weissman IL, Akashi K (2000) AML1/ETO-expressing nonleukemic stem cells in acute myelogenous leukemia with 8;21 chromosomal translocation. Proc Natl Acad Sci USA 97:7521–7526
73. Ellegast JM et al (2016) inv(16) and NPM1mut AMLs engraft human cytokine knock-in mice. Blood 128:2130–2134. https://doi.org/10.1182/blood-2015-12-689356
74. Patel S et al (2012) Successful xenografts of AML3 samples in immunodeficient NOD/shi-SCID IL2Rgamma(−)/(−) mice. Leukemia 26:2432–2435. https://doi.org/10.1038/leu.2012.154
75. Shlush LI (2018) Age-related clonal hematopoiesis. Blood 131:496–504. https://doi.org/10.1182/blood-2017-07-746453
76. Jan M, Ebert BL, Jaiswal S (2017) Clonal hematopoiesis. Semin Hematol 54:43–50. https://doi.org/10.1053/j.seminhematol.2016.10.002
77. Mossner M et al (2016) Mutational hierarchies in myelodysplastic syndromes dynamically adapt and evolve upon therapy response and failure. Blood 128:1246–1259. https://doi.org/10.1182/blood-2015-11-679167
78. Steensma DP et al (2015) Clonal hematopoiesis of indeterminate potential and its distinction from myelodysplastic syndromes. Blood 126:9–16. https://doi.org/10.1182/blood-2015-03-631747
79. Genovese G et al (2014) Clonal hematopoiesis and blood-cancer risk inferred from blood DNA sequence. N Engl J Med 371:2477–2487. https://doi.org/10.1056/NEJMoa1409405
80. Jaiswal S et al (2014) Age-related clonal hematopoiesis associated with adverse outcomes. N Engl J Med 371:2488–2498. https://doi.org/10.1056/NEJMoa1408617
81. Xie M et al (2014) Age-related mutations associated with clonal hematopoietic expansion and malignancies. Nat Med 20:1472–1478. https://doi.org/10.1038/nm.3733

82. Bejar R et al (2014) Somatic mutations predict poor outcome in patients with myelodysplastic syndrome after hematopoietic stem-cell transplantation. J Clin Oncol 32:2691–2698. https://doi.org/10.1200/JCO.2013.52.3381
83. Malcovati L et al (2014) Driver somatic mutations identify distinct disease entities within myeloid neoplasms with myelodysplasia. Blood 124:1513–1521. https://doi.org/10.1182/blood-2014-03-560227
84. Nazha A et al (2016) Incorporation of molecular data into the revised international prognostic scoring system in treated patients with myelodysplastic syndromes. Leukemia 30:2214–2220. https://doi.org/10.1038/leu.2016.138
85. Papaemmanuil E et al (2013) Clinical and biological implications of driver mutations in myelodysplastic syndromes. Blood 122:3616–3627.; quiz 3699. https://doi.org/10.1182/blood-2013-08-518886
86. Hou HA et al (2018) Incorporation of mutations in five genes in the revised international prognostic scoring system can improve risk stratification in the patients with myelodysplastic syndrome. Blood Cancer J 8:39. https://doi.org/10.1038/s41408-018-0074-7
87. Kennedy JA, Ebert BL (2017) Clinical implications of genetic mutations in myelodysplastic syndrome. J Clin Oncol 35:968–974. https://doi.org/10.1200/JCO.2016.71.0806
88. Lindsley RC et al (2015) Acute myeloid leukemia ontogeny is defined by distinct somatic mutations. Blood 125:1367–1376. https://doi.org/10.1182/blood-2014-11-610543
89. Takahashi K et al (2013) Dynamic acquisition of FLT3 or RAS alterations drive a subset of patients with lower risk MDS to secondary AML. Leukemia 27:2081–2083. https://doi.org/10.1038/leu.2013.165
90. Murphy DM et al (2013) NRAS mutations with low allele burden have independent prognostic significance for patients with lower risk myelodysplastic syndromes. Leukemia 27:2077–2081. https://doi.org/10.1038/leu.2013.160
91. Dan C, Chi J, Wang L (2015) Molecular mechanisms of the progression of myelodysplastic syndrome to secondary acute myeloid leukaemia and implication for therapy. Ann Med 47:1–9. https://doi.org/10.3109/07853890.2015.1009156
92. Abelson S et al (2018) Prediction of acute myeloid leukaemia risk in healthy individuals. Nature 559:400–404. https://doi.org/10.1038/s41586-018-0317-6
93. Shlush LI et al (2014) Identification of pre-leukaemic haematopoietic stem cells in acute leukaemia. Nature 506:328–333. https://doi.org/10.1038/nature13038
94. Sykes SM, Kokkaliaris KD, Milsom MD, Levine RL, Majeti R (2015) Clonal evolution of preleukemic hematopoietic stem cells in acute myeloid leukemia. Exp Hematol 43:989–992. https://doi.org/10.1016/j.exphem.2015.08.012
95. Corces-Zimmerman MR, Hong WJ, Weissman IL, Medeiros BC, Majeti R (2014) Preleukemic mutations in human acute myeloid leukemia affect epigenetic regulators and persist in remission. Proc Natl Acad Sci USA 111:2548–2553. https://doi.org/10.1073/pnas.1324297111
96. Stevens BM et al (2018) Characterization and targeting of malignant stem cells in patients with advanced myelodysplastic syndromes. Nat Commun 9:3694. https://doi.org/10.1038/s41467-018-05984-x
97. Woll PS et al (2014) Myelodysplastic syndromes are propagated by rare and distinct human cancer stem cells in vivo. Cancer Cell 25:794–808. https://doi.org/10.1016/j.ccr.2014.03.036
98. Goardon N et al (2011) Coexistence of LMPP-like and GMP-like leukemia stem cells in acute myeloid leukemia. Cancer Cell 19:138–152
99. Tremblay CS et al (2018) Restricted cell cycle is essential for clonal evolution and therapeutic resistance of pre-leukemic stem cells. Nat Commun 9:3535. https://doi.org/10.1038/s41467-018-06021-7
100. Shastri A, Will B, Steidl U, Verma A (2017) Stem and progenitor cell alterations in myelodysplastic syndromes. Blood 129:1586–1594. https://doi.org/10.1182/blood-2016-10-696062
101. Rubnitz JE et al (2010) Minimal residual disease-directed therapy for childhood acute myeloid leukaemia: results of the AML02 multicentre trial. Lancet Oncol 11:543–552. https://doi.org/10.1016/S1470-2045(10)70090-5

102. Shlush LI et al (2017) Tracing the origins of relapse in acute myeloid leukaemia to stem cells. Nature 547:104–108. https://doi.org/10.1038/nature22993
103. Jongen-Lavrencic M et al (2018) Molecular minimal residual disease in acute myeloid leukemia. N Engl J Med 378:1189–1199. https://doi.org/10.1056/NEJMoa1716863
104. Hollein A et al (2018) NPM1 mutated AML can relapse with wild-type NPM1: persistent clonal hematopoiesis can drive relapse. Blood Adv 2:3118–3125. https://doi.org/10.1182/bloodadvances.2018023432
105. Walter MJ et al (2012) Clonal architecture of secondary acute myeloid leukemia. N Engl J Med 366:1090–1098. https://doi.org/10.1056/NEJMoa1106968
106. Cancer Genome Atlas Research, N (2013) Genomic and epigenomic landscapes of adult de novo acute myeloid leukemia. N Engl J Med 368:2059–2074. https://doi.org/10.1056/NEJMoa1301689
107. Papaemmanuil E et al (2016) Genomic classification and prognosis in acute myeloid leukemia. N Engl J Med 374:2209–2221. https://doi.org/10.1056/NEJMoa1516192
108. Jan M et al (2012) Clonal evolution of preleukemic hematopoietic stem cells precedes human acute myeloid leukemia. Sci Transl Med 4:149ra118. https://doi.org/10.1126/scitranslmed.3004315
109. Wang L et al (2016) Transcriptomic characterization of SF3B1 mutation reveals its pleiotropic effects in chronic lymphocytic leukemia. Cancer Cell 30:750–763. https://doi.org/10.1016/j.ccell.2016.10.005
110. Lee SC et al (2018) Synthetic lethal and convergent biological effects of cancer-associated spliceosomal gene mutations. Cancer Cell 34:225–241. e228. https://doi.org/10.1016/j.ccell.2018.07.003
111. Will B et al (2012) Stem and progenitor cells in myelodysplastic syndromes show aberrant stage-specific expansion and harbor genetic and epigenetic alterations. Blood 120:2076–2086. https://doi.org/10.1182/blood-2011-12-399683
112. Pang WW et al (2013) Hematopoietic stem cell and progenitor cell mechanisms in myelodysplastic syndromes. Proc Natl Acad Sci USA 110:3011–3016. https://doi.org/10.1073/pnas.1222861110
113. Raza A, Galili N (2012) The genetic basis of phenotypic heterogeneity in myelodysplastic syndromes. Nat Rev Cancer 12:849–859. https://doi.org/10.1038/nrc3321
114. Haferlach T et al (2014) Landscape of genetic lesions in 944 patients with myelodysplastic syndromes. Leukemia 28:241–247. https://doi.org/10.1038/leu.2013.336
115. Lara-Astiaso D et al (2014) Immunogenetics. Chromatin state dynamics during blood formation. Science 345:943–949. https://doi.org/10.1126/science.1256271
116. Figueroa ME et al (2010) Leukemic IDH1 and IDH2 mutations result in a hypermethylation phenotype, disrupt TET2 function, and impair hematopoietic differentiation. Cancer Cell 18:553–567. https://doi.org/10.1016/j.ccr.2010.11.015
117. Abdel-Wahab O et al (2012) ASXL1 mutations promote myeloid transformation through loss of PRC2-mediated gene repression. Cancer Cell 22:180–193. https://doi.org/10.1016/j.ccr.2012.06.032
118. Issa JP (2012) DNA methylation as a clinical marker in oncology. J Clin Oncol 30:2566–2568. https://doi.org/10.1200/JCO.2012.42.1016
119. Bartholdy B et al (2014) HSC commitment-associated epigenetic signature is prognostic in acute myeloid leukemia. J Clin Invest 124:1158–1167. https://doi.org/10.1172/JCI71264
120. Jung N, Dai B, Gentles AJ, Majeti R, Feinberg AP (2015) An LSC epigenetic signature is largely mutation independent and implicates the HOXA cluster in AML pathogenesis. Nat Commun 6:8489. https://doi.org/10.1038/ncomms9489
121. Sarry JE et al (2011) Human acute myelogenous leukemia stem cells are rare and heterogeneous when assayed in NOD/SCID/IL2Rgammac-deficient mice. J Clin Invest 121:384–395. https://doi.org/10.1172/JCI41495

122. Ishikawa F et al (2005) Development of functional human blood and immune systems in NOD/SCID/IL2 receptor {gamma} chain(null) mice. Blood 106:1565–1573. https://doi.org/10.1182/blood-2005-02-0516
123. Shultz LD et al (2005) Human lymphoid and myeloid cell development in NOD/LtSz-scid IL2R gamma null mice engrafted with mobilized human hemopoietic stem cells. J Immunol 174:6477–6489
124. Taussig DC et al (2010) Leukemia-initiating cells from some acute myeloid leukemia patients with mutated nucleophosmin reside in the CD34(−) fraction. Blood 115:1976–1984. https://doi.org/10.1182/blood-2009-02-206565
125. Quek L et al (2016) Genetically distinct leukemic stem cells in human CD34- acute myeloid leukemia are arrested at a hemopoietic precursor-like stage. J Exp Med 213:1513–1535. https://doi.org/10.1084/jem.20151775
126. Jordan CT et al (2000) The interleukin-3 receptor alpha chain is a unique marker for human acute myelogenous leukemia stem cells. Leukemia 14:1777–1784
127. Bakker AB et al (2004) C-type lectin-like molecule-1: a novel myeloid cell surface marker associated with acute myeloid leukemia. Cancer Res 64:8443–8450. https://doi.org/10.1158/0008-5472.CAN-04-1659
128. Majeti R et al (2009) CD47 is an adverse prognostic factor and therapeutic antibody target on human acute myeloid leukemia stem cells. Cell 138:286–299
129. Jan M et al (2011) Prospective separation of normal and leukemic stem cells based on differential expression of TIM3, a human acute myeloid leukemia stem cell marker. Proc Natl Acad Sci USA 108:5009–5014
130. Hosen N et al (2007) CD96 is a leukemic stem cell-specific marker in human acute myeloid leukemia. Proc Natl Acad Sci USA 104:11008–11013
131. Jin L, Hope KJ, Zhai Q, Smadja-Joffe F, Dick JE (2006) Targeting of CD44 eradicates human acute myeloid leukemic stem cells. Nat Med 12:1167–1174
132. Kikushige Y, Miyamoto T (2015) Identification of TIM-3 as a leukemic stem cell surface molecule in primary acute myeloid leukemia. Oncology (Williston Park) 89(Suppl 1):28–32. https://doi.org/10.1159/000431062
133. Kikushige Y et al (2010) TIM-3 is a promising target to selectively kill acute myeloid leukemia stem cells. Cell Stem Cell 7:708–717
134. Jaiswal S et al (2009) CD47 is upregulated on circulating hematopoietic stem cells and leukemia cells to avoid phagocytosis. Cell 138:271–285
135. Bonardi F et al (2013) A proteomics and transcriptomics approach to identify leukemic stem cell (LSC) markers. Mol Cell Proteomics 12:626–637. https://doi.org/10.1074/mcp.M112.021931
136. van Rhenen A et al (2007) The novel AML stem cell associated antigen CLL-1 aids in discrimination between normal and leukemic stem cells. Blood 110:2659–2666
137. Moshaver B et al (2008) Identification of a small subpopulation of candidate leukemia-initiating cells in the side population of patients with acute myeloid leukemia. Stem Cells 26:3059–3067
138. Saito Y et al (2010) Induction of cell cycle entry eliminates human leukemia stem cells in a mouse model of AML. Nat Biotechnol 28:275–280
139. Askmyr M et al (2013) Selective killing of candidate AML stem cells by antibody targeting of IL1RAP. Blood 121:3709–3713. https://doi.org/10.1182/blood-2012-09-458935
140. Agerstam H et al (2015) Antibodies targeting human IL1RAP (IL1R3) show therapeutic effects in xenograft models of acute myeloid leukemia. Proc Natl Acad Sci USA 112:10786–10791. https://doi.org/10.1073/pnas.1422749112
141. Iwasaki M, Liedtke M, Gentles AJ, Cleary ML (2015) CD93 marks a non-quiescent human leukemia stem cell population and is required for development of MLL-rearranged acute myeloid leukemia. Cell Stem Cell 17:412–421. https://doi.org/10.1016/j.stem.2015.08.008

142. Pabst C et al (2016) GPR56 identifies primary human acute myeloid leukemia cells with high repopulating potential in vivo. Blood 127:2018–2027. https://doi.org/10.1182/blood-2015-11-683649
143. Al-Mawali A, Gillis D, Lewis I (2016) Immunoprofiling of leukemic stem cells CD34+/CD38-/CD123+ delineate FLT3/ITD-positive clones. J Hematol Oncol 9:61. https://doi.org/10.1186/s13045-016-0292-z
144. Zhang W et al (2015) Expressions of CD96 and CD123 in bone marrow cells of patients with myelodysplastic syndromes. Clin Lab 61:1429–1434
145. Testa U et al (2002) Elevated expression of IL-3Ralpha in acute myelogenous leukemia is associated with enhanced blast proliferation, increased cellularity, and poor prognosis. Blood 100:2980–2988. https://doi.org/10.1182/blood-2002-03-0852
146. Barve A et al (2018) Comparative utility of NRG and NRGS mice for the study of normal hematopoiesis, leukemogenesis, and therapeutic response. Exp Hematol 67:18–31. https://doi.org/10.1016/j.exphem.2018.08.004
147. Wunderlich M et al (2010) AML xenograft efficiency is significantly improved in NOD/SCID-IL2RG mice constitutively expressing human SCF, GM-CSF and IL-3. Leukemia 24:1785–1788. https://doi.org/10.1038/leu.2010.158
148. Wunderlich M, Mulloy JC (2016) MISTRG extends PDX modeling to favorable AMLs. Blood 128:2111–2112. https://doi.org/10.1182/blood-2016-09-738757
149. Ishikawa F et al (2007) Chemotherapy-resistant human AML stem cells home to and engraft within the bone-marrow endosteal region. Nat Biotechnol 25:1315–1321
150. Griessinger E et al (2016) Frequency and dynamics of leukemia-initiating cells during short-term ex vivo culture informs outcomes in acute myeloid leukemia patients. Cancer Res 76:2082–2086. https://doi.org/10.1158/0008-5472.CAN-15-2063
151. Vargaftig J et al (2012) Frequency of leukemic initiating cells does not depend on the xenotransplantation model used. Leukemia 26:858–860. https://doi.org/10.1038/leu.2011.250
152. Wang K et al (2017) Patient-derived xenotransplants can recapitulate the genetic driver landscape of acute leukemias. Leukemia 31:151–158. https://doi.org/10.1038/leu.2016.166
153. Vatakis DN et al. (2012) Using the BLT humanized mouse as a stem cell based gene therapy tumor model. J Visualized Exp:e4181. https://doi.org/10.3791/4181
154. Reinisch A et al (2016) A humanized bone marrow ossicle xenotransplantation model enables improved engraftment of healthy and leukemic human hematopoietic cells. Nat Med 22:812–821. https://doi.org/10.1038/nm.4103
155. Reinisch A, Hernandez DC, Schallmoser K, Majeti R (2017) Generation and use of a humanized bone-marrow-ossicle niche for hematopoietic xenotransplantation into mice. Nat Protoc 12:2169–2188. https://doi.org/10.1038/nprot.2017.088
156. Amir el AD et al (2013) viSNE enables visualization of high dimensional single-cell data and reveals phenotypic heterogeneity of leukemia. Nat Biotechnol 31:545–552. https://doi.org/10.1038/nbt.2594
157. Konopleva MY, Jordan CT (2011) Leukemia stem cells and microenvironment: biology and therapeutic targeting. J Clin Oncol 29:591–599. https://doi.org/10.1200/JCO.2010.31.0904
158. Becker MW, Jordan CT (2011) Leukemia stem cells in 2010: current understanding and future directions. Blood Rev 25:75–81
159. van Rhenen A et al (2005) High stem cell frequency in acute myeloid leukemia at diagnosis predicts high minimal residual disease and poor survival. Clin Cancer Res 11:6520–6527. https://doi.org/10.1158/1078-0432.CCR-05-0468
160. Vergez F et al (2011) High levels of CD34+CD38low/-CD123+ blasts are predictive of an adverse outcome in acute myeloid leukemia: a Groupe Ouest-Est des Leucemies Aigues et Maladies du Sang (GOELAMS) study. Haematologica 96:1792–1798. https://doi.org/10.3324/haematol.2011.047894
161. Eppert K et al (2011) Stem cell gene expression programs influence clinical outcome in human leukemia. Nat Med 17:1086–1093. https://doi.org/10.1038/nm.2415

162. Pearce DJ et al (2006) AML engraftment in the NOD/SCID assay reflects the outcome of AML: implications for our understanding of the heterogeneity of AML. Blood 107:1166–1173
163. Gentles AJ, Plevritis SK, Majeti R, Alizadeh AA (2010) Association of a leukemic stem cell gene expression signature with clinical outcomes in acute myeloid leukemia. JAMA 304:2706–2715
164. Bader P et al (2009) Prognostic value of minimal residual disease quantification before allogeneic stem-cell transplantation in relapsed childhood acute lymphoblastic leukemia: the ALL-REZ BFM study group. J Clin Oncol 27:377–384. https://doi.org/10.1200/JCO.2008.17.6065
165. Gerber JM et al (2012) A clinically relevant population of leukemic CD34(+)CD38(−) cells in acute myeloid leukemia. Blood 119:3571–3577. https://doi.org/10.1182/blood-2011-06-364182
166. Metzeler KH et al (2013) A stem cell-like gene expression signature associates with inferior outcomes and a distinct microRNA expression profile in adults with primary cytogenetically normal acute myeloid leukemia. Leukemia 27:2023–2031. https://doi.org/10.1038/leu.2013.181
167. Metzeler KH et al (2008) An 86-probe-set gene-expression signature predicts survival in cytogenetically normal acute myeloid leukemia. Blood 112:4193–4201. https://doi.org/10.1182/blood-2008-02-134411
168. Ng SW et al (2016) A 17-gene stemness score for rapid determination of risk in acute leukaemia. Nature 540:433–437. https://doi.org/10.1038/nature20598
169. Duployez N et al (2019) The stem cell-associated gene expression signature allows risk stratification in pediatric acute myeloid leukemia. Leukemia 33:348–357. https://doi.org/10.1038/s41375-018-0227-5
170. Ding L et al (2012) Clonal evolution in relapsed acute myeloid leukaemia revealed by whole-genome sequencing. Nature 481:506–510. https://doi.org/10.1038/nature10738
171. Kronke J et al (2013) Clonal evolution in relapsed NPM1-mutated acute myeloid leukemia. Blood 122:100–108. https://doi.org/10.1182/blood-2013-01-479188
172. Garg M et al (2015) Profiling of somatic mutations in acute myeloid leukemia with FLT3-ITD at diagnosis and relapse. Blood 126:2491–2501. https://doi.org/10.1182/blood-2015-05-646240
173. Kim YS, Kang MJ, Cho YM (2013) Low production of reactive oxygen species and high DNA repair: mechanism of radioresistance of prostate cancer stem cells. Anticancer Res 33:4469–4474
174. Diehn M et al (2009) Association of reactive oxygen species levels and radioresistance in cancer stem cells. Nature 458:780–783. https://doi.org/10.1038/nature07733
175. Simsek T et al (2010) The distinct metabolic profile of hematopoietic stem cells reflects their location in a hypoxic niche. Cell Stem Cell 7:380–390. https://doi.org/10.1016/j.stem.2010.07.011
176. Takubo K et al (2013) Regulation of glycolysis by pdk functions as a metabolic checkpoint for cell cycle quiescence in hematopoietic stem cells. Cell Stem Cell 12:49–61. https://doi.org/10.1016/j.stem.2012.10.011
177. Lewandowski D et al (2010) In vivo cellular imaging pinpoints the role of reactive oxygen species in the early steps of adult hematopoietic reconstitution. Blood 115:443–452. https://doi.org/10.1182/blood-2009-05-222711
178. Jang YY, Sharkis SJ (2007) A low level of reactive oxygen species selects for primitive hematopoietic stem cells that may reside in the low-oxygenic niche. Blood 110:3056–3063. https://doi.org/10.1182/blood-2007-05-087759
179. Mantel CR et al (2015) Enhancing hematopoietic stem cell transplantation efficacy by mitigating oxygen shock. Cell 161:1553–1565. https://doi.org/10.1016/j.cell.2015.04.054
180. Vannini N et al (2016) Specification of haematopoietic stem cell fate via modulation of mitochondrial activity. Nat Commun 7:13125. https://doi.org/10.1038/ncomms13125
181. Takubo K et al (2010) Regulation of the HIF-1alpha level is essential for hematopoietic stem cells. Cell Stem Cell 7:391–402. https://doi.org/10.1016/j.stem.2010.06.020

182. Ho TT et al (2017) Autophagy maintains the metabolism and function of young and old stem cells. Nature 543:205–210. https://doi.org/10.1038/nature21388
183. Ito K et al (2016) Self-renewal of a purified Tie2+ hematopoietic stem cell population relies on mitochondrial clearance. Science 354:1156–1160. https://doi.org/10.1126/science.aaf5530
184. Ito K et al (2012) A PML-PPAR-delta pathway for fatty acid oxidation regulates hematopoietic stem cell maintenance. Nat Med 18:1350–1358. https://doi.org/10.1038/nm.2882
185. Ward PS, Thompson CB (2012) Metabolic reprogramming: a cancer hallmark even Warburg did not anticipate. Cancer Cell 21:297–308. https://doi.org/10.1016/j.ccr.2012.02.014
186. Hanahan D, Weinberg RA (2011) Hallmarks of cancer: the next generation. Cell 144:646–674. https://doi.org/10.1016/j.cell.2011.02.013
187. Koppenol WH, Bounds PL, Dang CV (2011) Otto Warburg's contributions to current concepts of cancer metabolism. Nat Rev Cancer 11:325–337. https://doi.org/10.1038/nrc3038
188. Vlashi E et al (2011) Metabolic state of glioma stem cells and nontumorigenic cells. Proc Natl Acad Sci USA 108:16062–16067. https://doi.org/10.1073/pnas.1106704108
189. Janiszewska M et al (2012) Imp2 controls oxidative phosphorylation and is crucial for preserving glioblastoma cancer stem cells. Genes Dev 26:1926–1944. https://doi.org/10.1101/gad.188292.112
190. Lagadinou ED et al (2013) BCL-2 inhibition targets oxidative phosphorylation and selectively eradicates quiescent human leukemia stem cells. Cell Stem Cell 12:329–341. https://doi.org/10.1016/j.stem.2012.12.013
191. Roesch A et al (2013) Overcoming intrinsic multidrug resistance in melanoma by blocking the mitochondrial respiratory chain of slow-cycling JARID1B(high) cells. Cancer Cell 23:811–825. https://doi.org/10.1016/j.ccr.2013.05.003
192. Pei S et al (2018) AMPK/FIS1-mediated mitophagy is required for self-renewal of human AML stem cells. Cell Stem Cell 23:86–100 e106. https://doi.org/10.1016/j.stem.2018.05.021
193. Bossis G et al (2014) The ROS/SUMO axis contributes to the response of acute myeloid leukemia cells to chemotherapeutic drugs. Cell Rep 7:1815–1823. https://doi.org/10.1016/j.celrep.2014.05.016
194. Saito Y, Chapple RH, Lin A, Kitano A, Nakada D (2015) AMPK protects leukemia-initiating cells in myeloid leukemias from metabolic stress in the bone marrow. Cell Stem Cell 17:585–596. https://doi.org/10.1016/j.stem.2015.08.019
195. Carracedo A et al (2012) A metabolic prosurvival role for PML in breast cancer. J Clin Invest 122:3088–3100. https://doi.org/10.1172/JCI62129
196. Carracedo A, Cantley LC, Pandolfi PP (2013) Cancer metabolism: fatty acid oxidation in the limelight. Nat Rev Cancer 13:227–232. https://doi.org/10.1038/nrc3483
197. Shafat MS et al (2017) Leukemic blasts program bone marrow adipocytes to generate a protumoral microenvironment. Blood 129:1320–1332. https://doi.org/10.1182/blood-2016-08-734798
198. Tabe Y et al (2017) Bone marrow adipocytes facilitate fatty acid oxidation activating AMPK and a transcriptional network supporting survival of acute monocytic leukemia cells. Cancer Res 77:1453–1464. https://doi.org/10.1158/0008-5472.CAN-16-1645
199. Pike LS, Smift AL, Croteau NJ, Ferrick DA, Wu M (2011) Inhibition of fatty acid oxidation by etomoxir impairs NADPH production and increases reactive oxygen species resulting in ATP depletion and cell death in human glioblastoma cells. Biochim Biophys Acta 1807:726–734
200. Jeon SM, Chandel NS, Hay N (2012) AMPK regulates NADPH homeostasis to promote tumour cell survival during energy stress. Nature 485:661–665. https://doi.org/10.1038/nature11066
201. Tavitian S et al (2016) Impact of obesity in favorable-risk AML patients receiving intensive chemotherapy. Am J Hematol 91:193–198. https://doi.org/10.1002/ajh.24228
202. Jones CL et al (2018) Inhibition of amino acid metabolism selectively targets human leukemia stem cells. Cancer Cell 34:724–740 e724. https://doi.org/10.1016/j.ccell.2018.10.005

203. Guzman ML et al (2001) Nuclear factor-kappaB is constitutively activated in primitive human acute myelogenous leukemia cells. Blood 98:2301–2307
204. McCubrey JA et al (2014) Multifaceted roles of GSK-3 and Wnt/beta-catenin in hematopoiesis and leukemogenesis: opportunities for therapeutic intervention. Leukemia 28:15–33. https://doi.org/10.1038/leu.2013.184
205. Wang Y et al (2010) The Wnt/beta-catenin pathway is required for the development of leukemia stem cells in AML. Science 327:1650–1653
206. Jin L et al (2009) Monoclonal antibody-mediated targeting of CD123, IL-3 receptor alpha chain, eliminates human acute myeloid leukemic stem cells. Cell Stem Cell 5:31–42
207. Saito Y et al (2013) A pyrrolo-pyrimidine derivative targets human primary AML stem cells in vivo. Sci Transl Med 5:181ra152. https://doi.org/10.1126/scitranslmed.3004387
208. Majeti R (2011) Monoclonal antibody therapy directed against human acute myeloid leukemia stem cells. Oncogene 30:1009–1019
209. Zhang Y et al (2017) Targeting primary acute myeloid leukemia with a new CXCR4 antagonist IgG1 antibody (PF-06747143). Sci Rep 7:7305. https://doi.org/10.1038/s41598-017-07848-8
210. Wang Y, Liu Y, Malek SN, Zheng P, Liu Y (2011) Targeting HIF1alpha eliminates cancer stem cells in hematological malignancies. Cell Stem Cell 8:399–411. https://doi.org/10.1016/j.stem.2011.02.006
211. Schurch CM (2018) Therapeutic antibodies for myeloid neoplasms-current developments and future directions. Front Oncol 8:152. https://doi.org/10.3389/fonc.2018.00152
212. Tasian SK, Bornhauser M Rutella S (2018) Targeting leukemia stem cells in the bone marrow niche. Biomedicine 6. https://doi.org/10.3390/biomedicines6010022
213. Li F et al (2018) Characterization of SGN-CD123A, a potent CD123-directed antibody-drug conjugate for acute myeloid leukemia. Mol Cancer Ther 17:554–564. https://doi.org/10.1158/1535-7163.MCT-17-0742
214. Schepers K, Campbell TB, Passegue E (2015) Normal and leukemic stem cell niches: insights and therapeutic opportunities. Cell Stem Cell 16:254–267. https://doi.org/10.1016/j.stem.2015.02.014
215. Zhao C et al (2009) Hedgehog signalling is essential for maintenance of cancer stem cells in myeloid leukaemia. Nature 458:776–779
216. Sadarangani A et al (2015) GLI2 inhibition abrogates human leukemia stem cell dormancy. J Transl Med 13:98. https://doi.org/10.1186/s12967-015-0453-9
217. Queiroz KC et al (2010) Hedgehog signaling maintains chemoresistance in myeloid leukemic cells. Oncogene 29:6314–6322. https://doi.org/10.1038/onc.2010.375
218. Fukushima N et al (2016) Small-molecule hedgehog inhibitor attenuates the leukemia-initiation potential of acute myeloid leukemia cells. Cancer Sci 107:1422–1429. https://doi.org/10.1111/cas.13019
219. Savona MR et al (2018) Phase Ib study of glasdegib, a hedgehog pathway inhibitor, in combination with standard chemotherapy in patients with AML or high-risk MDS. Clin Cancer Res 24:2294–2303. https://doi.org/10.1158/1078-0432.CCR-17-2824
220. Cortes JE et al (2019) Randomized comparison of low dose cytarabine with or without glasdegib in patients with newly diagnosed acute myeloid leukemia or high-risk myelodysplastic syndrome. Leukemia 33:379–389. https://doi.org/10.1038/s41375-018-0312-9
221. Cortes JE et al (2018) Glasdegib in combination with cytarabine and daunorubicin in patients with AML or high-risk MDS: phase 2 study results. Am J Hematol 93:1301–1310. https://doi.org/10.1002/ajh.25238
222. Tibes R et al (2015) The hedgehog pathway as targetable vulnerability with 5-azacytidine in myelodysplastic syndrome and acute myeloid leukemia. J Hematol Oncol 8:114. https://doi.org/10.1186/s13045-015-0211-8
223. Radha G, Raghavan SC (2017) BCL2: a promising cancer therapeutic target. Biochim Biophys Acta Rev Cancer 1868:309–314. https://doi.org/10.1016/j.bbcan.2017.06.004

224. Hata AN, Engelman JA, Faber AC (2015) The BCL2 family: key mediators of the apoptotic response to targeted anticancer therapeutics. Cancer Discov 5:475–487. https://doi.org/10.1158/2159-8290.CD-15-0011
225. Opferman JT et al (2005) Obligate role of anti-apoptotic MCL-1 in the survival of hematopoietic stem cells. Science 307:1101–1104. https://doi.org/10.1126/science.1106114
226. DiNardo CD et al (2018) Clinical experience with the BCL2-inhibitor venetoclax in combination therapy for relapsed and refractory acute myeloid leukemia and related myeloid malignancies. Am J Hematol 93:401–407. https://doi.org/10.1002/ajh.25000
227. Naqvi K, Konopleva M, Ravandi F (2017) Targeted therapies in acute myeloid leukemia: a focus on FLT-3 inhibitors and ABT199. Expert Rev Hematol 10:863–874. https://doi.org/10.1080/17474086.2017.1366852
228. Leverson JD et al (2017) Found in translation: how preclinical research is guiding the clinical development of the BCL2-selective inhibitor venetoclax. Cancer Discov 7:1376–1393. https://doi.org/10.1158/2159-8290.CD-17-0797
229. Adams JM, Cory S (2018) The BCL-2 arbiters of apoptosis and their growing role as cancer targets. Cell Death Differ 25:27–36. https://doi.org/10.1038/cdd.2017.161
230. Oltersdorf T et al (2005) An inhibitor of Bcl-2 family proteins induces regression of solid tumours. Nature 435:677–681. https://doi.org/10.1038/nature03579
231. Tse C et al (2008) ABT-263: a potent and orally bioavailable Bcl-2 family inhibitor. Cancer Res 68:3421–3428. https://doi.org/10.1158/0008-5472.CAN-07-5836
232. Souers AJ et al (2013) ABT-199, a potent and selective BCL-2 inhibitor, achieves antitumor activity while sparing platelets. Nat Med 19:202–208. https://doi.org/10.1038/nm.3048
233. Schoenwaelder SM et al (2011) Bcl-xL-inhibitory BH3 mimetics can induce a transient thrombocytopathy that undermines the hemostatic function of platelets. Blood 118:1663–1674. https://doi.org/10.1182/blood-2011-04-347849
234. Kater AP et al (2018) Fixed duration of venetoclax-rituximab in relapsed/refractory chronic lymphocytic leukemia eradicates minimal residual disease and prolongs survival: post-treatment follow-up of the MURANO phase III study. J Clin Oncol:JCO1801580. https://doi.org/10.1200/JCO.18.01580
235. Roberts AW et al (2016) Targeting BCL2 with venetoclax in relapsed chronic lymphocytic leukemia. N Engl J Med 374:311–322. https://doi.org/10.1056/NEJMoa1513257
236. Anderson MA et al (2016) The BCL2 selective inhibitor venetoclax induces rapid onset apoptosis of CLL cells in patients via a TP53-independent mechanism. Blood 127:3215–3224. https://doi.org/10.1182/blood-2016-01-688796
237. Konopleva M et al (2016) Efficacy and biological correlates of response in a phase II study of venetoclax monotherapy in patients with acute myelogenous leukemia. Cancer Discov 6:1106–1117. https://doi.org/10.1158/2159-8290.CD-16-0313
238. Ramsey HE et al (2018) A novel MCL1 inhibitor combined with venetoclax rescues venetoclax-resistant acute myelogenous leukemia. Cancer Discov 8:1566–1581. https://doi.org/10.1158/2159-8290.CD-18-0140
239. Glaser SP et al (2012) Anti-apoptotic mcl-1 is essential for the development and sustained growth of acute myeloid leukemia. Genes Dev 26:120–125. https://doi.org/10.1101/gad.182980.111
240. Konopleva M et al (2006) Mechanisms of apoptosis sensitivity and resistance to the BH3 mimetic ABT-737 in acute myeloid leukemia. Cancer Cell 10:375–388
241. van Delft MF et al (2006) The BH3 mimetic ABT-737 targets selective Bcl-2 proteins and efficiently induces apoptosis via Bak/Bax if Mcl-1 is neutralized. Cancer Cell 10:389–399. https://doi.org/10.1016/j.ccr.2006.08.027
242. Teh TC et al (2018) Enhancing venetoclax activity in acute myeloid leukemia by co-targeting MCL1. Leukemia 32:303–312. https://doi.org/10.1038/leu.2017.243
243. Caenepeel S et al (2018) AMG 176, a selective MCL1 inhibitor, is effective in hematologic cancer models alone and in combination with established therapies. Cancer Discov 8:1582–1597. https://doi.org/10.1158/2159-8290.CD-18-0387

244. Konopleva M et al (2012) MEK inhibition enhances ABT-737-induced leukemia cell apoptosis via prevention of ERK-activated MCL-1 induction and modulation of MCL-1/BIM complex. Leukemia 26:778–787. https://doi.org/10.1038/leu.2011.287
245. Coloff JL et al (2011) Akt-dependent glucose metabolism promotes Mcl-1 synthesis to maintain cell survival and resistance to Bcl-2 inhibition. Cancer Res 71:5204–5213. https://doi.org/10.1158/0008-5472.CAN-10-4531
246. Bogenberger JM et al (2014) BCL-2 family proteins as 5-Azacytidine-sensitizing targets and determinants of response in myeloid malignancies. Leukemia 28:1657–1665. https://doi.org/10.1038/leu.2014.44
247. Niu X et al (2016) Binding of released Bim to Mcl-1 is a mechanism of intrinsic resistance to ABT-199 which can be overcome by combination with daunorubicin or cytarabine in AML cells. Clin Cancer Res 22:4440–4451. https://doi.org/10.1158/1078-0432.CCR-15-3057
248. Gupta N et al (2015) Comparison of epigenetic versus standard induction chemotherapy for newly diagnosed acute myeloid leukemia patients >/=60 years old. Am J Hematol 90:639–646. https://doi.org/10.1002/ajh.24016
249. Bogenberger JM et al (2015) Ex vivo activity of BCL-2 family inhibitors ABT-199 and ABT-737 combined with 5-azacytidine in myeloid malignancies. Leuk Lymphoma 56:226–229. https://doi.org/10.3109/10428194.2014.910657
250. Tsao T et al (2012) Concomitant inhibition of DNA methyltransferase and BCL-2 protein function synergistically induce mitochondrial apoptosis in acute myelogenous leukemia cells. Ann Hematol 91:1861–1870. https://doi.org/10.1007/s00277-012-1537-8
251. Pan R et al (2014) Selective BCL-2 inhibition by ABT-199 causes on-target cell death in acute myeloid leukemia. Cancer Discov 4:362–375. https://doi.org/10.1158/2159-8290.CD-13-0609
252. Chan SM et al (2015) Isocitrate dehydrogenase 1 and 2 mutations induce BCL-2 dependence in acute myeloid leukemia. Nat Med 21:178–184. https://doi.org/10.1038/nm.3788
253. Wei A, Strickland SA, Hou J-Z, Fiedler W, Lin TL, Walter RB, Enjeti AK, Hong W-J, Chyla B, Popovic R, Fakouhi K, Tu X, Hayslip J, Roboz GJ (2018) Venetoclax with low-dose cytarabine induces rapid, deep, and durable responses in previously untreated older adults with AML ineligible for intensive chemotherapy. Blood 132:284
254. Hope KJ, Jin L, Dick JE (2004) Acute myeloid leukemia originates from a hierarchy of leukemic stem cell classes that differ in self-renewal capacity. Nat Immunol 5:738–743
255. Samudio I et al (2010) Pharmacologic inhibition of fatty acid oxidation sensitizes human leukemia cells to apoptosis induction. J Clin Invest 120:142–156. https://doi.org/10.1172/JCI38942

Chapter 6
Tissue "Hypoxia" and the Maintenance of Leukemia Stem Cells

Persio Dello Sbarba and Giulia Cheloni

Contents

Abstract The relationship of the homing of normal hematopoietic stem cells (HSC) in the bone marrow to specific environmental conditions, referred to as the *stem cell niche* (SCN), has been intensively studied over the last three decades. These conditions include the action of a number of molecular and cellular players, as well as critical levels of nutrients, oxygen and glucose in particular, involved in energy production. These factors are likely to act also in leukemias, due to the strict analogy between the hierarchical structure of normal hematopoietic cell populations and that of leukemia cell populations. This led to propose that leukemic growth is fostered by cells endowed with stem cell properties, the leukemia stem cells (LSC), a concept readily extended to comprise the cancer stem cells (CSC) of solid tumors. Two alternative routes have been proposed for CSC generation, that is, the

P. Dello Sbarba (✉)
Department of Experimental and Clinical Biomedical Sciences, Università degli Studi di Firenze, Florence, Italy
e-mail: persio@unifi.it

G. Cheloni
Department of Experimental and Clinical Biomedical Sciences, Università degli Studi di Firenze, Florence, Italy

Cancer Research Institute, Department of Medicine and Pathology, Beth Israel Deaconess Medical Center, Harvard Medical School, Boston, Massachusetts, USA

H. Zhang, S. Li (eds.), *Leukemia Stem Cells in Hematologic Malignancies*, Advances in Experimental Medicine and Biology 1143,
https://doi.org/10.1007/978-981-13-7342-8_6

oncogenic staminalization (acquisition of self-renewal) of a normal progenitor cell (the "CSC in normal progenitor cell" model) and the oncogenic transformation of a normal (self-renewing) stem cell (the "CSC in normal stem cell" model). The latter mechanism, in the hematological context, makes LSC derive from HSC, suggesting that LSC share SCN homing with HSC. This chapter is focused on the availability of oxygen and glucose in the regulation of LSC maintenance within the SCN. In this respect, the most critical aspect in view of the outcome of therapy is the long-term maintenance of the LSC subset capable to sustain minimal residual disease and the related risk of relapse of disease.

Keywords Leukemia stem cell maintenance · Hypoxia · Oxygen · Stem cell niche

6.1 Introduction

The relationship of the homing of normal hematopoietic stem cells (HSC) in the bone marrow (BM) to specific environmental conditions, referred to as the "stem cell niche" (SCN), has been the object of intensive study over the last three decades. These conditions include the action of a number of molecular and cellular players as well as critical levels of nutrients, oxygen and glucose in particular, involved in energy production [1, 2]. These factors are likely to act also in leukemias. A strict analogy indeed emerged between the hierarchical structure of normal hematopoietic cell populations and that of leukemia cell populations [3]. This led to propose [4] that leukemic growth is fostered by cells endowed with stem cell properties, the leukemia stem cells (LSC), a concept readily extended to comprise the cancer stem cells (CSC) of solid tumors [5, 6]. More than one model exists to explain the relationship of CSC to their normal counterpart. Two alternative routes have been proposed for CSC generation [6], that is, the oncogenic staminalization (acquisition of self-renewal) of a normal progenitor cell (the "CSC in normal progenitor cell" model) and the oncogenic transformation of a normal (self-renewing) stem cell (the "CSC in normal stem cell" model). The latter mechanism, in the hematological context, makes LSC derive from HSC, an assumption which suggests that LSC share with HSC the homing behavior within SCN of BM [7, 8]. This chapter is focused on the role of the availability of oxygen and glucose in the regulation of LSC maintenance. In this respect, the most critical aspect in view of the outcome of therapy is the long-term maintenance of the LSC subset capable to sustain minimal residual disease (MRD) and the related risk of relapse of disease following successful induction of remission. The description of aspects of the regulation of LSC maintenance related to the symbiosis with stromal cells, to the action of soluble or cell surface/matrix-bound cytokines or growth factors, and to the modulation of intracellular signaling is left to other chapters of this book and/or to independent reviews.

6.2 Why Hypoxia Is Written "Hypoxia"

Tissue zones where stem cell potential is maintained are commonly referred to as "hypoxic" SCN [9], being "hypoxia" an inappropriate term in this context, as it is used to define in vitro or in vivo situations where O_2 concentration is lower than that of sea-level atmosphere (21% O_2). Indeed, O_2 concentration in normal tissues, even in the lung, is always significantly lower than 21% [10–13]. The papers we are referring to here provided convincing arguments for the use of the term "relatively (to 21%) low O_2 concentration" (to be shortened to "low O_2"), instead of "hypoxia," to define conditions corresponding to physiological tissue O_2 concentrations. Accordingly, any O_2 concentration typical of any tissue under physiological conditions should be considered as "*in situ* normoxia" [13]. On the other hand, the term hypoxia can be correctly used to refer to pathological situations where O_2 concentration is lower than that of the corresponding normal tissue. Therefore, it is important to point out that the concept of in situ normoxia applies to significantly different O_2 concentrations, between different tissues and different sites within the same tissue [13]. In BM, for instance, the normal overall O_2 concentration is lower than in most other tissues. BM cells are indeed physiologically distributed along a gradient of partial O_2 pressure (ppO_2) that ranges from around 5% O_2 close to blood vessels to about 0 ("anoxia") in the most distant regions. Thus, within BM, "normoxic" conditions correspond to O_2 concentrations as different as 5% from 0.1%, being even anoxia to be considered as in situ normoxia in some zones of BM [13].

6.3 The Low-Oxygen Stem Cell Niche in the Bone Marrow

The key feature of SCN is to enable HSC to proliferate without losing (or losing as slowly as possible) stem cell potential, i.e., to undergo the so-called self-renewal, which is a defining property of stem cells. Such a pattern of HSC proliferation is what sustains long-term maintenance of HSC, yet allowing their contribution to active hematopoiesis. The SCN concept emerged as a theoretical model [14], yet based on experimental ex vivo data indicating an uneven compartmentalization in BM of HSC and less immature hematopoietic progenitor cells (HPC), being HSC located preferentially close to the bone surface and HPC instead in proximity of the central sinus. This concept evolved later from the HSC/HPC contraposition to a model where different HSC subsets are distributed between "endosteal SCN," where the stem cell potential of HSC is maintained, and "vascular SCN," where HSC commitment to clonal expansion and differentiation is driven [15–17]. The relationship of SCN function to proximity to blood vessels is in keeping with a regulation based on nutrient supply, O_2 and glucose in particular [18, 19]. It emerged that SCN where HSC are physiologically long-term maintained are placed in BM zones where O_2 concentration is lower than in the surrounding tissue and that in situ normoxic conditions for HSC maintenance are ensured in sites at the lower extreme of ppO_2

gradient in BM [20]. It is this HSC-maintaining SCN that will be referred to hereafter with the acronym SCN, according to the "restrictive" original Schofield's definition [14].

The SCN model started to be enriched with mechanistic details in the early 1990s, on the basis of results of in vitro studies [21]. In murine BM cell cultures incubated in atmosphere at 1% O_2, HSC maintenance was found enhanced, while HPC were suppressed and the overall hematopoietic output markedly reduced. Thus, low ppO_2 emerged as a regulatory aspect of SCN function, and the concept of "hypoxic" SCN was introduced. An interesting aspect of this study was that pooled data obtained using a number of different progenitor/stem cell assays revealed a precise hierarchical "gradient" of resistance to low O_2, being the progenitor the more resistant, the higher its hierarchical level [21]. This means that an environment which is "hypoxic" for the bulk of hematopoietic cells and HPC is actually "normoxic" for HSC. Thus, the first mechanistic feature of SCN to emerge pointed to metabolic issues, and to the role of ppO_2 in particular, in the regulation of HSC maintenance. Similar data were then collected for human hematopoiesis and long-term repopulating HSC [22–24], and the existence of "hypoxic" SCN in vivo was later confirmed experimentally [9], leading to a complete systematization of the issue [13]. That said, it is necessary here to underscore a point relative to the abovementioned "gradient" of resistance to low ppO_2. To home selectively within SCN placed in tissue zones at the lowest ppO_2, HSC need to exhibit a metabolic profile that is not shared by HPC. This profile cannot simply consist of the compatibility with a generic tissue "hypoxia," which refers to a wide range of tissue O_2 concentrations. Either HSC or HPC exhibit indeed a "hypoxic" metabolic profile [25]. Thus, what qualifies HSC from HPC is that HSC (but not HPC) are capable to stand the lowest physiological ppO_2 and for extended times. The question invariably slides toward the upregulation of HIFα signaling, the best known driver of "cell adaptation to hypoxia." However, HIFα stabilization threshold is around 3% O_2 [26], and HIFα signaling is active in both HSC and HPC, so that it cannot sustain alone the critical metabolic differences between HSC and HPC or confer upon HSC all the features enabling their selective homing in low ppO_2 SCN. Only a small minority of cells where HIFα is stabilized is indeed capable to stand the low ppO_2 typical of SCN [9]. Thus, HIFα stabilization is a necessary, although not sufficient, condition for HSC maintenance, in keeping with what demonstrated for the maintenance of LSC [27] of chronic myeloid leukemia (CML).

One of the mechanisms operating within the SCN to sustain hematopoiesis while making HSC lose stem cell potential as slowly as possible consists in the regulation of balance between HSC quiescence and cycling. It was therefore crucial to establish whether low O_2 modulates this balance. The issue was addressed using murine BM cell cultures incubated at 1% O_2 and two different strategies, i.e., determining the suppression of cycling HSC following treatment with 5-fluoro-uracil (5FU) [28] and the coupling between mitotic history and maintenance of stem cell potential [29]. It was found that, after 5 days in low O_2, 1/3 of HSC are induced to quiescence (5FU-resistant), while 2/3 are 5FU-sensitive, suggesting that low ppO_2 of SCN in vivo is compatible with the cycling of a substantial proportion of HSC [28]. It also emerged that one replication cycle at 1% O_2 boosts stem cell potential and that this effect is lost when cycling is sustained for more than one cycle at 1% O_2 or cells are incubated in "normoxia" or in the presence of interleukin-3. This indicates that HSC self-renewal occurs immediately upon HSC recruitment from quiescence to cycling, provided this happens in low O_2, suggesting that low ppO_2 of SCN in vivo is a crucial factor for the maintenance of stem cell potential of HSC, helping HSC to proliferate as stem cells (self-renewal), and preventing their commitment to clonal expansion. On the contrary, stem cell potential is typically lost when proliferation is extensively stimulated, such as in the presence of IL-3 or relatively high ppO_2 [29].

While all the earlier studies of the effects of low O_2 on HSC maintenance were carried out incubating cells at 1% O_2 (or even higher O_2%), the question emerged of whether O_2 concentrations lower than 1%, detectable in vivo [10, 11, 13, 20], should be considered to play a regulatory role in SCN. An affirmative answer was obtained in vitro, showing that at 0.1% O_2 not only HSC survival but also HSC cycling, a crucial aspect of SCN function, is maintained, although induction of quiescence is favored [30, 31]. Actually, O_2 concentrations ranging between 0.1% and 1.0% were classified even as "moderate hypoxia" [26]. A direct measurement of O_2 concentration in BM showed that ppO_2 is around 10 mmHg, equivalent to 1.3% O_2, within the vascular SCN [32], the SCN type where ppO_2 is believed to be relatively high. This suggests that ppO_2 of endosteal SCN should be well below the 1% O_2 concentration used for the earlier studies in vitro of the effects of low O_2 on HSC maintenance. The compatibility of O_2 concentrations as low as 0.1–0.2% with SCN function led to the use these concentrations for the studies of leukemias summarized below.

6.4 Low Oxygen and "Oncogene Suppression" in Leukemias

Working with CML stabilized cell lines or primary explants, it was shown that the culture in atmosphere at 0.1% O_2 for relatively long incubation times (7–10 days) markedly reduced cell bulk with respect to time zero, while in the residual population stem cell potential was integrally maintained on a per-cell basis. This parallels strictly what observed for normal hematopoiesis. A crucial outcome of these experiments was that incubation in low O_2 also completely suppressed $BCR/Abl_{protein}$, the product of the fusion oncogene responsible for CML pathogenesis, but not BCR/Abl_{mRNA} [33]. Consequently, the CML cell subset capable to stand incubation in low O_2 is independent of BCR/Abl for persistence in culture but remains genetically leukemic (Figs. 6.1 and 6.2). Thus, when BCR/Abl-independent cells were transferred to growth-permissive incubation conditions ("normoxia"), their stem cell potential was exploited, as expected, via $BCR/Abl_{protein}$ re-expression, which is capable to ensure a maximal expansion of cell population. It is extremely relevant that the suppression of driver oncogenic proteins ("oncogene suppression") following cell incubation in low O_2 is not restricted to CML (Fig. 6.1) but is extended to other types of leukemias [34, 35], such as murine Friend's erythroleukemia (MEL; erythropoietin receptor and glycoprotein 55), human acute myeloblastic leukemia (AML; AML1/ETO), and human acute promyelocytic leukemia (APL; PML/

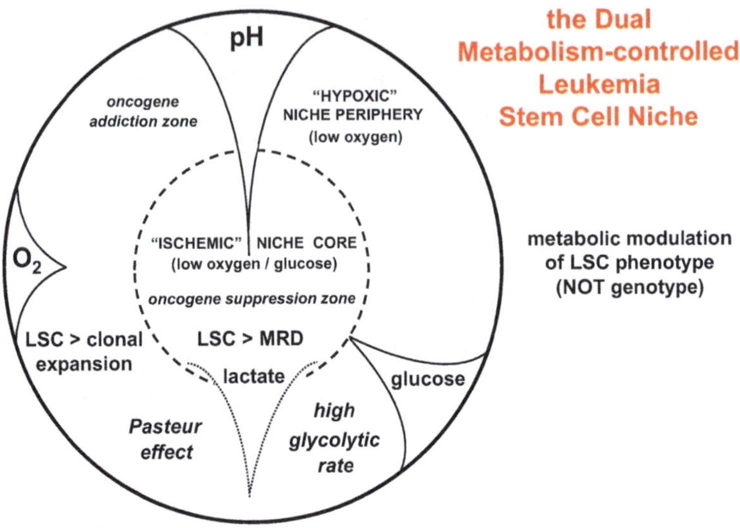

Fig. 6.1 Schematic representation of the "Dual Metabolism-controlled CML Stem Cell Niche," as hypothesized on the basis of published data and hypotheses [41, 49–51, 66, 86]. The "hypoxic" stem cell niche (SCN) periphery, where ppO_2 is low but glucose concentration is relatively high, allows the expression of the oncogenic protein (maintenance of oncogene addiction), steering leukemia stem cells (LSC) to clonal expansion. In the "ischemic" SCN core, characterized by both O_2 and glucose shortage, the oncogenic protein is suppressed, boosting LSC self-renewal and maintanance of minimal residual disease (MRD)

RARα). In all leukemias tested, oncogene suppression is relatively slow (3–7 days after the beginning of incubation in low O_2), indicating that the phenomenon cannot be considered as a direct consequence of HIFα stabilization, which is triggered within few minutes of cell transfer to low O_2 [9]. Nevertheless, HIFα activity has been shown necessary for the maintenance of LSC of CML [27]. As for the downstream, effector mechanisms of oncogene suppression in low O_2, they belong to a complex panel which includes transcriptional, translational, and posttranslational events [36–39]. It is important to point out that the downregulation of these processes occurring in low O_2 is not a generalized phenomenon and that the block of protein synthesis does not apply to a number of factors, many of which are apparently important for cell maintenance under metabolic restriction [39, 40].

A key point of the scenario summarized above is that oncogene suppression occurring in low O_2 does not correlate with a loss of stem cell potential, at least in the case of CML [33, 41] and MEL [34] cells. This points to the existence of a leukemia cell subset endowed with LSC properties and capable to survive and cycle (34 and Tanturli M, Cheloni G, Bono S and Poteti M, unpublished data) in the absence of oncogenic signaling (Fig. 6.1). The selection of such an LSC subset under metabolic pressure suggests that maintained oncogenic stimulation makes LSC unfit to home the SCN, due to the detrimental effects of growth-promoting stimuli in a growth-limiting environment. On the contrary, oncogene suppression would enable LSC to persist as such within the SCN, without risking to be pushed by oncogenic signaling to commitment to clonal expansion and differentiation. Thus, the loss of "oncogene addiction" [42–44] would be a necessary step of LSC adaptation to SCN (Fig. 6.1). In other words, in order to earn their long-term maintenance within the SCN, LSC are actually required to revert to a normal, HSC-like phenotype. On the other hand, when the homeostatic balance within the LSC compartment shifts from the maintenance of stem cell potential to commitment to clonal expansion and then differentiation, LSC would shift from the core to the periphery SCN environment, which is permissive for the expression of the oncogenic protein and the boosting of clonal expansion by oncogenic signaling. Under this perspective, two points are worth being underscored: (a) oncogene suppression is not a genetically blocked event but a fully reversible phenotypical adaptation, like the one proposed to describe the relationship between HSC and HPC [45]; (b) the withdrawal of oncogenic signaling in LSC appears functionally equivalent to that of cytokine signaling in HSC [3]. With respect to (b), it is to note that BCR/Abl-induced growth factor independence of hematopoietic cells is mediated at least in part by IL3 signaling [46], which indeed impairs HSC maintenance in low O_2 [29].

6.5 The Dual "Hypoxic/Ischemic" LSC Niche in Chronic Myeloid Leukemia

When the metabolic consequences of incubation of CML cells in low O_2 were addressed, it was glucose consumption from culture medium to emerge as the crucial condition for the reduction of total cell number with respect to time zero, as well as for $BCR/Abl_{protein}$ suppression. By varying time zero cell density and glucose concentration in cultures incubated in low O_2, it was established that the kinetics of this suppression is strictly related to that of glucose exhaustion [41]. Deepening the issue, it was shown that the lack of glucose is capable to drive $BCR/Abl_{protein}$ suppression independently of O_2 shortage [40]. Thus, glucose availability emerged as a converging trait of the metabolic control of $BCR/Abl_{protein}$ expression, suggesting that it is severe energy restriction in general, rather than low O_2 in particular, to trigger the $BCR/Abl_{protein}$ suppression machinery that qualifies the SCN environment (Figs. 6.1 and 6.2). If this is the case, why do we focus on low O_2 environments? For two reasons: (a) because the lower solubility of O_2 when compared to glucose, affecting their diffusion in tissues, makes O_2 shortage precede glucose shortage physiologically while getting away from blood vessel [47]; (b) because the enhanced glucose consumption rate (the Pasteur effect) typical of cell metabolism at low ppO_2 makes of low O_2 a powerful modulator of glucose availability. The time lapse between the onset of glucose shortage and that of O_2 shortage (a) well explains the markedly delayed kinetics of $BCR/Abl_{protein}$ suppression with respect to that of HIFα stabilization. Point (b) is affected by HIFα stabilization itself, glucose exhaustion being more

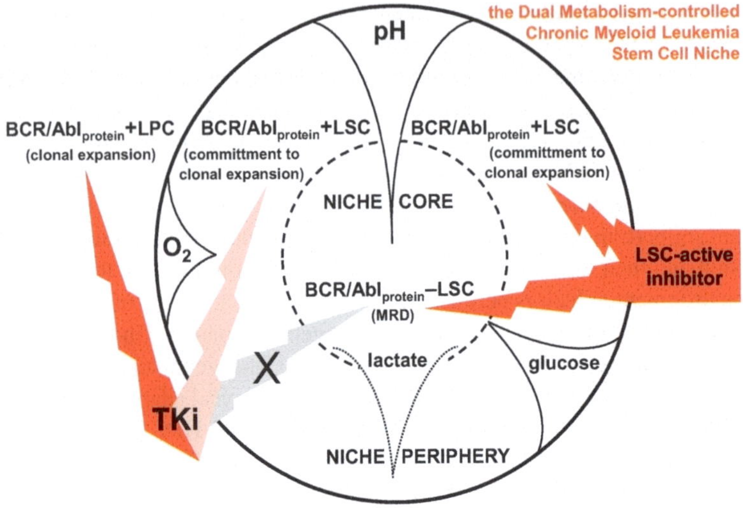

Fig. 6.2 Schematic representation of the "Dual Metabolism-controlled CML Stem Cell Niche," as hypothesized on the basis of published data and hypotheses [41, 49–51, 66, 86]. The "hypoxic" stem cell niche (SCN) periphery, where ppO_2 is low but glucose concentration is relatively high, allows BCR/Abl$_{protein}$ expression, making BCR/Abl-dependent self-renewal of leukemia stem cells (LSC) sensitive to tyrosine kinase inhibitors (TKi) active on BCR/Abl signaling (pink bolt). In the "ischemic" SCN core, characterized by both O_2 and glucose shortage, BCR/Abl$_{protein}$ suppression and BCR/Abl-independent self-renewal of LSC maintain minimal residual disease (MRD) protected from TKi action (grey bolt/X). On the other hand, TKi reach maximal activity outside SCN (red bolt on the left) against leukemia progenitor cells (LPC) engaged in CML clonal expansion (100% in replication). When BCR/Abl$_{protein}$ is suppressed, the question of the presence or the absence of primary or secondary *BCR/abl* mutations determining BCR/Abl resistance to TKI is irrelevant. Many non-TKi, LSC-active inhibitors act irrespective of BCR/Abl$_{protein}$ expression or suppression, thus succeeding in targeting both MRD and clonal expansion (red bolts on the right)

efficiently driven in low ppO_2 zones than in zones at higher ppO_2 where HIFα is not stabilized. The role of HIFα in the control of leukemic growth is reviewed in [48].

What summarized above led to define two different nutrient-restricted SCN environments, the low O_2 and the low O_2/glucose, playing distinct roles in the biology of leukemias (Fig. 6.1). These two SCN environments can be referred to, respectively, as the "hypoxic" and the "ischemic" SCN zones [41, 49–51]. The term "ischemic" sounds as inappropriate as "hypoxic" in the context of SCN. Indeed, "ischemia" refers to in vivo (but not in vitro) pathological occurrences where the tissue is damaged as a consequence of nutrient shortage, whereas in the SCN this shortage has a precise regulative role. A permanently insufficient glucose diffusion, affecting cell viability, has been shown in solid neoplasias [47] and is likely to characterize also the hyperplastic BM of leukemias, CML in particular. A simple model drawing the relationship between the two SCN zones predicts the onset of ischemia-like conditions within the core of hypoxic SCN [49–51]. In the SCN core, glucose would get close to exhaustion due to the combined effects of the high rate of consumption (in the SCN periphery) and the scarce diffusion (higher distance from blood vessels of SCN core than SCN periphery). While HIFα is surely stabilized in both the SCN core and SCN periphery, it is most probably the HIFα-dependent metabolic reprogramming of the latter which conditions the asset and function of SCN core. It is worth pointing out here that the existence of ischemia-like conditions emerged in normal BM from studies of the role of blood perfusion, rather than O_2 diffusion, in the control of hematopoiesis and in the hierarchical distribution of HSC in BM. These studies showed that the most primitive HSC reside in BM zones where blood perfusion is lowest, implying that HSC maintenance within the SCN is supported by a strongly reduced supply of nutrients as well as of blood-borne soluble factors [52, 53]. The capacity of LSC to nest in HSC niches has been also demonstrated [54, 55].

The dual "hypoxic/ischemic" SCN delineated above impacts LSC modeling in CML. As mentioned in the *Introduction*, Reya et al. summarized two alternative models for CSC generation, the cancerization of a normal stem cell or the staminalization of a normal progenitor cell, meaning that cells exhibiting either the stem or the progenitor cell phenotype may long-term sustain cancer growth [6]. The identification of two possible "metabolism-driven" phenotypes for LSC of CML led us to propose that, short of being alternative to each other, *both* models fit with CML biology, i.e., that two LSC subsets with different functional roles coexist in CML (Fig. 6.2). The "stem cell LSC model" seems indeed adequate to describe a BCR/$Abl_{protein}$-negative LSC subset capable of BCR/Abl-independent self-renewal, while the "progenitor cell LSC model" suits an LSC subset where BCR/$Abl_{protein}$ is expressed and self-renewal is sustained by BCR/Abl-dependent signaling [49–51]. It is worth reminding here that CML and MEL include an LSC subset capable to cycle in the absence of oncogenic signaling (Tanturli et al. unpublished data; [34]). It is reasonable to think that this subset cycles more slowly than that where oncogenic signaling is active, as it is deprived of the growth advantage oncogenic signaling confers. Such an advantage, however, while being useful to maximize clonal expansion whenever possible, would be highly detrimental to the long-term maintenance of LSC. The dual CML stem cell compartment described above is well in keeping with the view of CML as a stem cell-derived but progenitor cell-driven disease [56], if one transposes the concepts of "stem" and "progenitor" cells to the two different LSC subsets described above.

6.6 The LSC Niche and the Refractoriness of MRD to Treatment in CML

The studies where the response of CML to low O_2 was initially characterized also addressed the effects of treatment with imatinib mesylate (IM; Gleevec®), the inhibitor of the constitutive enzymatic activity of BCR/Abl which is the prototype of tyrosine kinase inhibitors (TKi) used for CML therapy. As expected on the basis of what explained above, cells where BCR/$Abl_{protein}$ had been reversibly suppressed following incubation at 0.1% O_2 were resistant to treatment, obviously because devoid of the molecular target of IM [33, 41]. The crucial aspect of such an outcome is that the maintenance of CML stem cell potential was completely insensitive to IM (Fig. 6.2). This insensitivity is appropriately referred to as "refractoriness," while the term "primary resistance" should be reserved to genetically determined (and therefore irreversible) insensitivity to IM already present before the treatment starts, and the term "secondary resistance" to the irreversible insensitivity acquired by a cell subset via mutations under treatment pressure. It is worth discussing resistance to IM under the light of the SCN model delineated in the previous section.

Metabolic pressure within the SCN controls LSC sensitivity or refractoriness to IM as a phenotypical adaptation, rather than a genetically blocked event, depending on whether LSC reside within the "hypoxic" or the "ischemic" SCN zones where $BCR/Abl_{protein}$ is expressed or suppressed, respectively (Fig. 6.2). The concept of resistance to IM as a phenotypical change is in keeping with the marked phenotypical heterogeneity of CML cells as for BCR/Abl expression and sensitivity to IM [57, 58] and with the fact that relapse of disease upon discontinuation of IM most often occurs without signs of development of mutation-driven secondary resistance to IM, so that the relapsed patient responds well to the reintroduction of IM [59]. Accordingly, the TKi-resistant CML progenitor cells shown to be BCR/Abl-positive by FISH or PCR [60–62] are likely to be BCR/Abl_{mRNA}-positive/$BCR/Abl_{protein}$-negative. This is a conceptually simple way to resolve the controversy of whether TKi inhibit or not BCR/Abl kinase in LSC [60–63], by concluding that TKi are always effective but LSC do not respond due to the lack of the TKi molecular target. The presence of $BCR/Abl_{protein}$-negative LSC also explains easily the lack of LSC-suppressive effects of not only IM but also "second-generation" BCR/Abl-active TKi, despite their enhanced action on CML cell bulk [64, 65]. The ineffectiveness of IM as well as dasatinib on LSC was confirmed using CML cells, from stabilized lines or explanted from patients, kept at 0.1% O_2 for incubation times sufficient to drive glucose exhaustion from culture medium [66]. In this scenario, it was predicted that even next generations of BCR/Abl-active TKi will be useless to suppress LSC [35, 50, 51].

SCN zones where LSC are capable to cycle slowly and self-renew independently of BCR/Abl signaling and are therefore protected from TKi action represent ideal sites for the long-term maintenance of treatment-resistant MRD of CML (Fig. 6.2). Such an outline facilitates MRD modeling, because it prevents us from restricting MRD to quiescent LSC, although there is no doubt that quiescence boosts resistance to treatment when drugs generically active against cycling cells are used. What we call a "dynamic maintenance" of MRD [49–51] appears indeed better suited than LSC quiescence to explain the combination of refractoriness to TKi with liability to neoplastic progression. First, the finding that cycling of HSC in low O_2 fosters self-renewal rather than clonal expansion [29] suggests that the same may apply to LSC. A relatively high stem cell potential of the mutated CML subclone stored in the SCN before relapse of disease reaches the clinical level may be behind the often very rapid and aggressive course of relapse once it is triggered. Second, neoplastic progression while disease remains at a subclinical level is best explained if LSC cycle, because cell cycling is necessary to transmit mutations to progeny, which are undoubtedly frequent in the genetically unstable CML cells.

The process leading from MRD to relapse of disease in CML can be summarized hypothetically as follows: (a) BCR/Abl-independent, IM-insensitive self-renewal within the "ischemic" SCN core would maintain (MRD) non-mutated or mutated LSC; (b) migration out of the SCN core or (transient?) increase of glucose supply therein would make LSC switch to conditions typical of the "hypoxic" SCN periphery, where ppO_2 is still very low but glucose concentration is higher, which allow the rescue of $BCR/Abl_{protein}$ expression; (c) BCR/Abl-dependent signaling in the SCN periphery would enhance LSC self-renewal and enable commitment to clonal expansion; and (d) a move of LSC to better oxygenated BM zones (the "vascular" niches) would drive clonal expansion and differentiation (if any), making relapse of disease emerge at the clinical level (Fig. 6.2). Of course, from phase (b) on, CML growth returns to be TKi-sensitive in the majority of cases, so that there is risk of relapse only in the absence of therapy [59]. The majority of relapsed patients indeed does not present mutations, or, if it does, the mutations are incapable of conferring secondary TKi resistance. It is likely that the shift from chronic to accelerated to blast clinical phase of CML is founded on the above biological frame.

6.7 To Exploit the Metabolic Control of LSC Compartment for Therapeutic Purposes

The long-standing use for cancer therapy of drugs most of which are cell cycle-specific antiblastic agents obviously leads to trace back treatment-resistant MRD to quiescence, to quiescent stem cells in particular. Furthermore, especially in the hematopoietic field, quiescence is a substantial component within the SCN of the mechanisms protecting stem cells from premature commitment to clonal expansion and differentiation [14]. Therefore, the combination of treatments capable of inducing (stem) cell exit from quiescence with cycle-specific antiblastic agents has been proposed to try to suppress MRD [67]. A more advanced approach to the therapy of neoplasias is based on the use, instead of drugs acting generically against cycling cells, of selective inhibitors of disease-specific targets, such as BCR/Abl in CML, with obvious advantages in terms of reduced toxicity on normal cells (therapeutic index). In theory, such an approach should be also active on quiescent cells sustaining MRD, as neoplastic cells may depend on oncogenic signaling not only for proliferation but also survival (oncogene-addicted cells). An opposite, less optimistic perspective springs from the SCN model we proposed for CML, which also applies to MEL. In these leukemias, the loss of oncogene addiction driven under metabolic pressure makes not only survival but also self-renewal independent of oncogenic signaling (see above). Thus, the pool of cells resistant to drugs targeting the oncogenic protein may be larger than the quiescent LSC subset, to include LSC capable to cycle in the absence of oncogenic signaling. Therefore, to push LSC to exit from quiescence would not be equivalent to make them sensitive to treatment. In this case, a possible alternative strategy would be to combine the targeting of the oncogenic protein with treatments capable to prevent or revert its suppression. In CML, either the treatment with specific drugs [68] or the interference with the metabolic adaptation of LSC to the SCN (Poteti M et al., manuscript in preparation) was shown to prevent $BCR/Abl_{protein}$ suppression following incubation at 0.1% O_2.

A perhaps more straightforward approach to suppress MRD maintained under conditions where oncogenic signaling is suppressed is represented by the targeting of mechanisms driving the overall LSC adaptation to these conditions (rather than those specifically related to oncogene suppression, discussed in the previous paragraph). As those conditions are likely to consist of a low O_2/glucose environment, HIFα appears an obvious potential therapeutic target (see text above and, as examples, [69–71]), and HIFα targeting was tested. Echinomycin was the first HIFα inhibitor to be used to treat leukemias. The drug, capable to block the DNA-binding activity of HIFα, was found to target human AML and LSC in particular in a xenogeneic model in vivo, reducing leukemia burden and extending mouse survival, as well as, more importantly, suppressing disease development following transplantation into secondary recipients. Equally important is the finding that echinomycin does not interfere with HSC self-renewal, thus exhibiting a high therapeutic index [72, 73]. In murine APL models driven by both PML-RARα and PLZF-RARα, the HIF-1α transcription antagonist EZN-2968 or the HIF-1α inhibitor EZN-2208 are capable alone of debulking leukemia and prolonging mouse survival and, when administered in combination with all-trans retinoic acid, suppressing LSC and eradicating leukemia [71, 74]. Furthermore, L-ascorbic acid inhibits HIF-1α transcription and reduces growth of APL and CML cell lines [75]. Finally, acriflavine, which decreases HIF transcriptional activity by inhibiting α/β dimerization, reduces growth and LSC maintenance in cultures incubated at low O_2 of CML cell lines and cells explanted from a number of patients as well as in *BCR/abl*-induced mice used as a stem cell-driven disease model [66].

An alternative strategy to the block of HIF-dependent LSC adaptation to SCN consists of taking advantage of low O_2 to activate prodrugs which then target LSC already adapted to SCN. TH-302 is a "hypoxia"-activated prodrug that is specifically cytotoxic in low O_2 via multiple mechanisms and is capable to reduce leukemia burden and to prolong recipient survival in murine models of human AML [76]. Similar results were obtained by treating acute lymphoblastic leukemia (ALL) cells with the "hypoxia"-activated DNA cross-linking agent PR-104; in a phase I/II trial study, PR-104 was shown effective in the treatment of AML or ALL patients, even refractory to standard therapy or in relapse [77, 78].

As explained above, HIFα is a necessary, but not sufficient, mediator of LSC maintenance. Indeed, other critical regulators of stem cell compartment unrelated to HIFα signaling, or more in general to the metabolic control of LSC homing within the SCN, have been identified as potential therapeutic targets. The product of the arachidonate-5-lipoxygenase (5-LO) gene (alox5) is one of these regulators, and Zileuton, a specific 5-LO inhibitor, induces depletion of short-term LSC and multipotent progenitor cells of CML, likely by blocking the commitment of long-term

LSC, in the murine CML model mentioned above [79, 80]. Zileuton also impairs stem cell function in polycythemia vera [81]. Likewise, alox15 inhibition by PD146176 causes LSC depletion and prevents CML initiation [82]. Omacetaxine (formerly homoharringtonine) markedly reduces both LSC and leukemia bulk in vitro and in vivo, also when *BCR/abl* is mutated [83], exemplifying a general advantage of non-BCR/Abl-targeting drugs. The hedgehog signaling inhibitors cyclopamine or GDC-0449 target LSC in CML [84]. BMS-214662 induces apoptosis of both proliferating and quiescent LSC of CML via mechanism not associated with the farnesyltransferase-inhibitory activity of the drug [85]. Finally, the inhibition of MEK5/ERK5 pathway reduces growth and LSC maintenance in cultures incubated at low O_2 of CML cell lines and cells explanted from a number of patients and in the murine CML model mentioned above [86]. Very often, normal HSC were found much less sensitive than LSC to these treatments. The vast array of new options for leukemia therapy summarized above, once its translatability to clinical practice is verified, induces a reasonable hope to succeed in suppressing LSC and MRD in patients affected by a number of different leukemias, thus aiming at cure rather than care.

References

1. Hoggatt J, Scadden DT (2012) The stem cell niche: tissue physiology at a single cell level. J Clin Invest 122:3029–3034. https://doi.org/10.1172/JCI60238.2
2. Rafalski VA, Mancini E, Brunet A (2012) Energy metabolism and energy-sensing pathways in mammalian embryonic and adult stem cell fate. J Cell Sci 125:5597–5608. https://doi.org/10.1242/jcs.114827
3. Bonnet D, Dick JE (1997) Human acute myeloid leukemia is organized as a hierarchy that originates from a primitive hematopoietic cell. Nat Med 7:730–737. https://doi.org/10.1038/nm0797-730
4. Lapidot T, Sirard C, Vormoor J, Murdoch B, Hoang T, Caceres-Cortes J et al (1994) A cell initiating human acute myeloid leukaemia after transplantation into SCID mice. Nature 367:645–648. https://doi.org/10.1038/367645a0
5. Zagozdzon R, Golab J (2015) Cancer stem cells in haematological malignancies. Contemp Oncol 19:A1–A6. https://doi.org/10.5114/wo.2014.47127
6. Reya T, Morrison SJ, Clarke MF, Weissman IL (2001) Stem cells, cancer, and cancer stem cells. Stem Cells 414:105–111. https://doi.org/10.1007/978-1-60327-933-8
7. Giuntoli S, Rovida E, Barbetti V, Cipolleschi M-G, Olivotto M, Dello Sbarba P (2006) Hypoxia suppresses BCR/Abl and selects Imatinib-insensitive progenitors within clonal CML population. Leukemia 20:1291–1293. PMID: 16710305
8. Schepers K, Campbell TB, Passegué E (2015) Normal and leukemic stem cell niches: insights and therapeutic opportunities. Cell Stem Cell 16:254–267. https://doi.org/10.1016/j.stem.2015.02.0147
9. Simsek T, Kocabas F, Zheng J, Deberardinis RJ, Mahmoud AI, Olson EN et al (2010) The distinct metabolic profile of hematopoietic stem cells reflects their location in a hypoxic niche. Cell Stem Cell 7:380–390
10. Chow DC, Wenning LA, Miller WM, Papoutsakis ET (2001) Modeling pO(2) distributions in the bone marrow hematopoietic compartment. I. Krogh's model. Biophys J 81:675–684. https://doi.org/10.1016/S0006-3495(01)75733-5

11. Chow DC, Wenning LA, Miller WM, Papoutsakis ET (2001) Modeling pO(2) distributions in the bone marrow hematopoietic compartment. II. Modified Kroghian models. Biophys J 81:685–696. https://doi.org/10.1016/S0006-3495(01)75733-5
12. Tondevold E, Eriksen J, Jansen E (1979) Observations on long bone medullary pressures in relation to arterial PO2, PCO2 and pH in the anaesthetized dog. Acta Orthop Scand 50:645–651
13. Ivanović Z (2009) Hypoxia or in situ normoxia: the stem cell paradigm. J Cell Physiol 219:271–275
14. Schofield R (1978) The relationship between the spleen colony-forming cell and the haemopoietic stem cell. Blood Cells 4:7–25; PMID:747780
15. Nilsson SK, Johnston HM, Coverdale JA (2001) Spatial localization of transplanted hemopoietic stem cells: inferences for the localization of stem cell niches. Blood 97:2293–2299
16. Zhang J, Niu C, Ye L, Huang H, He X, Tong WG et al (2003) Identification of the hematopoietic stem cell niche and control of the niche size. Nature 425:836–841
17. Calvi LM, Adams GB, Weibrecht KW, Weber JM, Olson DP, Knight MC et al (2003) Osteoblastic cells regulate the hematopoietic stem cell niche. Nature 425:841–846
18. Parmar K, Mauch P, Vergilio JA, Sackstein R, Down JD (2007) Distribution of hematopoietic stem cells in the bone marrow according to regional hypoxia. Proc Natl Acad Sci U S A 104:5431–5436
19. Jing D, Wobus M, Poitz DM, Bornhauser M, Ehninger G, Ordemann R (2012) Oxygen tension plays a critical role in the hematopoietic microenvironment in vitro. Haematologica 97:331–339. https://doi.org/10.3324/haematol.2011.050815
20. Guitart AV, Hammoud M, Dello Sbarba P, Ivanović Z, Praloran V (2010) Slow-cycling/quiescence balance of hematopoietic stem cells is related to physiological gradient of oxygen. Exp Hematol 38:847–851. PMID:20547202. https://doi.org/10.1016/j.exphem.2010.06.002
21. Cipolleschi MG, Dello Sbarba P, Olivotto M (1993) The role of hypoxia in the maintenance of hematopoietic stem cells. Blood 82:2031–2037
22. Cipolleschi MG, D'Ippolito G, Bernabei PA, Caporale R, Nannini R, Mariani M et al (1997) Severe hypoxia enhances the formation of erythroid bursts from human cord blood cells and the maintenance of BFU-E in vitro. Exp Hematol 25:1187–1194
23. Ivanović Z, Dello Sbarba P, Trimoreau F, Faucher JL, Praloran V (2000) Primitive human HPCs are better maintained and expanded in vitro at 1 percent oxygen than at 20 percent. Transfusion 40:1482–1488
24. Danet GH, Pan Y, Luongo JL, Bonnet DA, Simon MC (2003) Expansion of human SCID-repopulating cells under hypoxic conditions. J Clin Invest 112:126–135
25. Broxmeyer HE, O'Leary HA, Huang X, Mantel C (2015) The importance of hypoxia and EPHOSS for collection and processing of stem and progenitor cells to understand true physiology/pathology of these cells ex-vivo. Curr Opin Hematol 22:273–278
26. Koumenis C, Wouters BG (2006) "Translating" tumor hypoxia: unfolded protein response (UPR)-dependent and UPR-independent pathways. Mol Cancer Res 4:423–436
27. Zhang H, Li H, Xi HS, Li S (2012) HIF1alpha is required for survival maintenance of chronic myeloid leukemia stem cells. Blood 119:2595–2607
28. Cipolleschi MG, Rovida E, Ivanović Z, Praloran V, Olivotto M, Dello Sbarba P (2000) The expansion of murine bone marrow cells preincubated in hypoxia as an in vitro indicator of their marrow-repopulating ability. Leukemia 14:735–739. PMID: 10764163
29. Ivanović Z, Belloc F, Faucher JL, Cipolleschi MG, Praloran V, Dello Sbarba P (2002) Hypoxia maintains and interleukin-3 reduces the pre-colony-forming cell potential of dividing CD34(+) murine bone marrow cells. Exp Hematol 30:67–73
30. Hermitte F, Brunet de la Grange P, Belloc F, Praloran V, Ivanovic Z (2006) Very low O2 concentration (0.1%) favors G0 return of dividing CD34+ cells. Stem Cells 24:65–73
31. Guitart AV, Debeissat C, Hermitte F, Villacreces A, Ivanović Z, Boeuf H, Praloran V (2011) Very low oxygen concentration (0.1%) reveals two FDCP-Mix cell subpopulations that differ by their cell cycling, differentiation and p27KIP1 expression. Cell Death Differ 18:174–182
32. Spencer JA, Ferraro F, Roussakis E, Klein A, Wu J, Runnels JM et al (2014) Direct measurement of local oxygen concentration in the bone marrow of live animals. Nature 508:269–273. https://doi.org/10.1038/nature13034

33. Giuntoli S, Rovida E, Barbetti V, Cipolleschi MG, Olivotto M, Dello Sbarba P (2006) Hypoxia suppresses BCR/Abl and selects imatinib-insensitive progenitors within clonal CML populations. Leukemia 20:1291–1293
34. Giuntoli S, Rovida E, Gozzini A, Barbetti V, Cipolleschi MG, Olivotto M, Dello Sbarba P (2007) Severe hypoxia defines heterogeneity and selects highly immature progenitors within clonal erythroleukemia cells. Stem Cells 25:1119–1125
35. Cheloni G, Poteti M, Bono S, Masala E, Mazure NM, Rovida E et al (2017) The leukemic stem cell niche: adaptation to "hypoxia" versus oncogene addiction. Stem Cells Int 2017:4979474. https://doi.org/10.1155/2017/4979474
36. Hochachka PW, Buck LT, Doll CJ, Land SC (1996) Unifying theory of hypoxia tolerance: molecular/metabolic defense and rescue mechanisms for surviving oxygen lack. Proc Natl Acad Sci U S A 93:9493–9498
37. Pettersen EO, Juul NO, Rønning OW (1986) Regulation of protein metabolism of human cells during and after acute hypoxia. Cancer Res 46:4346–4351
38. Liu L, Simon MC (2004) Regulation of transcription and translation by hypoxia. Cancer Biol Ther 3:492–497
39. Stein I, Itin A, Einat P, Skaliter R, Grossman Z, Keshet E (1998) Translation of vascular endothelial growth factor mRNA by internal ribosome entry: implications for translation under hypoxia. Mol Cell Biol 18:3112–3119
40. Bono S, Lulli M, D'Agostino VG, Di Gesualdo F, Loffredo R, Cipolleschi MG et al (2016) Different BCR/Abl protein suppression patterns as a converging trait of chronic myeloid leukemia cell adaptation to energy restriction. Oncotarget 7:84810–84825
41. Giuntoli S, Tanturli M, Di Gesualdo F, Barbetti V, Rovida E, Dello Sbarba P (2011) Glucose availability in hypoxia regulates the selection of Chronic Myeloid Leukaemia progenitor subsets with different resistance to Imatinib-mesylate. Haematologica 96:204–212. PMID:21071498. https://doi.org/10.3324/haematol.2010.029082
42. Sharma SV, Settleman J (2010) Exploiting the balance between life and death: targeted cancer therapy and "oncogenic shock". Biochem Pharmacol 80:666–673
43. Meyer N, Kim SS, Penn LZ (2006) The Oscar-worthy role of Myc in apoptosis. Semin Cancer Biol 16:275–287
44. Li B, Simon MC (2013) Molecular pathways: targeting MYC-induced metabolic reprogramming and oncogenic stress in cancer. Clin Cancer Res 19:5835–5841. https://doi.org/10.1158/1078-0432.CCR-12-3629
45. Quesenberry PJ, Colvin GA, Lambert JF (2002) The chiaroscuro stem cell: a unified stem cell theory. Blood 100:4266–4271
46. Jiang X, Ng E, Yip C, Eisterer W, Chalandon Y, Stuible M et al (2002) Primitive interleukin-3-null hematopoietic cells transduced with BCR-ABL show accelerated loss after culture of factor-independence in vitro and leukemogenic activity in vivo. Blood 100:3731–3740
47. Tannock IF (1968) The relation between cell proliferation and the vascular system in a transplanted mouse mammary tumour. Br J Cancer 22:258–273
48. Deynoux M, Sunter N, Hérault O, Mazurier F (2016) Hypoxia and hypoxia-inducible factors in leukemias. Front Oncol 6:41. https://doi.org/10.3389/fonc.2016.00041
49. Cipolleschi MG, Rovida E, Dello Sbarba P (2013) The culture-repopulating ability assays and incubation in low oxygen: a simple way to test drugs on leukemia stem or progenitor cells. Curr Pharm Des 19:5374–5383. PMID: 23394087
50. Rovida E, Marzi I, Cipolleschi MG, Dello Sbarba P (2014) One more stem cell niche: how the sensitivity of chronic myeloid leukemia cells to imatinib-mesylate is modulated within a "hypoxic" environment. Hypoxia 2:1–10. PMID: 27774462
51. Rovida E, Peppicelli S, Bono S, Bianchini F, Tusa I, Cheloni G et al (2014) The metabolically-modulated stem cell niche: a dynamic scenario regulating cancer cell phenotype and resistance to therapy. Cell Cycle 13:3169–3175. PMID: 25485495. https://doi.org/10.4161/15384101.2014.964107
52. Winkler IG, Barbier V, Wadley R, Zannettino AC, Williams S, Lévesque JP (2010) Positioning of bone marrow hematopoietic and stromal cells relative to blood flow in vivo: serially reconstituting hematopoietic stem cells reside in distinct nonperfused niches. Blood 116:375–385

53. Lévesque JP, Winkler IG (2011) Hierarchy of immature hematopoietic cells related to blood flow and niche. Curr Opin Hematol 18:220–225
54. Ishikawa F, Yoshida S, Saito Y, Hijikata A, Kitamura H, Tanaka S et al (2007) Chemotherapy-resistant human AML stem cells home to and engraft within the bone-marrow endosteal region. Nat Biotech 25:1315–1321
55. Colmone A, Amorim M, Pontier AL, Wang S, Jablonski E, Sipkins DA (2008) Leukemic cells create bone marrow niches that disrupt the behavior of normal hematopoietic progenitor cells. Science 322:1861–1865
56. Savona M, Talpaz M (2008) Getting to the stem of chronic myeloid leukaemia. Nat Rev Cancer 8:341–350
57. Modi H, McDonald T, Chu S, Yee JK, Forman SJ, Bhatia R (2007) Role of BCR/ABL gene-expression levels in determining the phenotype and imatinib sensitivity of transformed human hematopoietic cells. Blood 109:5411–5421
58. Kumari A, Brendel C, Hochhaus A, Neubauer A, Burchert A (2012) Low BCR-ABL expression levels in hematopoietic precursor cells enable persistence of chronic myeloid leukemia under imatinib. Blood 119:530–539
59. Mahon F-X, Rea D, Guilhot J, Guilhot F, Huguet F, Nicolini F et al (2010) Intergroupe Français des Leucémies Myéloïdes Chroniques. Discontinuation of imatinib in patients with chronic myeloid leukaemia who have maintained complete molecular remission for at least 2 years: the prospective, multicentre STop IMatinib (STIM) trial. Lancet Oncol 11:1029–1035
60. Graham SM, Jørgensen HG, Allan E, Pearson C, Alcorn MJ, Richmond L, Holyoake TL (2002) Primitive, quiescent, Philadelphia-positive stem cells from patients with chronic myeloid leukemia are insensitive to STI571 in vitro. Blood 99:319–325
61. Corbin AS, Agarwal A, Loriaux M, Cortes J, Deininger MW, Druker BJ (2010) Human chronic myeloid leukemia stem cells are insensitive to imatinib despite inhibition of BCR-ABL activity. J Clin Invest 121:396–409
62. Perl A, Carroll M (2011) BCR-ABL kinase is dead; long live the CML stem cell. J Clin Invest 121:22–25
63. Donato NJ, Wu JY, Stapley J, Lin H, Arlinghaus R, Aggarwal BB et al (2004) Imatinib mesylate resistance through BCR-ABL independence in chronic myelogenous leukemia. Cancer Res 64:672–677, erratum in: Cancer Res. 2004;64:2306
64. Copland M, Hamilton A, Elrick LJ, Baird JW, Allan EK, Jordanides N et al (2006) Dasatinib (BMS-354825) targets an earlier progenitor population than imatinib in primary CML but does not eliminate the quiescent fraction. Blood 107:4532–4539
65. Konig H, Holtz M, Modi H, Manley P, Holyoake TL, Forman SJ, Bhatia R (2008) Enhanced BCR-ABL kinase inhibition does not result in increased inhibition of downstream signaling pathways or increased growth suppression in CML progenitors. Leukemia 22:748–755
66. Cheloni G, Tanturli M, Tusa I, DeSouza NH, Shan Y, Gozzini A et al (2017) Targeting chronic myeloid leukemia stem cells with the hypoxia-inducible factor inhibitor acriflavine. Blood 130:655–665. PMID: 28576876. https://doi.org/10.1182/blood-2016-10-745588
67. Pettit K, Stock W, Walter RB (2016) Incorporating measurable ('minimal') residual disease-directed treatment strategies to optimize outcomes in adults with acute myeloid leukemia. Leuk Lymphoma 57:1527–1533. https://doi.org/10.3109/10428194.2016.1160085
68. Del Poggetto E, Tanturli M, Ben-Califa N, Gozzini A, Tusa I, Cheloni G et al (2015) Salarin C inhibits the maintenance of chronic myeloid leukemia progenitor cells. Cell Cycle 14:3146–3154. PMID: 26291130. https://doi.org/10.1080/15384101.2015.1078029
69. Matsunaga T, Imataki O, Torii E, Kameda T, Shide K, Shimoda H et al (2012) Elevated HIF-1α expression of acute myelogenous leukemia stem cells in the endosteal hypoxic zone may be a cause of minimal residual disease in bone marrow after chemotherapy. Leuk Res 36:e122–e124. https://doi.org/10.1016/j.leukres.2012.02.028
70. Drolle H, Wagner M, Vasold J, Kütt A, Deniffel C, Sotlar K et al (2015) Hypoxia regulates proliferation of acute myeloid leukemia and sensitivity against chemotherapy. Leuk Res 39:779–785. https://doi.org/10.1016/j.leukres.2015.04.019

71. Coltella N, Percio S, Valsecchi R, Cuttano R, Guarnerio J, Ponzoni M et al (2014) HIF factors cooperate with PML-RARα to promote acute promyelocytic leukemia progression and relapse. EMBO Mol Med 6:640–650. https://doi.org/10.1002/emmm.201303065
72. Wang Y, Liu Y, Malek SN, Zheng P, Liu Y (2011) Targeting HIF1α eliminates cancer stem cells in hematological malignancies. Cell Stem Cell 8:399–411. https://doi.org/10.1016/j.stem.2011.02.006
73. Wang Y, Liu Y, Tang F, Bernot KM, Schore R, Marcucci G et al (2014) Echinomycin protects mice against relapsed acute myeloid leukemia without adverse effects on hematopoietic stem cells. Blood 124:1127–1136. https://doi.org/10.1182/blood-2013-12-544221
74. Coltella N, Valsecchi R, Ponente M, Ponzoni M, Bernardi R (2015) Synergistic leukemia eradication by combined treatment with retinoic acid and HIF inhibition by EZN-2208 (PEG-SN38) in preclinical models of PML-RAR and PLZF-RAR-driven leukemia. Clin Cancer Res 21:3685–3694. https://doi.org/10.1158/1078-0432.CCR-14-3022
75. Kawada H, Kaneko M, Sawanobori M, Uno T, Matsuzawa H, Nakamura Y et al (2013) High concentrations of l-ascorbic acid specifically inhibit the growth of human leukemic cells via downregulation of HIF-1α transcription. PLoS One 8:e62717. https://doi.org/10.1371/journal.pone.0062717
76. Portwood S, Lal D, Hsu YC, Vargas R, Johnson MK, Wetzler M et al (2013) Activity of the hypoxia-activated prodrug, TH-302, in preclinical human acute myeloid leukemia models. Clin Cancer Res 19:6506–6519. https://doi.org/10.1158/1078-0432.CCR-13-0674
77. Patterson AV, Ferry DM, Edmunds SJ, Gu Y, Singleton RS, Patel K et al (2007) Mechanism of action and preclinical antitumor activity of the novel hypoxia-activated DNA cross-linking agent PR-104. Clin Cancer Res 13:3922–3932. https://doi.org/10.1158/1078-0432.CCR-07-0478
78. Konopleva M, Thall PF, Yi CA, Borthakur G, Coveler A, Bueso-Ramos C et al (2015) Phase I/II study of the hypoxia-activated prodrug PR104 in refractory/relapsed acute myeloid leukemia and acute lymphoblastic leukemia. Haematologica 100:927–934. https://doi.org/10.3324/haematol.2014.118455
79. Chen Y, Hu Y, Zhang H, Peng C, Li S (2009) Loss of the Alox5 gene impairs leukemia stem cells and prevents chronic myeloid leukemia. Nat Genet 41:783–792
80. Chen Y, Li D, Li S (2009) The Alox5 gene is a novel therapeutic target in cancer stem cells of chronic myeloid leukemia. Cell Cycle 8:3488–3492
81. Chen Y, Shan Y, Lu M, DeSouza NH, Guo Z, Hoffman R, Liang A, Li S (2017) Alox5 Blockade Eradicates JAK2V617F-induced polycythemia Vera in Mice. Cancer Res 77:164–174.https://doi.org/10.1158/0008-5472.CAN-15-2933
82. Chen Y, Peng C, Abraham SA, Shan Y, Guo Z, Desouza N et al (2014) Arachidonate 15-lipoxygenase is required for chronic myeloid leukemia stem cell survival. J Clin Invest 124:3847–3862. https://doi.org/10.1172/JCI66129
83. Chen Y, Hu Y, Michaels S, Segal D, Brown D, Li S (2009) Inhibitory effects of omacetaxine on leukemic stem cells and BCR-ABL-induced chronic myeloid leukemia and acute lymphoblastic leukemia in mice. Leukemia 23:1446–1454. https://doi.org/10.1038/leu.2009.52
84. Zhao C, Chen A, Jamieson CH, Fereshteh M, Abrahamsson A, Blum J et al (2009) Hedgehog signalling is essential for maintenance of cancer stem cells in myeloid leukaemia. Nature 458:776–779
85. Copland M, Pellicano F, Richmond L, Allan EK, Hamilton A, Lee FY, Weinmann R, Holyoake TL (2008) BMS-214662 potently induces apoptosis of chronic myeloid leukemia stem and progenitor cells and synergizes with tyrosine kinase inhibitors. Blood 111:2843–2853. PMID: 18156496
86. Tusa I, Cheloni G, Poteti M, Gozzini A, DeSouza NH, Shan Y et al (2018) Targeting the extracellular signal-regulated kinase 5 pathway to suppress human chronic myeloid leukemia stem cells. Stem Cell Reports 11:929–943

Chapter 7
DNA Damage Response in Quiescent Hematopoietic Stem Cells and Leukemia Stem Cells

Wenjun Zhang, Guangming Wang, and Aibin Liang

Contents

Abstract In humans, hematopoietic stem cells (HSCs) adopt unique responsive pathways counteracting with the DNA-damaging assaults to weigh the balance between the maintenance of normal stem cell poor for whole-life blood regeneration and the transformation to leukemia stem cells (LSCs) for leukemia initiation. LSCs also take actions of combating with the attack launched by externally therapeutic drugs that can kill most leukemic cells, to avoid extermination and promote

Authors Wenjun Zhang and Guangming Wang have been contributed equally to this chapter.

W. Zhang · G. Wang · A. Liang (✉)
Department of Hematology, Tongji Hospital, Tongji University School of Medicine, Shanghai, China
e-mail: lab7182@tongji.edu.cn

H. Zhang, S. Li (eds.), *Leukemia Stem Cells in Hematologic Malignancies*, Advances in Experimental Medicine and Biology 1143,
https://doi.org/10.1007/978-981-13-7342-8_7

disease relapse. Therefore, the collection of knowledge about all these underlined mechanisms would present a preponderance for later studies. In this chapter, the universal DNA damage response (DDR) mechanisms were firstly introduced, and then DDR of HSCs were presented focusing on the DNA double-strand breaks in the quiescent state of HSCs, which poses a big advantage in promoting its transformation into preleukemic HSCs. Lastly, the DDR of LSCs were summarized based on the major outcomes triggered by different pathways in specific leukemia, upon which some aspects for future investigations were envisioned under our currently limited scope of knowledge.

Keywords DNA damage · Hematopoietic stem cell · Leukemia stem cell

7.1 Introduction

As the most important material in a cell of the whole organism, DNA is very fragile and constantly under attack of various, endogenous, and exogenous assaults which cause different kinds of DNA lesions [11, 66]. To counteract the deleterious action of DNA damage, organisms have developed multiple and robust DNA damage response (DDR) mechanisms through which the DNA defects are either removed or maintained to induce differential cellular outcomes [10, 68].

Hematopoietic stem cells (HSCs), which are crucial for the maintenance of homeostasis and regeneration of the blood system over the lifetime of an individual, inevitably suffer damaging factors leading to genome instability. Accelerating evidence highlights the occurrence of DDR and the presence of DNA lesions being the underlined factors contributing to the functional decline and the leukemogenic accumulation of HSCs [9, 31, 75]. More strikingly, an emerging model proposes mutations in HSCs giving rise to the emergence of aberrant preleukemic HSCs that, with the additional oncogenic/environmental events, are finally transformed into leukemic stem cells (LSCs); in this way, the mutations in HSCs seem to serve as the time bomb for the leukemia [38]. LSCs, being the rare populations in leukemia with the capacity of self-renewal to drive the expansion of leukemia and taking the responsibility for the relapse of leukemia, also employ various means to cope with DNA damage, which now is garnering widespread interest of investigators [34].

7.2 The DNA Damage Response

All living cells suffer the DNA lesions constantly caused by endogenous and exogenous resources and combat with them through a well-organized process called DDR. Different DNA-damaging insults generate specific type of DNA lesions that are processed by distinct DDR leading to quite different outcomes for a cell, which depends on, at the cellular level, the developmental stage, e.g., stem or mutual cell;

cell cycle state, e.g., dividing or nondividing cell; and tissue origin, e.g., blood or neural cell. Here, we firstly give a brief description of DNA damage induction, and then we make a summary of the universally reactive DDR based on current understanding.

7.2.1 *Various Insults Generate Different Types of DNA Lesions*

DNA damage is normally classified based on the lesions that occurred at the base or strand level of the DNA resulting from various endogenous or exogenous factors [11, 66] (Fig. 7.1). At the DNA base level, the bases within the normal sequences of the DNA can be base-base mismatched, dimerized (e.g., pyrimidine dimers), deleted (e.g., abasic sites), and abnormally modified (e.g., Uracil, 8-oxoguanine, O-6-methylguanine) [3]. At the strand level, the damage may take place within one single strand, such as single-strand nick (SSN) or single-strand break (SSB), or between two individual strands, such as interstrand cross-link (ICL) and double-strand break (DSB).

DNA base-base mismatches are endogenously generated during the leading- and lagging-strand replication of DNA by DNA polymerase picking up the incorrect dNTPs while posing a very low frequency due to the high selectivity of substrates by the DNA polymerase and the proofreading mechanism in vivo [53, 60]. The pyrimidine dimers are produced between adjacent pyrimidine residues in one strand of the DNA upon ultraviolet (UV) radiation [74, 92]. Abasic sites, also known as AP sites (apurinic/apyrimidinic site), form spontaneously when the glycosidic linkage between the deoxyribose and its nitrogenous base is broken due to the enzymatic reaction induced by chemical compounds [67]. The abnormally modified bases, e.g., oxidized/reduced bases, alkylated bases, and deaminated bases, are majorly introduced directly by agents of external origin such as chemotherapeutic drugs, and reactive oxygen species (ROS) are generated internally by metabolic processes or externally by ionizing radiation and UV [19, 36].

DNA SSNs and SSBs which predominantly resulted from the oxidative attack by ROS are produced endogenously by metabolisms or exogenously upon radiation, also generated by cellular enzymes such as DNA topoisomerase 1 (TOP1) that poses a DNA nick to relax DNA for transcription and replication, or produced as intermediates containing a 1–30 nucleotide break during base and nucleotide excision repair [20]. The DNA ICLs are mainly created by the involvement of agents covalently binding to two base sites in the DNA duplex arising from endogenous metabolites such as nitrous acid and bifunctional aldehydes or exogenous chemicals such as cisplatin, nitrogen mustard, and mitomycin D and also oxidatively generated by free oxygen radicals from internal and external sources [30, 77]. As the most severe lesions to DNA, DSBs are broadly induced by various factors under physiologic condition such as V(D)J recombination and class switch during early lymphocytic development and pathologic situation such as ionizing radiation, oxidative stress, replication fork, inadvertent enzymatic action, topoisomerase failures, and mechanical stress, which was well summarized by Michael R. Lieber [65].

In a cell, DDR is a multicomponent, well-organized biological process employed by cells when encountering with DNA lesions. Different DNA lesions will be primarily repaired by different means (Fig. 7.1) and trigger other mechanistically distinct DDR, composed of signaling pathways in a hierarchical manner, which similarly involves in the following steps: sensors recognize the damage; mediators and transducers process the signal; and effectors regulate the cellular outcome [10, 89]. As the exact pathway of DDR between prokaryotes and eukaryotes, on the other hand, lower and higher eukaryotes, represents some difference in details, we mainly take higher eukaryotes for the following illustration under current understanding.

7.2.2 DNA Base Damage Is Primarily Repaired

DNA base-base mismatches are predominantly processed by DNA mismatch repair (MMR), a highly conserved biological process between prokaryotes and eukaryotes, which involves in the collaboration of MMR-specific proteins and replication and/or recombination components [60, 63]. In eukaryotic cells, MutSα recognizes a mismatch, followed by recruitment of MutLα in an ATP-dependent manner, which is then activated by PCNA to incise at both sides of the mismatched base giving rise

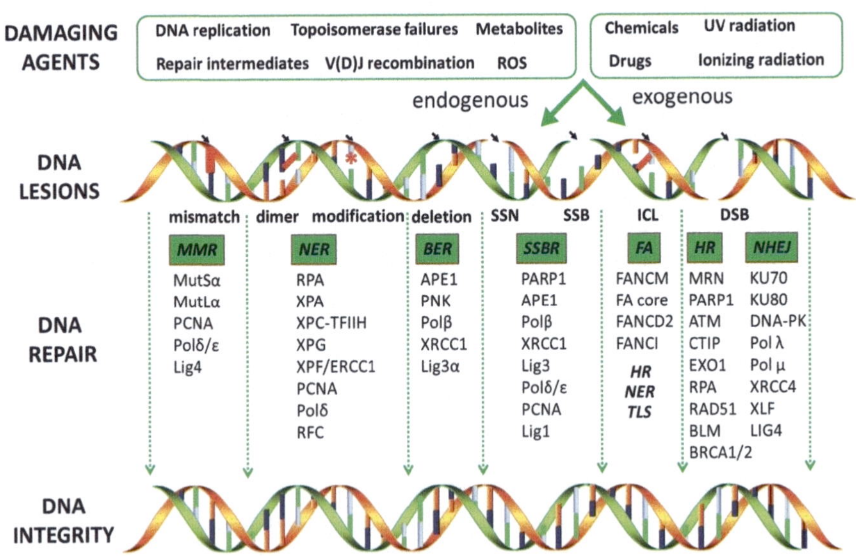

Fig. 7.1 The direct induction and major repair pathway of DNA damage. Various factors from endogenous and exogenous stimulus can cause different types of DNA lesions that are primarily repaired by distinct pathways to sustain the DNA integrity

to the error-containing strand, and then the repair is completed following error removal, DNA resynthesis, and ligation [60, 63]. The pyrimidine dimers are directly recognized and repaired by photolyase in *E. coli*, which is consisted of a photoantenna cofactor in the form of 5,10-methenyltetrahydrofolate (5,10-MTHF) and a catalytic cofactor in the form of $FADH^-$. Photolyase interacts with the backbone of DNA duplex and forms specific contacts with pyrimidine dimers, initiated by light; 5,10-MTHF absorbs and transfers energy to $FADH^-$, which then transfers the electron to dimer to split it into individual pyrimidines [78, 88].

In higher eukaryotes, the AP sites are majorly repaired by base excision repair (BER), during which AP-endonucleases (APE1) firstly recognize and incise the AP sites producing SSBs containing a 5′-deoxyribosephosphate (5′-dRP) end, which are then removed by DNA polymerase β or 5′-flap endonuclease resulting in gaps of one or few nucleotides that are finally processed by DNA polymerase and ligase [55]. Other responses to AP sites may involve nucleotide excision repair (NER), recombination, DNA translesion synthesis (TLS), or cell death, which is revealed by studies based on BER deficiency [12]. The abnormally modified bases are primarily repaired by specific process depending on the type of modification. For example, uracil bases and 8-oxoguanine are efficiently repaired by BER and O-6-methylguanine by methylguanine DNA methyltransferase in a direct way [59, 89]. NER is the major system to deal with bulky DNA adducts in which for human cells, RPA, XPA, and XPC-TFIIH forming a complex recognizes the DNA lesion and makes successively preincision complex 1, 2 and 3 through ATP hydrolysis of damaged duplex, replacement of XPC with XPG and recruitment of XPF/ERCC1, respectively. Then the damaged strand is incised by XPG and XPF/ERCC1 leaving a gap that is taken cared by replication machinery including Polδ, PCNA, and RFC [95].

7.2.3 DNA Strand Damage Impacts Cell Fate

Most SSBs in a cell are rapidly repaired by global SSB repair process [20, 21]. In the SSB repair procedure, PARP1 recognizing SSBs from oxidative reaction or abortive TOP1 activity or APE1 remaining bound to the cleaved SSB intermediate during BER stimulates the recruitment of XRCC1 protein complexes that are composed of constitutive Polβ and Lig3 and in addition either PNKP, APTX, or APLF for DNA end processing. Then Polβ replaces the single missing nucleotide, and XRCC1/Lig3α complex ligates the patch in a more commonly short-patch repair model. While in a less frequent long-patch model, the gap usually 2–12 nt is filled by Polβ and/or Polδ/ε leaving a single-strand flap that is removed by flap endonuclease-1 (FEN1), and this patch is then ligated by PCNA/Lig1 complexes. Aside from direct repair, SSBs may lead to paused replication forks or DSBs during S phase that are processed by PARP1 augmented homologous recombination (HR) in cycling cells [15, 61, 98] or stalled transcription resulted in altered transcription profiles and/or cell death in nonproliferating cells [54, 73].

The DNA ICLs have very bad impacts on cells, in the worst case, leading to cell death. The unrepaired ICLs can block replication by preventing the progression of the replisome or stall transcription through impeding the interaction between proteins and DNA resulting in p53-dependent apoptosis; moreover, accumulated ICLs directly cause cell death due to the loss of vital DNA integrity [30]. Nevertheless, cells employ sophisticated repair mechanisms, which involves the coordination of several repair processes including homologous recombination (HR), NER, and TLS, to protect themselves from the death fate [30, 56, 77]. The repair of ICLs is majorly a concomitant Fanconi anemia (FA) pathway in which FANCM and associated proteins recognize ICL followed by the recruitment of the FA core complex, the monoubiquitylation of FANCD2-FANCI, and the subsequently coordinative action of repairing components including nucleases and polymerases, such as BRCA2 and RAD51C of HR and FAN1 and Polν of TLS. This procedure is recently found to be aided by a non-erythroid alpha spectrin (αIISp) serving as a scaffold for the recruitment of repair proteins to sites of damage [62].

The responses to DSBs that are particularly well understood due to its great importance and accordingly deep investigation of researchers in the past few decades are still the hot area drawing attention of the followers for further exploration [41, 106]. DSBs processed by distinct pathways can lead to various cellular outcomes including error-prone repair, error-free repair, cell cycle arrest, apoptosis, senescence, and specifically differentiation for stem cells, depending majorly upon the cell cycle status and the activity of p53 [24, 35, 68, 103, 106] (Fig. 7.2). DSBs recognized by the KU dimer (KU70–KU80) are repaired by nonhomologous end joining (NHEJ) which is an error-prone repair system employed by proliferating and nonproliferating cells [64]. In its classically simplest way, the DNA-dependent protein kinase (DNA-PK) catalytic subunits (DNA-PKcs) are recruited by KU70–KU80 to activate DNA-PK that conducts end processing and resection with the participation of the nuclease Artemis, which is followed by gap filling with Pol λ and Pol μ and subsequent ends ligation performed by a ligase complex XRCC4/XLF/LIG4 (Fig. 7.3).

The detection of DSBs by PARP and MRE11-RAD50-NBS1 (MRN) complex leads to a higher-fidelity repair named HR or other consequences including cell cycle arrest, apoptosis, and senescence in cycling cells [10, 68]. In this complicated process, briefly the protein kinase ataxia-telangiectasia mutated (ATM), recruited by MRN complex and activated by autophosphorylation, phosphorylates and activates downstream targets that can operate repair or define the cell fate [10, 68]. The cell fate determinants include checkpoint kinase 2 (CHK2) and p53 leading to cell cycle arrest through p21 transactivation, bcl2 gene family members resulting in apoptosis, and p16/Ink4a and p19/ARF causing senescence. In the procedure of HR, ATM phosphorylates histone H2AX and other members of the DDR leading to the following repair [7] (Fig. 7.3), in which the two short stretches of single-stranded DNA (ssDNA) at DSBs are processed by nucleases including CTIP, EXO1, and

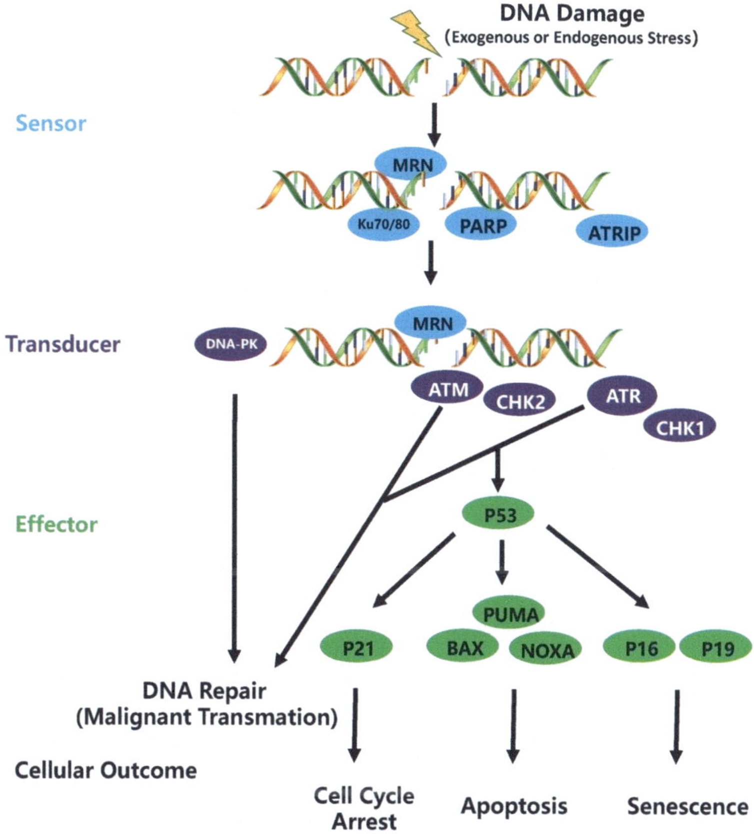

Fig. 7.2 DNA damage response pathways. DNA damage is caused by a variety of sources. There are distinct sensors, transducers, and effectors in the DNA damage response pathways. This DNA damage response converges upon p53 which, depending on the target genes activated, regulates different cellular outcomes, which may involve the cell cycle arrest, senescence, execution of DNA repair, or, when the damage is severe, initiation of apoptosis

BLM leaving the long ssDNA overhangs bound by RPA. RPA is then replaced by RAD51 recombinase, with the assistance of BRCA1, BRCA2, and PALB2, forming multimers to produce the nucleoprotein filament that invades the sister chromatid and promotes the formation of Holliday junctions through the generation of a D-loop intermediate and primed DNA synthesis of the invading strand when met with the homology. The Holliday junctions are lastly processed by some enzyme complex such as BLM, MUS81, EME1, and GEN1 resulting in the final crossover or noncrossover products.

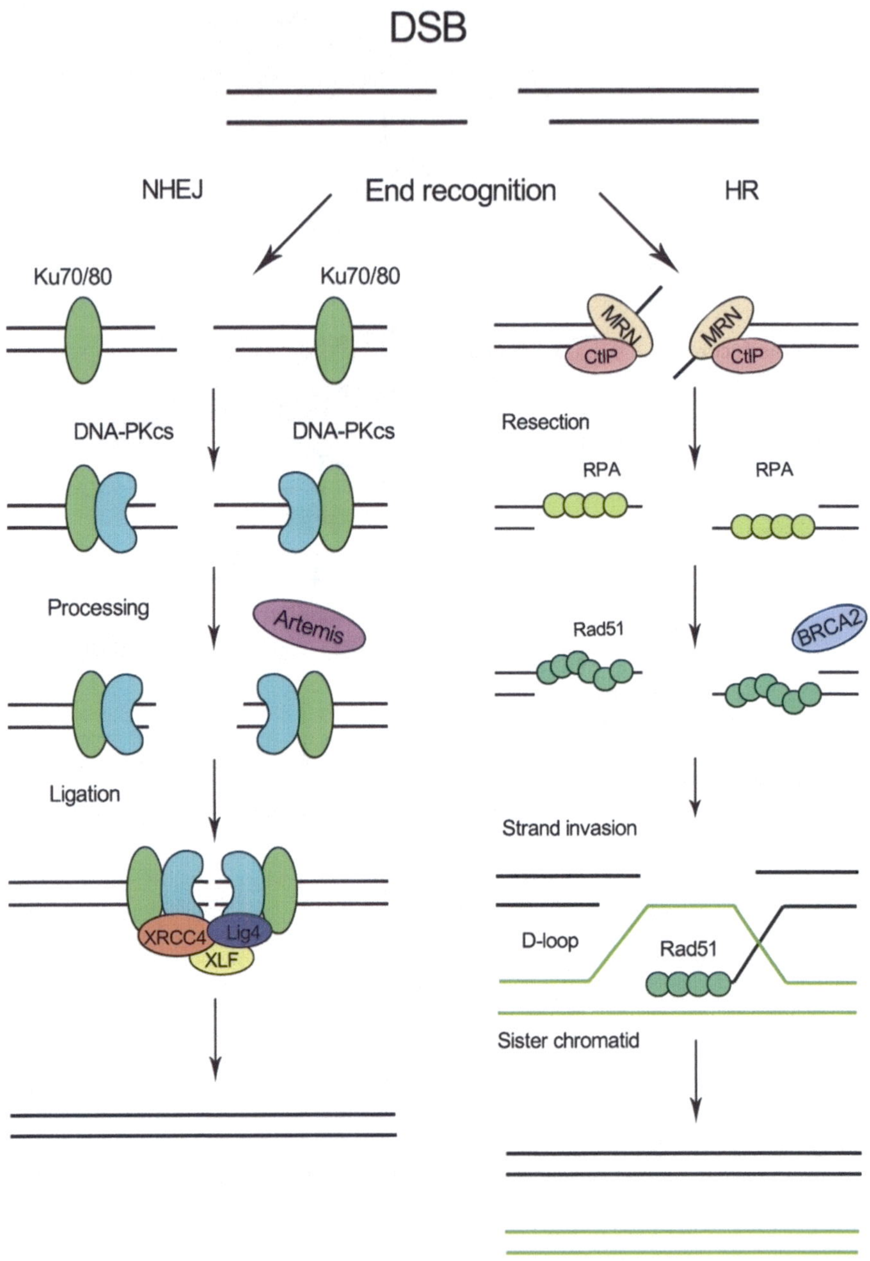

Fig. 7.3 NHEJ and HR repair pathway. In NHEJ pathway, the KU70-KU80 heterodimer firstly recognizes the DNA ends and subsequently recruits DNA-PKcs. If the ends are incompatible, nucleases such as Artemis can trim the ends. The XRCC4-DNA Ligase IV-XLF ligation complex seals the break. In HR pathway, the MRN-CtIP complex resects the DNA breaks to generate single-stranded DNA (ssDNA). After resection the break can no longer be repaired by NHEJ. RPA first coats the ssDNA and subsequently replaces Rad51 with the help of BRCA2. These Rad51 nucleoprotein filaments mediate strand invasion on the homologous template. Extension of the D-loop and capture of the second end lead to repair

7.3 DNA Damage Response in Hematopoietic Stem Cells

Lifelong blood production depends on HSCs-a subset of primitive hematopoietic cells endowed with high self-renewal potential. HSCs give rise to committed progenitors (CPs) with limited or no self-renewal, which, in turn, differentiate into various mature blood cells. Mounted evidence have revealed that, aside from DSBs, DDR of various DNA lesion is involved in leukemia [32] and quiescence serves as the most prominent property of adult stem cells [23]; here we take DSBs in quiescent HSCs for detail illustration.

7.3.1 DNA Damage Response in Quiescent Hematopoietic Stem Cells

Given that HSCs maintain the homeostasis and regenerate the blood system over the lifetime of an individual, any exogenous or endogenous factor leading to genome instability would lead to shortening of organism survival. In order to preserve HSCs' genomic integrity to ensure lifelong function, some strategies were proposed to focus on minimizing endogenous DNA damage. Firstly, endogenous DNA damages were mostly from ROS of oxidative stress and DNA replication stress. In order to be protected from these DNA damages, a quiescent phase of the cell cycle mostly kept in the human and mouse contributes to their overall maintenance and protection of their genomic integrity by minimizing DNA damage associated with ROS production, cellular respiration, and cell division [107]. Compared to quiescent HSCs, activated HSCs showed a higher levels of ROS and ROS-induced DNA lesions [104]. Secondly, the distinct niches where adult HSCs reside helped to reduce the level of ROS, as exampled by hypoxic niches in the bone marrow [45] where active ROS transportation occurs in adjacent mesenchymal stromal cells [101]. Despite reducing the endogenous damages to capacity, accumulation of DNA damage and formation of γH2AX foci and comet tail lengths as the markers for SSBs and DSBs are increased in old HSCs from both mice and humans [6]. In contrast with HSCs, the number of γH2AX foci is similar between progenitors in old animals and those in young animals [86]. The follow-up researches indicate that DNA damage accumulation in HSCs was associated with broad attenuation of DNA repair and response pathways that were dependent upon HSC quiescence. Accordingly, cycling fetal HSCs and adult HSCs driven into cycle upregulated these pathways leading to repair of strand breaks [6, 8]. Therefore, HSC quiescence and concomitant attenuation of DNA repair and response pathways underlie DNA damage accumulation in HSCs (Fig. 7.4).

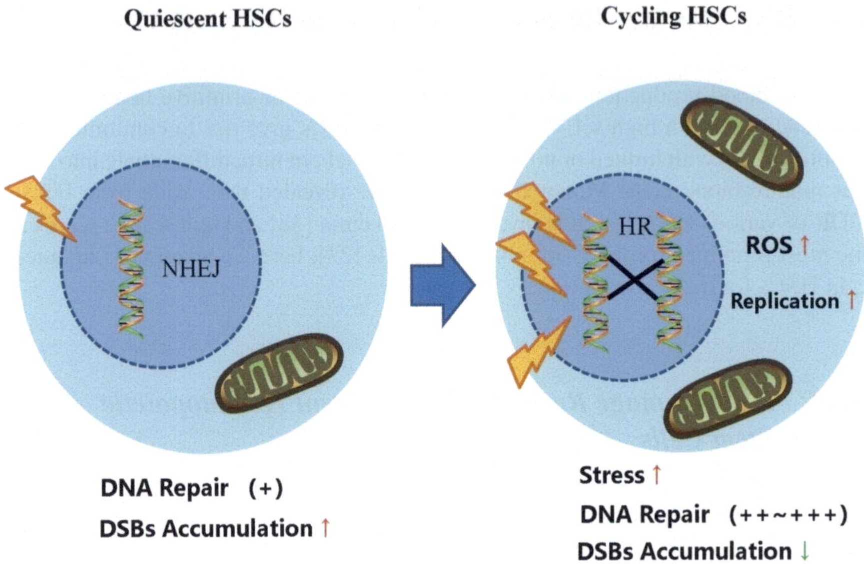

Fig. 7.4 Differences between quiescent and cycling HSCs in DNA repair. Although quiescent HSCs are more likely to utilize NHEJ as a repair mechanism, the attenuation of DNA repair and response pathways underlies DNA damage accumulation in these cells. Compared to quiescent HSCs, elevated ROS and ROS-induced DNA lesions were observed in those cycling HSCs; the enhanced HR could still reduce the DNA damage accumulation in these activated cells

The continuous DNA damage accumulations in HSCs are bound to activate the cellular DNA damage machinery and lead to the corresponding DDR outcome. Two studies uncover different outcomes of irradiation-induced DDR between mouse HSCs and human HSCs (Fig. 7.5). Murine hematopoietic stem and progenitor cells (HSPCs) were more resistant to apoptosis than differentiated progenitor cells [71]. The unique DNA damage response of mouse HSPCs involves the tumor suppressor protein p53 and is lost when stem cells are forced out of quiescence and into the cell cycle by treatment with chemotherapy or cytokines. In human umbilical cord blood cells, an enhanced sensitivity to apoptosis was observed in contrast with differentiated cells. Survival and the clonogenic and reconstitutive capacity of the irradiated human HSPCs were rescued by blocking p53 expression or by overexpression of the antiapoptotic factor Bcl-2. However, irradiated human HSPCs lacking p53 were unable to sustain hematopoiesis and showed evidence of persistent DNA double-strand breaks when serially transplanted into recipient mice. Therefore, a short-term gain in survival could be achieved by human hematopoietic stem cells as found in the mouse, but the default setting for irradiated human HSPCs is an increase in p53 expression resulting in apoptosis [70].

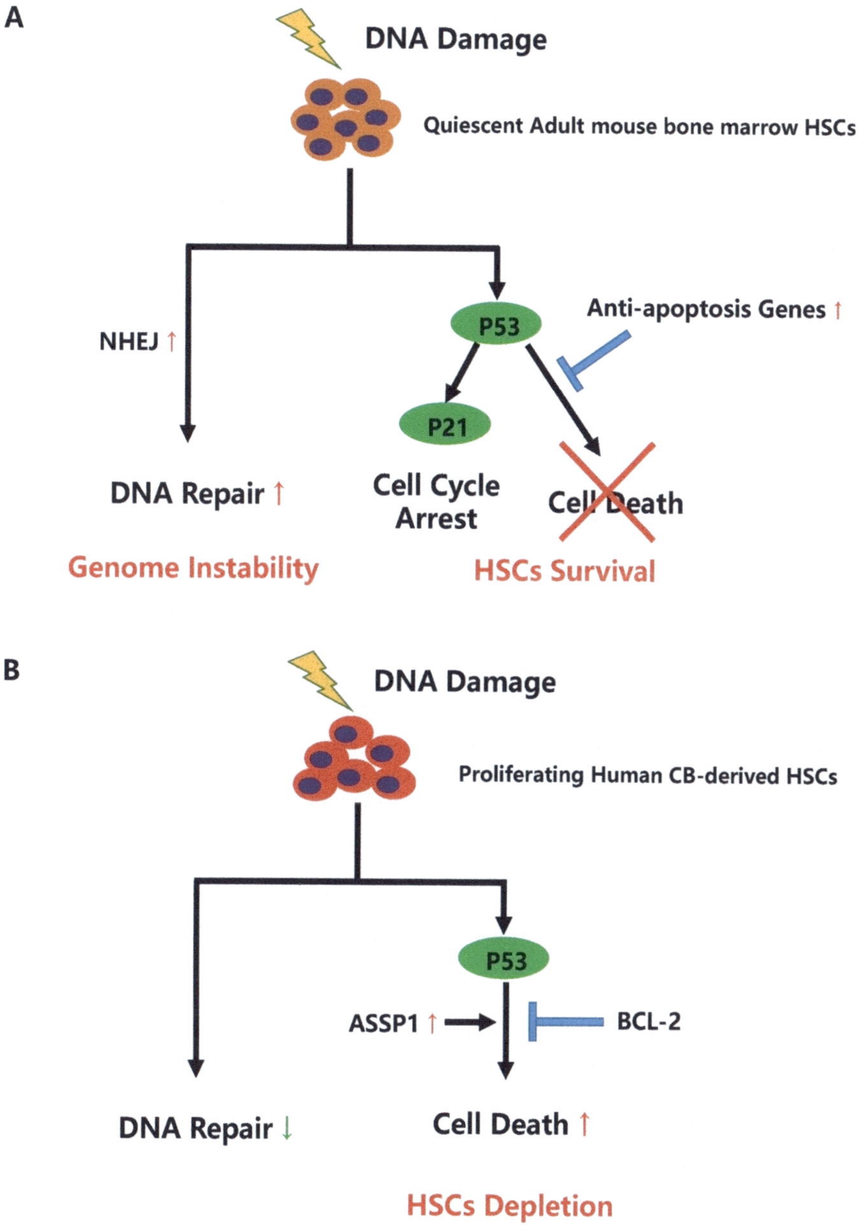

Fig. 7.5 DNA damage response in murine and human HSCs. Quiescent murine HSCs (**a**) and human CB-derived HSCs (**b**) exhibit opposite outcomes following DNA damage induced by low-level ionizing radiation, with different consequences for their overall maintenance and genomic integrity

7.3.2 NHEJ Pathway Was Responsible for the Mutagenesis of Quiescent HSCs

DSBs, which are considered the most lethal form of DNA damage, can be repaired by two major pathways, homology-directed repair (HR) and nonhomologous end joining (NHEJ) [42]. HR is active during the late S and G2 phases of the cell cycle and uses the intact sister chromatid as the template for repair. And NHEJ, which joins DNA ends of DSBs directly and occurs throughout the cell cycle (G0, G1, and early S phase), is the major DSB repair pathway in mammalian cells. However, this error-prone repair pathway could lead to accumulation of mutation in cells. On the one hand, in order to maintain their genome integrity and survival, HSCs must employ the repair pathway to reduce the accumulated DNA damages caused by exogenous or endogenous factors. Due to the cell cycle status of quiescence, repair of DNA damage through homologous recombination (which has a lower error rate than NHEJ) requires that cells enter the cell cycle; thus, quiescent HSCs must rely on NHEJ as an alternative to repair the DNA. In one study, it was revealed that irradiated quiescent HSCs display high levels of chromosomal abnormalities after the rapid DNA repair through the error-prone NHEJ pathway and their progeny show persistent genomic instability associated with some of the same abnormal cytogenetic findings and engraftment defects in secondary mouse recipients. These events would pose an obvious risk to a long-lived organism as serial exposure of stem cells to genotoxic agents could readily result in leukemia or aplasia. As mentioned above, NHEJ appears to be the most commonly used DNA repair pathway in quiescent HSCs, and this result is in line with the fact that the deficiency of functional components of DSB recognition and repair pathways affected the function of HSCs upon genotoxic stress in mouse models. Recurrent chromosome translocations occur frequently in both de novo and therapy-related in acute myeloid leukemia (AML) and myelodysplastic syndromes (MDS). Cloning of the genomic breakpoints in the common chromosome translocations in leukemia reveals that most of the genomic breakpoints tend to cluster in a restricted intronic region. In addition, sequencing of the translocation junctions identified regions of microhomology, strongly indicative of the involvement of ALT-NHEJ in the repair of DSBs and the generation of these chromosomal abnormalities [111]. It may explain why cancer patients treated with radiotherapy or chemotherapy may develop blood cancer or myelodysplasias bone marrow failure because of the use of error-prone DNA repair in quiescent HSCs.

On the other hand, in order to reduce the risk of mutagenesis, quiescent HSPCs exhibit reduced NHEJ activities in comparison with cycling progenitors. In contrast with the two- to sixfold difference in total NHEJ capacity, Alt-EJ activity was more moderately reduced by 30% in HSPCs compared with cycling progenitors, thus supporting the notion that canonical NHEJ capacity rises with the degree of differentiation in hematopoietic cells [8]. In line with these results, the global transcriptional analysis showed decreased expression of multiple DNA damage checkpoint and DNA repair pathways which are involved in DNA DSB repair and can be detected in highly purified murine and human quiescent HSCs with much less in their cycling progeny [6, 8, 18].Secondly, there might be a link between quiescence status and DNA repair gene expression. For example, NF-kB and PI3K/AKT/mTOR signaling pathways have recently been implicated in the transcriptional regulation of a group of DNA repair genes that mediate HR and NHEJ in human primary and cancer cells [22, 57]. Collectively, for one thing, in order to reduce the accumulated DNA damage and survive, quiescent HSCs rely on NHEJ repair pathway that has the potential to cause mutations for malignant hematology; on the other hand, high rigidity of DSB repair in HSCs reduces the risk of mutagenesis.

7.3.3 *Preleukemic HSC Emergence: The Cost of Long-Term Genomic Integrity*

The mutagenesis caused by the error-prone NHEJ pathway is a short-term survival strategy at an expense of their long-term genomic integrity in quiescent HSCs. Misrepair or unrepair of DNA damage in these cells can disrupt normal hematopoiesis and promote leukemogenesis. In some studies, attempts were made to conduct next-generation sequencing on leukemia cells and highly purified HSCs from the same AML patients with a goal of identifying the initiating lesions and determining the order of their acquisition of the lesions and their impact on the function of HSCs [27, 47, 94]. In human HSCs, a set of leukemia-related recurrent mutations could be detected prior to the onset of leukemia for years in the same patient. These HSCs with mutations are called preleukemic HSCs and retained normal stem cell activity, showing a multilineage repopulation advantage over non-mutated HSCs in xenografts [94]. Preleukemic HSCs preferentially display mutations in "landscaping" genes involved in DNA methylation, chromatin modification, and cohesin complex (such as IDH1, IDH2, IKZF1, DNMT3A, and ASXL1), indicating that these functional mutations represent initiating events in leukemogenesis [37, 46, 69, 109]. The existence of preL-HSCs with certain frequencies that meet the definition of clonal hematopoiesis poses a high risk for developing hematologic cancers (hazard ratio, 11.1–12.9) [37, 46]. In sum, preleukemic mutations that are essential for the initial development of hematologic malignancies, such as AML, are critical for the early stages of leukemogenesis.

7.4 DNA Damage Response in Leukemia Stem Cells

It was proposed that LSCs are some rare cells possessing the ability to self-renew to drive the expansion of leukemia from the entire population of leukemia cells [34]. Both as stem cells, LSCs are different from HSCs in that they can only differentiate into specific lineage. Being the two most outstanding characterizations of LSCs, including other CSCs, self-renewal is unlimited [25, 79], and quiescent state is a major barrier for regular therapeutic approaches such as chemotherapy and radiotherapy [17, 58, 99]. Although DNA damage can be various for one cell and is processed by very different responding, most studies focused on DSBs as which are the most severe lesions for DNA, and the cell employs the most sophisticated means to deal with them [10, 89]. On the other hand, the general response of leukemia cells to DNA damage was detailedly documented in a recent nice review [76], so here we summarized the major DDR to particular DSBs in stem cell compartment of leukemia.

7.4.1 Leukemia Stem Cells Adopt a Quiescent State

LSCs are now thought to be in a quiescent state that, on the one hand, protect them from exhaustion and responsible for continuous activation to maintain leukemic progression under normal leukemogenesis and, on the other hand, render them resistance to cell cycle-dependent chemotherapeutic drugs and relapse [33, 39, 44]. Many studies have suggested that cell cycle restriction could be one of the distinctly protective mechanisms that LSCs utilize to their advantage.

In an early study, through the isolation of a highly quiescent subpopulation of primitive leukemic cells in CML, Holyoake et al. proved that LSCs, which continuously overproduces myeloid progenitors and mature cells, are maintained as a quiescent, self-renewing, nonexpanding population [43]. Since then, accelerating evidence support the quiescence of LSCs being a prominent mechanism underlying resistance to cell cycle-dependent cytotoxic therapy [51]. Recently, using the humanized mouse models, the researchers have revealed that the transplanted human LSCs in immunodeficient mice pose a quiescent state rendering them anti-chemotherapy, and based on which they developed the valuable research tools for LSC specific therapeutic schedule [44, 87]. In these studies, Ishikawa and colleagues firstly developed a human AML xenotransplantation model with NOD/SCID/IL2r-gamma (null) mice and then demonstrated that it was the transplanted LSCs which remained the self-renewal capacity and homed to and engrafted within the osteoblast-rich area of the bone marrow, contributed to the escape of chemotherapy-induced apoptosis [44]. Next, they proved that these LSCs harboring in the endosteal region are predominantly quiescent and can be stimulated to proliferation following in vivo cytokine treatment such as granulocyte colony-stimulating factor (G-CSF), upon which they can be effectively eliminated by treatment with chemotherapeutic agents such as cytarabine (AraC) [87]. Here, particularly worth mentioning is that the NOD/SCID/IL2r-gamma (null) mice have become the most commonly used research models for leukemia studies nowadays.

7.4.2 Cancer Stem Cells Are More Resistant to DNA Damage Induction

In a cell, DDR is a multicomponent joined and fine-regulated processing, if damage not being repaired, leading to distinctly cellular outcomes such as transient cell cycle arrest, apoptosis, and senescence depending upon DNA damage extent, cell type, cell cycle stage, and p53 activity [10, 89]. On the other hand, DNA damage being the hallmark of cancer is less well processed in cancer cells compared to normal ones [40, 48]. Consequently, the components of DDR has long been proposed as a relatively ideal therapeutic target for almost all kinds of cancers with the combination of DDR inhibitors and cytotoxic agents, as well as the more promising drugs based on synthetic lethality [13, 29, 80]. Even with these promising therapeutic strategies, cancer is still one of the most refractory diseases with which we are combating, probably due to the existing CSCs that are responsible for relapse which are thought to give rise to therapy resistance through various mechanisms including drug efflux, DDR enhancement, quiescence status, and so on [26, 96, 97, 110]. The enhanced DDR in CSCs was majorly investigated in solid tumors under radiotherapy, such as carcinoma from glioma, prostate, and breast [1, 2], which could be achieved by inhibition of cell cycle progressing, promotion of DNA repair, and scavenging of reactive oxygen species [26, 81]. Based on these studies, it was suggested that CSCs majorly being a quiescent state are less sensitive to radiation-induced DNA damage reasoning probably from that the fast dividing cells were effectively eradicated, while the slowly growing ones remained undamaged upon irradiation [72, 97]. As quiescence is not limited to CSCs, which also resides in other normal tissues [23], the cellular outcome and related molecular mechanism of DDR in different tissue compartments are distinctly various, so as to that of normal stem cells [10, 103]; it will be of great value to identify the specific DDR of any kinds of stem cells, including LSCs.

7.4.3 Cell Cycle Arrests Upon DNA Damage in Leukemia Stem Cells

When DNA damage occurs in a cell, it can be sensed, transduced, and mediated to the ultimate effectors to induce the cell cycle arrest for later repair, not only in cycling stem cells but also in quiescent ones. Although LSCs are primarily in quiescent state, the induction of cell cycle arrest is their powerful weapons to defend the attack of anticancer drugs as discussed below.

Many studies have revealed that DNA damage-induced cell cycle arrest is p53 mediated and p21 dependent. For example, two decades ago, Wilson et al. had proved that induction of DNA damage with irradiation upregulating p53 and p21 expression suppressed cell proliferation in cycling stem cells of intestinal tissue through the detection of p53-/p21-positive cells along with [^{3}H] thymidine incorporation [108]. The work form Takubo et al. manifested that the undifferentiated spermatogonia, as the stem cell compartment in spermatogenesis, lost self-renewal and gained arrested cell cycle due to accumulated DNA damage under ATM deficiency, which could be restored by suppression of p21 [100]. Also in HSCs, DNA damage induced by iron radiation caused transient growth arrest through p53-mediated induction of p21 expression for the following DNA repair, which protected them from apoptosis compared to the progenitors [71]. So far, in these stem cells, the p21-dependent cell cycle arrest upon DNA damaging may serve as a protective mechanism employed by them to cope with DNA damage. However, this kind of protection could be utilized by LSCs in AML to get themselves more comfortable to survive, which was well illustrated by the study from Viale et al. [102]. The abovementioned response emphasized the fact that p21-dependent cell cycle arrest was induced by DNA damage induction exogenously; on the contrary, as revealed by Viale and colleagues, p21 loss caused increased DNA damage in HSCs. In preleukemic HSCs, high levels of DNA damage were introduced by expression of PML-RAR or AML1-ETO fusion oncoproteins endogenously, which resulted in upregulation of p21-induced cell cycle restriction, which is also applied to LSCs. These findings highlighted that LSCs can respond to DNA damage endogenously through the induction of p21-dependent cell cycle arrest limiting DNA damage accumulation and preventing exhaustion of LSCs, which indicates some novel antileukemic strategies based on DNA repair; however, studies about the cell cycle progressing response of LSCs upon direct DNA-damaging agents are lacking.

7.4.4 Differentiation Is Induced by DNA Damage in Leukemia Stem Cells

Differentiating into descendent cells being one of the hallmarks of stem cells can also be induced by DNA damage, which serves as a DDR specific to stem cells [93]. DNA damage-induced differentiation of stem cells can have a bad effect on the normal stem cell compartment by accelerating the exhaustion of stem cell pool or reducing the product of proper cell types for homeostasis; on the other hand, it may serve a positive role in anticancer treatment to exhaust CSCs. In neural stem cell (NSC) compartment, Schneider and colleagues found that ionizing radiation-induced DNA damage promotes NSCs into astrocytic differentiation in normal brain tissue while through which the tumorigenic potential of CSCs in glioblastoma is affected [91]. An interesting phenomenon was observed in muscle stem cells in which a differentiation checkpoint exists whose target, MyoD, is controlled by DNA-damaging induction that causes a reversible inhibition of myogenic differentiation [83]. In HSCs, previous studies seemed to indicate that DNA damage may lead to myeloid differentiation of HSCs, since DNA damage, including DSBs, accumulates with age [84] and aging HSCs prefer to differentiate into myeloid lineage rather than lymphoid descendants [85]. Now the direct connection between DNA damage and differentiation in HSCs was clearly revealed by the study from Wang et al., who proved that DNA damage induces lymphoid differentiation of HSCs through a basic leucine zipper transcription factor, BATF, in a G-CSF-/STAT3-dependent manner [105].

More recently, the link between differentiation and DNA damage in LSCs was also well characterized by Santos and colleagues [90]. In this outstanding work, upon the elucidation of histone methyltransferase MLL4 playing the suppressing role in AML by regulating antioxidant response, Santos et al. demonstrated that expression of MLL-AF9 oncogene in LSCs induces DNA damage directly leading to cell cycle exit and terminal differentiation through the activation of p21 in AML. Besides, a balance between the differentiation and self-renewal may exist in LSCs to account for the final decision of cellular outcomes (Fig. 7.6).

7.4.5 The Damages Are Repaired in Leukemia Stem Cells

DNA repair is a major molecular process responding to DNA damage for all cell types, including LSCs, which were deeply studied particularly in CML. The progress of CML can be divided into two stages, the chronic phase (CML-CP) and the blastic phase (CML-BP). The generally accepted fact is that BCR-ABL1 emerging in one single HSC initiates the genesis of CML-CP, in which LSCs or LPCs (leukemia progenitor cells) progress to CML-BP upon the acquisition of mutations and chromosomal aberrations [14, 28, 82]. LSCs in CML-CP, defined as $Lin^{-}CD34^{+}CD38^{-}$ population [52], can survive from the treatment with tyrosine kinase inhibitor (TKI)-based drugs, such as imatinib, by adopting a quiescence, high expression of BCR-ABL1 and dysregulated DNA repair, which is rooted in the progression of CML-BP [4, 50, 82].

BCR-ABL1 can cause enhanced DNA damage but, on the other hand, unfaithful and inefficient DNA repair universally in leukemia cells [16], which is also applied to LSCs in CML-CP to promote CML progression. The treasonable DNA repair in LSCs can be well illustrated by the prominent genomic instability of LSCs or LPCs in CML. BCR-ABL1 of LSCs itself may gain mutation in CML-CP, as described by Jiang et al., who revealed that LSCs from patients with CML-CP displayed massive mutation corresponding to frameshift or premature stop codon at the kinase domain [49]. Also, LPCs from patients in CML-CP displayed ABL kinase domain mutations preceding the CML-BP [5], so it was suggested that genomic instability in CML may occur at LSC/LPC level which can be passed on to the successive generation promoting the CML progression, furthermore contributing to the drug resistance [82].

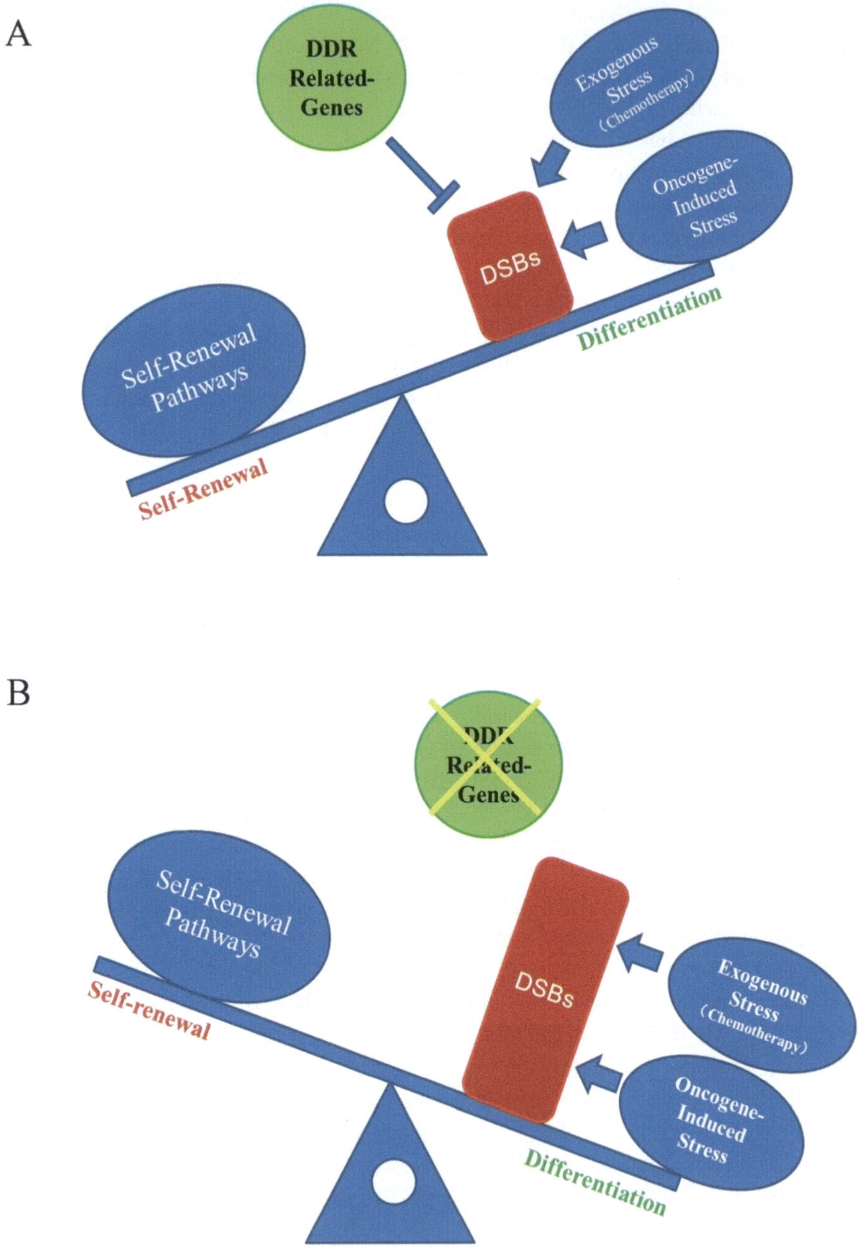

Fig. 7.6 Model showing the balance between the self-renewal and differentiation in leukemia stem cells. (**a**) Exogenous (chemotherapy et al.) or endogenous (oncogene-induced replication et al) stresses underlie the accumulation of DSBs in LSCs. In order to maintain the capabilities of self-renewal, DNA damage response machinery is used to reduce the DSBs. (**b**) Disruptions of DNA damage response of LSCs impair the above balance and lead to the genome instabilities, which are sufficient to induce the differentiation of LSCs

7.5 Perspective

In the past few decades, the studies about DDR of somatic cells, tissue stem cells, cancer cells, and cancer stem cells have made great progress, with which many novel therapeutic drugs aiming to cancers have been developed. However, the systemic understanding of DDR in LSCs hurdled by various factors is still ill defined to us. The first is the sophistication of DDR system in which one specific DNA lesion could be processed by more than one pathway and different repair processes may share enzymes and reaction intermediates. The second difficulty is the complexity of leukemia that poses so many types; the exact DDR of one specific LSC of leukemia may not be applied to another. The third and perhaps most challenging obstacle is the real modeling of leukemia in humans; so far, so many studies are based on humanized mouse model with expression of oncogenes, which can hardly capture the exact process in human leukemia. So even with the big achievement on DDR in organisms, the direct DDR of specific LSCs upon DNA damage has less been elucidated, which needs the endeavor from current and following researchers, with which we can better understand the regulation of DDR in LSCs and efficiently create novel weapons to exterminate leukemia.

Acknowledgements This work was supported by Ministry of Science and Technology of China (Grant No. 2016YFE0107200) and the National Natural Science Foundation of China (Grant No. 81770151).

References

1. Arnold CR, Abdelmoez A, Thurner G, Debbage P, Lukas P, Skvortsov S, Skvortsova II (2014) Rac1 as a multifunctional therapeutic target to prevent and combat cancer metastasis. Oncoscience 1(8):513–521
2. Bao SD, Wu QL, McLendon RE, Hao YL, Shi Q, Hjelmeland AB et al (2006) Glioma stem cells promote radioresistance by preferential activation of the DNA damage response. Nature 444(7120):756–760
3. Barnes DE, Lindahl T (2004) Repair and genetic consequences of endogenous DNA base damage in mammalian cells. Annu Rev Genet 49(38):445–476. https://doi.org/10.1146/annurev.genet.38.072902.092448
4. Barnes DJ, Melo JV (2006) Primitive, quiescent and difficult to kill: the role of non-proliferating stem cells in chronic myeloid leukemia. Cell Cycle 5(24):2862–2866. https://doi.org/10.4161/cc.5.24.3573
5. Barnes DJ, Palaiologou D, Panousopoulou E, Schultheis B, Yong ASM, Wong A et al (2005) Bcr-Abl expression levels determine the rate of development of resistance to imatinib mesylate in chronic myeloid leukemia. Cancer Res 65(19):8912–8919
6. Beerman I, Seita J, Inlay MA, Weissman IL, Rossi DJ (2014) Quiescent hematopoietic stem cells accumulate DNA damage during aging that is repaired upon entry into cell cycle. Cell Stem Cell 15(1):37–50. https://doi.org/10.1016/j.stem.2014.04.016
7. Bernstein KA, Rothstein R (2009) At loose ends: resecting a double-strand break. Cell 137(5):807–810

8. Biechonski S, Olender L, Zipin-Roitman A, Yassin M, Aqaqe N, Marcu-Malina V et al (2018) Attenuated DNA damage responses and increased apoptosis characterize human hematopoietic stem cells exposed to irradiation. Sci Rep 8(1):6071. https://doi.org/10.1038/s41598-018-24440-w
9. Biechonski S, Yassin M, Milyavsky M (2017) DNA-damage response in hematopoietic stem cells: an evolutionary trade-off between blood regeneration and leukemia suppression. Carcinogenesis 38(4):367–377
10. Blanpain C, Mohrin M, Sotiropoulou PA, Passegué E (2011) DNA-damage response in tissue-specific and cancer stem cells. Cell Stem Cell 8(1):16–29. https://doi.org/10.1016/j.stem.2010.12.012
11. Bohlander SK, Kakadia PM (2015) DNA repair and chromosomal translocations. Recent Results Cancer Res 200:1–37
12. Boiteux S, Guillet M (2004) Abasic sites in DNA: repair and biological consequences in *Saccharomyces cerevisiae*. DNA Repair (Amst) 3(1):1–12. https://doi.org/10.1016/j.dnarep.2003.10.002
13. Brown JS, O'Carrigan B, Jackson SP, Yap TA (2017) Targeting DNA repair in cancer: beyond PARP inhibitors. Cancer Discov 7(1):20–37. https://doi.org/10.1158/2159-8290.CD-16-0860
14. Bruns I, Czibere A, Fischer JC, Roels F, Cadeddu RP, Buest S et al (2009) The hematopoietic stem cell in chronic phase CML is characterized by a transcriptional profile resembling normal myeloid progenitor cells and reflecting loss of quiescence. Leukemia 23(5):892–899. https://doi.org/10.1038/leu.2008.392
15. Bryant HE, Petermann E, Schultz N, Jemth AS, Loseva O, Issaeva N et al (2009) PARP is activated at stalled forks to mediate Mre11-dependent replication restart and recombination. EMBO J 28(17):2601–2615. https://doi.org/10.1038/emboj.2009.206
16. Burke BA, Carroll M (2010) BCR-ABL: a multi-faceted promoter of DNA mutation in chronic myelogeneous leukemia. Leukemia 24(6):1105–1112
17. Cabezas-Wallscheid N, Eichwald V, de Graaf J, Lower M, Lehr HA, Kreft A et al (2013) Instruction of haematopoietic lineage choices, evolution of transcriptional landscapes and cancer stem cell hierarchies derived from an AML1-ETO mouse model. EMBO Mol Med 5(12):1804–1820. https://doi.org/10.1002/emmm.201302661
18. Cabezas-Wallscheid N, Klimmeck D, Hansson J, Lipka DB, Reyes A, Wang Q et al (2014) Identification of regulatory networks in HSCs and their immediate progeny via integrated proteome, transcriptome, and DNA methylome analysis. Cell Stem Cell 15(4):507–522. https://doi.org/10.1016/j.stem.2014.07.005
19. Cadet J, Wagner JR (2013) DNA base damage by reactive oxygen species, oxidizing agents, and UV radiation. Cold Spring Harb Perspect Biol 5(2). doi: https://doi.org/10.1101/cshperspect.a012559
20. Caldecott KW (2008) Single-strand break repair and genetic disease. Nat Rev Genet 9(8):619–631. https://doi.org/10.1038/nrg2380
21. Caldecott KW (2014) DNA single-strand break repair. Exp Cell Res 329(1):2–8. https://doi.org/10.1016/j.yexcr.2014.08.027
22. Chang L, Graham PH, Hao J, Ni J, Bucci J, Cozzi PJ et al (2014) PI3K/Akt/mTOR pathway inhibitors enhance radiosensitivity in radioresistant prostate cancer cells through inducing apoptosis, reducing autophagy, suppressing NHEJ and HR repair pathways. Cell Death Dis 5:e1437. https://doi.org/10.1038/cddis.2014.415
23. Cheung TH, Rando TA (2013) Molecular regulation of stem cell quiescence. Nat Rev Mol Cell Biol 14(6):329–340
24. Ciccia A, Elledge SJ (2010) The DNA damage response: making it safe to play with knives. Mol Cell 40(2):179–204
25. Clarke MF, Dick JE, Dirks PB, Eaves CJ, Jamieson CH, Jones DL et al (2006) Cancer stem cells – perspectives on current status and future directions: AACR workshop on cancer stem cells. Cancer Res 66(19):9339–9344. https://doi.org/10.1158/0008-5472.can-06-3126

26. Cojoc M, Mäbert K, Muders MH, Dubrovska A (2015) A role for cancer stem cells in therapy resistance: cellular and molecular mechanisms. Semin Cancer Biol 31:16–27. https://doi.org/10.1016/j.semcancer.2014.06.004
27. Corces-Zimmerman MR, Hong WJ, Weissman IL, Medeiros BC, Majeti R (2014) Preleukemic mutations in human acute myeloid leukemia affect epigenetic regulators and persist in remission. Proc Natl Acad Sci U S A 111(7):2548–2553. https://doi.org/10.1073/pnas.1324297111
28. Cross NC, Daley GQ, Green AR, Hughes TP, Jamieson C, Manley P et al (2008) BCR-ABL1-positive CML and BCR-ABL1-negative chronic myeloproliferative disorders: some common and contrasting features. Leukemia 22(11):1975–1989. https://doi.org/10.1038/leu.2008.231
29. Curtin NJ (2012) DNA repair dysregulation from cancer driver to therapeutic target. Nat Rev Cancer 12(12):801–817. https://doi.org/10.1038/nrc3399
30. Deans AJ, West SC (2011) DNA interstrand crosslink repair and cancer. Nat Rev Cancer 11(7):467–480
31. Delia D, Mizutani S (2017) The DNA damage response pathway in normal hematopoiesis and malignancies. Int J Hematol 106(3):328–334
32. Economopoulou P, Pappa V, Papageorgiou S, Dervenoulas J, Economopoulos T (2011) DNA repair deficiency associated with hematological neoplasms. InTech
33. Elrick LJ, Jorgensen HG, Mountford JC, Holyoake TL (2005) Punish the parent not the progeny. Blood 105(5):1862–1866. https://doi.org/10.1182/blood-2004-08-3373
34. Estrov Z (2010) The leukemia stem cell. In: Nagarajan L (ed) Acute myelogenous leukemia: genetics, biology and therapy. Springer New York, New York, pp 1–17
35. Fortini P, Ferretti C, Dogliotti E (2013) The response to DNA damage during differentiation: pathways and consequences. Mutat Res-Fundam Mol Mech Mutagen 743:160–168
36. Fu D, Calvo JA, Samson LD (2012) Balancing repair and tolerance of DNA damage caused by alkylating agents. Nat Rev Cancer 12(2):104–120
37. Genovese G, Kahler AK, Handsaker RE, Lindberg J, Rose SA, Bakhoum SF et al (2014) Clonal hematopoiesis and blood-cancer risk inferred from blood DNA sequence. N Engl J Med 371(26):2477–2487. https://doi.org/10.1056/NEJMoa1409405
38. Gonzalez-Herrero I, Rodriguez-Hernandez G, Luengas-Martinez A, Isidro-Hernandez M, Jimenez R, Garcia-Cenador MB, … Vicente-Duenas C (2018) The making of leukemia. Int J Mol Sci 19(5)
39. Guan YH, Gerhard B, Hogge DE (2003) Detection, isolation, and stimulation of quiescent primitive leukemic progenitor cells from patients with acute myeloid leukemia (AML). Blood 101(8):3142–3149. https://doi.org/10.1182/blood-2002-10-3062
40. Hanahan D, Weinberg RA (2011) Hallmarks of cancer: the next generation. Cell 144(5):646–674. https://doi.org/10.1016/j.cell.2011.02.013
41. Harper JW, Elledge SJ (2007) The DNA damage response: ten years after. Mol Cell 28(5):739–745. https://doi.org/10.1016/j.molcel.2007.11.015
42. Helleday T, Lo J, van Gent DC, Engelward BP (2007) DNA double-strand break repair: from mechanistic understanding to cancer treatment. DNA Repair (Amst) 6(7):923–935. https://doi.org/10.1016/j.dnarep.2007.02.006
43. Holyoake T, Jiang X, Eaves C, Eaves A (1999) Isolation of a highly quiescent subpopulation of primitive leukemic cells in chronic myeloid leukemia. Blood 94(6):2056–2064
44. Ishikawa F, Yoshida S, Saito Y, Hijikata A, Kitamura H, Tanaka S et al (2007) Chemotherapy-resistant human AML stem cells home to and engraft within the bone-marrow endosteal region. Nat Biotechnol 25(11):1315–1321. https://doi.org/10.1038/nbt1350
45. Itkin T, Gur-Cohen S, Spencer JA, Schajnovitz A, Ramasamy SK, Kusumbe AP et al (2016) Distinct bone marrow blood vessels differentially regulate haematopoiesis. Nature 532(7599):323–328. https://doi.org/10.1038/nature17624
46. Jaiswal S, Fontanillas P, Flannick J, Manning A, Grauman PV, Mar BG et al (2014) Age-related clonal hematopoiesis associated with adverse outcomes. N Engl J Med 371(26):2488–2498. https://doi.org/10.1056/NEJMoa1408617

47. Jan M, Snyder TM, Corces-Zimmerman MR, Vyas P, Weissman IL, Quake SR, Majeti R (2012) Clonal evolution of preleukemic hematopoietic stem cells precedes human acute myeloid leukemia. Sci Transl Med 4(149):149ra118. https://doi.org/10.1126/scitranslmed.3004315
48. Jeggo PA, Lobrich M (2015) How cancer cells hijack DNA double-strand break repair pathways to gain genomic instability. Biochem J 471(1):1–11. https://doi.org/10.1042/bj20150582
49. Jiang X, Saw KM, Eaves A, Eaves C (2007a) Instability of BCR-ABL gene in primary and cultured chronic myeloid leukemia stem cells. J Natl Cancer Inst 99(9):680–693. https://doi.org/10.1093/jnci/djk150
50. Jiang X, Zhao Y, Smith C, Gasparetto M, Turhan A, Eaves A, Eaves C (2007b) Chronic myeloid leukemia stem cells possess multiple unique features of resistance to BCR-ABL targeted therapies. Leukemia 21(5):926–935. https://doi.org/10.1038/sj.leu.2404609
51. Jordan CT, Guzman ML, Noble M (2006) Cancer stem cells. N Engl J Med 355(12):1253–1261. https://doi.org/10.1056/NEJMra061808
52. Jorgensen HG, Holyoake TL (2007) Characterization of cancer stem cells in chronic myeloid leukaemia. Biochem Soc Trans 35(Pt 5):1347–1351. https://doi.org/10.1042/bst0351347
53. Josef J (2006) The multifaceted mismatch-repair system. Nat Rev Mol Cell Biol 7(5):335–346
54. Kathe SD, Shen GP, Wallace SS (2004) Single-stranded breaks in DNA but not oxidative DNA base damages block transcriptional elongation by RNA polymerase II in HeLa cell nuclear extracts. J Biol Chem 279(18):18511–18520. https://doi.org/10.1074/jbc.M313598200
55. Kelley MR, Kow YW, Wilson DM 3rd. (2003) Disparity between DNA base excision repair in yeast and mammals: translational implications. Cancer Res 63(3):549–554
56. Kottemann MC, Smogorzewska A (2013) Fanconi anaemia and the repair of Watson and Crick DNA crosslinks. Nature 493(7432):356–363
57. Kraft D, Rall M, Volcic M, Metzler E, Groo A, Stahl A et al (2015) NF-kappaB-dependent DNA damage-signaling differentially regulates DNA double-strand break repair mechanisms in immature and mature human hematopoietic cells. Leukemia 29(7):1543–1554. https://doi.org/10.1038/leu.2015.28
58. Kreso A, Dick JE (2014) Evolution of the cancer stem cell model. Cell Stem Cell 14(3):275–291. https://doi.org/10.1016/j.stem.2014.02.006
59. Krokan HE, Bjoras M (2013) Base excision repair. Cold Spring Harb Perspect Biol 5(4):a012583. https://doi.org/10.1101/cshperspect.a012583
60. Kunkel TA, Erie DA (2015) Eukaryotic mismatch repair in relation to DNA replication. Annu Rev Genet 49(49):291–313
61. Kuzminov A (2001) Single-strand interruptions in replicating chromosomes cause double-strand breaks. Proc Natl Acad Sci U S A 98(15):8241–8246. https://doi.org/10.1073/pnas.131009198
62. Lambert MW (2016) Nuclear alpha spectrin: critical roles in DNA interstrand cross-link repair and genomic stability. Exp Biol Med 241(15):1621–1638
63. Li GM (2008) Mechanisms and functions of DNA mismatch repair. Cell Res 18(1):85–98. https://doi.org/10.1038/cr.2007.115
64. Lieber MR (2010a) The mechanism of double-strand DNA break repair by the nonhomologous DNA end-joining pathway. Annu Rev Biochem 79(79):181–211
65. Lieber MR (2010b) The mechanism of double-strand DNA break repair by the nonhomologous DNA end-joining pathway. [Review] [164 refs]. Annu Rev Biochem 79(1):181–211
66. Linn S (2003) Molecular mechanisms of mammalian DNA repair and the DNA damage checkpoints. Annu Rev Biochem 73(73):39–85
67. Talpaert-Borlé M (1987) Formation, detection and repair of AP sites. Mutat Res/Fundam Mol Mech Mutagen 181(1):45–56
68. Mandal PK, Blanpain C, Rossi DJ (2011) DNA damage response in adult stem cells: pathways and consequences. Nat Rev Mol Cell Biol 12(3):198–202

69. McKerrell T, Park N, Moreno T, Grove CS, Ponstingl H, Stephens J et al (2015) Leukemia-associated somatic mutations drive distinct patterns of age-related clonal hemopoiesis. Cell Rep 10(8):1239–1245. https://doi.org/10.1016/j.celrep.2015.02.005
70. Milyavsky M, Gan OI, Trottier M, Komosa M, Tabach O, Notta F et al (2010) A distinctive DNA damage response in human hematopoietic stem cells reveals an apoptosis-independent role for p53 in self-renewal. Cell Stem Cell 7(2):186–197. https://doi.org/10.1016/j.stem.2010.05.016
71. Mohrin M, Bourke E, Alexander D, Warr MR, Barry-Holson K, Le Beau MM et al (2010a) Hematopoietic stem cell quiescence promotes error-prone DNA repair and mutagenesis. Cell Stem Cell 7(2):174–185. https://doi.org/10.1016/j.stem.2010.06.014
72. Moore N, Lyle S (2011) Quiescent, slow-cycling stem cell populations in cancer: a review of the evidence and discussion of significance. J Oncol 2011:396076. https://doi.org/10.1155/2011/396076
73. Moroni F (2008) Poly(ADP-ribose)polymerase 1 (PARP-1) and postischemic brain damage. Curr Opin Pharmacol 8(1):96–103. https://doi.org/10.1016/j.coph.2007.10.005
74. Mouret S, Baudouin C, Charveron M, Favier A, Cadet J, Douki T (2006) Cyclobutane pyrimidine dimers are predominant DNA lesions in whole human skin exposed to UVA radiation. Proc Natl Acad Sci U S A 103(37):13765–13770
75. Niedernhofer LJ (2008) DNA repair is crucial for maintaining hematopoietic stem cell function. DNA Repair (Amst) 7(3):523–529
76. Nilles N, Fahrenkrog B (2017) Taking a bad turn: compromised DNA damage response in leukemia. Cells 6(2). doi: Artn 11783390/Cells6020011
77. Noll DM, Mason TM, Miller PS (2006) Formation and repair of interstrand cross-links in DNA. Chem Rev 106(2):277–301
78. Park H, Zhang K, Ren Y, Nadji S, Sinha N, Taylor JS, Kang C (2002) Crystal structure of a DNA decamer containing a cis-syn thymine dimer. Proc Natl Acad Sci U S A 99(25):15965–15970. https://doi.org/10.1073/pnas.242422699
79. Passegue E (2005) Hematopoietic stem cells, leukemic stem cells and chronic myelogenous leukemia. Cell Cycle 4(2):266–268
80. Pearl LH, Schierz AC, Ward SE, Al-Lazikani B, Pearl FMG (2015) Therapeutic opportunities within the DNA damage response. Nat Rev Cancer 15(3):166–180. https://doi.org/10.1038/nrc3891
81. Peitzsch C, Kurth I, Kunz-Schughart L, Baumann M, Dubrovska A (2013) Discovery of the cancer stem cell related determinants of radioresistance. Radiother Oncol 108(3):378–387. https://doi.org/10.1016/j.radonc.2013.06.003
82. Perrotti D, Jamieson C, Goldman J, Skorski T (2010) Chronic myeloid leukemia: mechanisms of blastic transformation. J Clin Invest 120(7):2254–2264. https://doi.org/10.1172/jci41246
83. Puri PL, Bhakta K, Wood LD, Costanzo A, Zhu JY, Wang JYJ (2002) A myogenic differentiation checkpoint activated by genotoxic stress. Nat Genet 32(4):585–593
84. Rossi DJ, Bryder D, Seita J, Nussenzweig A, Hoeijmakers J, Weissman IL (2007a) Deficiencies in DNA damage repair limit the function of haematopoietic stem cells with age. Nature 447(7145):725-U715
85. Rossi DJ, Bryder D, Zahn JM, Ahlenius H, Sonu R, Wagers AJ, Weissman IL (2005) Cell intrinsic alterations underlie hematopoietic stem cell aging. Proc Natl Acad Sci USA 102(26):9194–9199
86. Rossi DJ, Seita J, Czechowicz A, Bhattacharya D, Bryder D, Weissman IL (2007b) Hematopoietic stem cell quiescence attenuates DNA damage response and permits DNA damage accumulation during aging. Cell Cycle 6(19):2371–2376. https://doi.org/10.4161/cc.6.19.4759
87. Saito Y, Uchida N, Tanaka S, Suzuki N, Tomizawa-Murasawa M, Sone A et al (2010) Induction of cell cycle entry eliminates human leukemia stem cells in a mouse model of AML. Nat Biotechnol 28(3):275-U133

88. Sancar A (2003) Structure and function of DNA photolyase and cryptochrome blue-light photoreceptors. Chem Rev 103(6):2203–2237. https://doi.org/10.1021/cr0204348
89. Sancar A, Lindsey-Boltz LA, Unsal-Kacmaz K, Linn S (2004) Molecular mechanisms of mammalian DNA repair and the DNA damage checkpoints. Annu Rev Biochem 73:39–85. https://doi.org/10.1146/annurev.biochem.73.011303.073723
90. Santos MA, Faryabi RB, Ergen AV, Day AM, Malhowski A, Canela A, … Nussenzweig A (2014) DNA-damage-induced differentiation of leukaemic cells as an anti-cancer barrier. Nature, 514(7520):107.–+
91. Schneider L, Pellegatta S, Favaro R, Pisati F, Roncaglia P, Testa G et al (2013) DNA damage in mammalian neural stem cells leads to astrocytic differentiation mediated by BMP2 signaling through JAK-STAT. Stem Cell Rep 1(2):123–138
92. Setlow RB, Carrier WL (1966) Pyrimidine dimers in ultraviolet-irradiated DNA's. J Mol Biol 17(1):237–254
93. Sherman MH, Bassing CH, Teitell MA (2011) Regulation of cell differentiation by the DNA damage response. Trends Cell Biol 21(5):312–319
94. Shlush LI, Zandi S, Mitchell A, Chen WC, Brandwein JM, Gupta V et al (2014) Identification of pre-leukaemic haematopoietic stem cells in acute leukaemia. Nature 506(7488):328–333. https://doi.org/10.1038/nature13038
95. Shuck SC, Short EA, Turchi JJ (2008) Eukaryotic nucleotide excision repair: from understanding mechanisms to influencing biology. Cell Res 18(1):64–72. https://doi.org/10.1038/cr.2008.2
96. Skvortsov S, Debbage P, Lukas P, Skvortsova I (2015) Crosstalk between DNA repair and cancer stem cell (CSC) associated intracellular pathways. Semin Cancer Biol 31:36–42. https://doi.org/10.1016/j.semcancer.2014.06.002
97. Skvortsova I, Debbage P, Kumar V, Skvortsov S (2015) Radiation resistance: cancer stem cells (CSCs) and their enigmatic pro-survival signaling. Semin Cancer Biol 35:39–44. https://doi.org/10.1016/j.semcancer.2015.09.009
98. Sugimura K, Takebayashi S, Taguchi H, Takeda S, Okumura K (2008) PARP-1 ensures regulation of replication fork progression by homologous recombination on damaged DNA. J Cell Biol 183(7):1203–1212. https://doi.org/10.1083/jcb.200806068
99. Takeishi S, Nakayama KI (2016) To wake up cancer stem cells, or to let them sleep, that is the question. Cancer Sci 107(7):875–881. https://doi.org/10.1111/cas.12958
100. Takubo K, Ohmura M, Azuma M, Nagamatsu G, Yamada W, Arai F et al (2008) Stem cell defects in ATM-deficient undifferentiated spermatogonia through DNA damage-induced cell-cycle arrest. Cell Stem Cell 2(2):170–182. https://doi.org/10.1016/j.stem.2007.10.023
101. Taniguchi Ishikawa E, Gonzalez-Nieto D, Ghiaur G, Dunn SK, Ficker AM, Murali B et al (2012) Connexin-43 prevents hematopoietic stem cell senescence through transfer of reactive oxygen species to bone marrow stromal cells. Proc Natl Acad Sci U S A 109(23):9071–9076. https://doi.org/10.1073/pnas.1120358109
102. Viale A, De Franco F, Orleth A, Cambiaghi V, Giuliani V, Bossi D et al (2009) Cell-cycle restriction limits DNA damage and maintains self-renewal of leukaemia stem cells. Nature 457(7225):51-U52. https://doi.org/10.1038/nature07618
103. Vitale I, Manic G, De Maria R, Kroemer G, Galluzzi L (2017) DNA damage in stem cells. Mol Cell 66(3):306–319. https://doi.org/10.1016/j.molcel.2017.04.006
104. Walter D, Lier A, Geiselhart A, Thalheimer FB, Huntscha S, Sobotta MC et al (2015) Exit from dormancy provokes DNA-damage-induced attrition in haematopoietic stem cells. Nature 520(7548):549–552. https://doi.org/10.1038/nature14131
105. Wang J, Sun Q, Morita Y, Jiang H, Groß A, Lechel A et al (2012) A differentiation checkpoint limits hematopoietic stem cell self-renewal in response to DNA damage. Cell 148(5):1001–1014. https://doi.org/10.1016/j.cell.2012.01.040
106. Weeden CE, Asselin-Labat M-L (2018) Mechanisms of DNA damage repair in adult stem cells and implications for cancer formation. Biochim Biophys Acta (BBA) – Mol Basis Dis 1864(1):89–101. https://doi.org/10.1016/j.bbadis.2017.10.015

107. Wilson A, Laurenti E, Oser G, van der Wath RC, Blanco-Bose W, Jaworski M et al (2008) Hematopoietic stem cells reversibly switch from dormancy to self-renewal during homeostasis and repair. Cell 135(6):1118–1129. https://doi.org/10.1016/j.cell.2008.10.048
108. Wilson JW, Pritchard DM, Hickman JA, Potten CS (1998) Radiation-induced p53 and p21WAF–1/CIP1 expression in the murine intestinal epithelium: apoptosis and cell cycle arrest. Am J Pathol 153(3):899–909. https://doi.org/10.1016/S0002-9440(10)65631-3
109. Xie M, Lu C, Wang J, McLellan MD, Johnson KJ, Wendl MC et al (2014) Age-related mutations associated with clonal hematopoietic expansion and malignancies. Nat Med 20(12):1472–1478. https://doi.org/10.1038/nm.3733
110. Yoshida GJ, Saya H (2016) Therapeutic strategies targeting cancer stem cells. Cancer Sci 107(1):5–11. https://doi.org/10.1111/cas.12817
111. Zhang Y, Rowley JD (2006) Chromatin structural elements and chromosomal translocations in leukemia. DNA Repair (Amst) 5(9–10):1282–1297. https://doi.org/10.1016/j.dnarep.2006.05.020

Chapter 8
Epigenetic Abnormalities in Acute Myeloid Leukemia and Leukemia Stem Cells

Jing Xu, Xiaohang Hang, Baohong Wu, Chong Chen, and Yu Liu

Contents

Abstract Recently advances in cancer genomics revealed the unexpected high frequencies of epigenetic abnormalities in human acute myeloid leukemia (AML). Accumulating data suggest that these leukemia-associated epigenetic factors play critical roles in both normal hematopoietic stem cells (HSCs) and leukemia stem cells (LSCs). In turn, these abnormalities result in susceptibilities of LSC and related diseases to epigenetic inhibitors. In this chapter, we will focus on the mutations of epigenetic factors in AML, their functional roles and mechanisms in normal hematopoiesis and leukemia genesis, especially in LSC, and potential treatment opportunities specifically for AML with epigenetic dysregulations.

Keywords Epigenetic changes · Leukemia stem cell · AML

J. Xu · X. Hang · B. Wu · C. Chen · Y. Liu (✉)
Department of Hematology, State Key Laboratory of Biotherapy, West China Hospital, Sichuan University, Chengdu, Sichuan, China
e-mail: yuliuscu@scu.edu.cn

H. Zhang, S. Li (eds.), *Leukemia Stem Cells in Hematologic Malignancies*, Advances in Experimental Medicine and Biology 1143,
https://doi.org/10.1007/978-981-13-7342-8_8

8.1 Introduction

Epigenetics is heritable phenotype changes beyond the DNA sequence. Waddington coined the term "epigenetics" in 1942 and proposed the famous Waddington's epigenetic landscape for embryonic development [1]. These changes include DNA modifications and histone modifications, called epigenetic marks. These epigenetic markers are added to specific DNA and histones by a variety of enzymes, called epigenetic writers. Another group of proteins, called epigenetic readers, can recognize and bind these modified DNA and histones to regulate the expressions of involved genes. The third group of proteins, called erasers, is able to remove these epigenetic modifications [2].

Histones help DNA to fold into structural units called nucleosomes. Posttranslational modifications on histones can change chromosomal folding and regulate gene expressions. Each nucleosome contains two copies of core histones H2A, H2B, H3 and H4, and histone H1 is the linker histone between nucleosomes. Frequent histone modifications include methylation, acetylation, phosphorylation, and ubiquitination. Histone methylation are catalyzed by histone methyltransferases (HMTs), including two major types, lysine-specific and arginine-specific HMTs. There are many lysine-specific HMTs, such as SET domain-containing ASH1L, EZH1/2 and MLL family members (MLL1–5), SET family and NSD1, and non-SET domain containing DOT1L [3, 4]. Arginine-specific HMTs can be divided into three subtypes, dependent on methylation form on arginine. H3 and H4 lysine methylations can either activate or repress the expressions of related genes, dependent on the lysine sites and also gene regions. Lysine methylations can be mono-, di-, or trimethylations, which can have different effects on gene expressions. H3K4 methylations are generally active marks. Histone methylations are removed by histone demethylases, which can have different catalytic domains with various functions. There are two major types of histone demethylases. One is lysine demethylases (KDMs), and the other one is the Jumonji C domain-containing family. The first discovered KDM is LSD1. LSD1 can remove methyl groups from H3K4 or H3K9 [5]. The level of histone acetylation is mainly determined by histone acetyltransferase (HATs) and histone deacetylase (HDAC). Both HATs and HDACs can form protein complexes with transcription factors to coordinate the fate of DNA transcripts.

DNA methylation is the most studied epigenetic modification on DNA. DNA methyltransferase enzymes (DNMTs) are the writers of DNA methylations. There are three DNMTs, DNMT1, DNMT3A, and DNMT3B, in mammals. They catalyze the additions of methyl group to the 5-cytosine to form 5-methylcytosine (5mC) in CpG. DNMT3A and DNMT3B are the major de novo methyltransferases, while DNMT1 is mainly responsible for methylation maintenance during DNA replication. In contrast, TET genes (TET1, TET2, and TET3) can successively oxidize 5mC to 5-hydroxymethylcytosine (5hmC), 5-formylcytosine (5fC), and 5-carboxylcytosine (5caC). TDG and BER can remove carboxyl group from 5caC and then complete the "erasion" of cytosine methylation. 5mC at CpG sites on a gene promoter generally leads to the expression silencing, while the functions of other methylation modifications or 5mC on other regions are less clear.

It has been shown that there is critical crosstalk between DNA methylation and histone modifications, especially histone methylations, to regulate the expressions of associated genes. First discovered in plant, there is a self-reinforcing loop between DNA and histone methylation mediated by DNA methyltransferase CMT3 and histone methyltransferase KYP [6]. DNA methylation can inhibit H3K4 methylation [7], and, in turn, H3K9 methyltransferase can also reduce DNA methylation [8].

Numerous studies have showed that these epigenetic regulators are essential for normal development and healthy hemostasis. Mutations and dysregulations of these regulators are involved in multiple human diseases, including metabolic disorders, autoimmune diseases, and various cancers. In recent years, with the rapid advances and applications in high-throughput sequencing, amounting data have described a more and more clear picture of human cancer genomics. Genomic studies have identified novel numerous recurrent mutations in hematopoietic malignancies. Among them, the most interesting ones are a group of genes which can affect the epigenetic landscapes or the readouts of the abnormal epigenetic modifications in leukemia and lymphoma cells. The landmark studies by The Cancer Genome Atlas (TCGA) research group revealed unexpected high frequent mutations of DNA methylation-related genes and chromatin-modifying genes in adult de novo AML [9]. Forty percent AML contained mutations in DNA methylation-related genes, including DNMT3A, DNMT3B, DNMT1, TET1, TET2, IDH1, and IDH2. And more than 30% AML had mutations on chromatin modifiers, such as MLL fusions, ASXL1, EZH2, and KDM6A. To add up, more than 70% AML, at least in this cohort, had epigenetic-related mutations. Similarly, epigenetic genes are also involved in lymphoid neoplasms. About 50% of early T-cell precursor acute lymphoblastic leukemia (ALL) had mutations on histone-modifying genes, such as EZH2, EED, SUZ12, SETD2, and EP300 [10]. Thus, understanding the functional roles and underlying molecular mechanisms of these epigenetic factors in both normal hematopoiesis and hematopoietic malignancies would have great impact on understanding the pathology of these diseases and exploring novel treatment options.

8.2 Epigenetic Regulations of Normal Hematopoietic Stem Cells

Hematopoietic stem cells can self-renew to maintain the hemostasis of the HSC pool and also differentiate into all blood and immune cell lineages. Dysregulations of HSC have been linked to numerous human diseases, including multiple hematopoietic malignancies. Epigenetic factors play critical roles in the tightly regulated self-renewal and gradual differentiation of HSC.

It has been shown that there are dramatic DNA methylation patterns in HSC, lineage-specific progenitors, and differentiated cells [11]. DNA methylation maintenance is critical in these processes. DNMT1 mutant HSCs have severe defects in their functions, including self-renewal, niche retention, and differentiation [12]. DNA de novo methylation is important for gradually silencing of self-renewal genes during HSC differentiation. DNMT3A deficiency leads to impaired HSC differentiation and enhanced self-renewal [13]. Loss of both DNMT3A and DNMT3B had more dramatic phenotype, possibly through regulating the DNA methylations on multiple self-renewal genes, including b-catenin [14]. Interestingly, erasers of DNA methylation also play significant roles in HSC self-renewal. TET2 loss mice have progressive enlargement of their HSC pool, associated with both enhanced HSC self-renewal and block of differentiation. Differentiation block by TET2 deficiency can also happen at multiple stages besides HSC, including myeloid progenitors [15]. Beyond gene expression regulations, DNA methylation seems to play roles in maintaining the genome integrity. It has been shown that TET2-mutant HSC and progenitors display hypermutagenicity [16].

The functions of HSC and the hematopoietic system decline during aging, which might contribute to the hematopoietic pathophysiology in old people. It seems that the global DNA methylation levels don't significantly change in aged HSC. However, site-specific DNA methylation can vary a lot, especially on genes involving HSC self-renewal, which are largely affected by the proliferation history of HSC over time [17]. Consistent with the association of DNA methylation patterns with HSC aging, importantly, recurrent TET2 and DNMT3A mutations have been observed in normal elder with clonal hematopoiesis, a syndrome associated with abnormal self-renewal of HSC [18–20].

Among histone modifiers, HMTs are the best studied ones in hematopoietic system, given their frequent involvement in hematopoietic malignancies. One example is the polycomb repressive complex 1 and 2 (PRC1 and PRC2). The catalytic components in PRC1 and PRC2 are EZH1 and EZH2, respectively. PRC1 can ubiquitylates H2AK119 in mammalian cells and thus promotes chromatin folding. While PRC2 is responsible for H3K27 methylation, BMI1 is a unique component in PRC1. It has been shown that BMI1 is critical for the self-renewal of HSCs in adult largely through repressing the expression of p16INK4A and p19ARF [21] and balancing the symmetric and asymmetric divisions of HSCs [22]. BMI1 also represses the production of reactive oxidative species to prevent the premature of HSCs and plays roles [23]. Interestingly, EZH1 can repress primitive and EMP-biased hematopoiesis in early embryos [24]. The roles of PRC2 in HSCs are more complicated. EZH2 loss in adult HSCs doesn't significantly affect HSC self-renewal or hematopoiesis, except lymphoid differentiation. However, deletion of EZH2 in embryonic HSCs dramatically reduces their self-renewal and functions, associated with H3K27 methylation reduction [25]. However, knockout of Eed, a component of PRC2, has minimal effect on fetal liver HSCs but depletes both neonatal and adult HSCs [26]. MLL genes are H3K4 mono- and di-methyltransferases. MLL1, the founding member of this gene family, is frequently involved in chromosome translocations in both lymphoid and myeloid leukemia. MLL1 is essential for the hemostasis of adult HSCs. MLL1 knockout drives quiescent HSCs into cell cycle, and thus these mutant HSCs lose their long-term self-renewal and repopulation capacity. Interestingly, MLL1-deficient progenitor cells show reduced proliferation [27].

Histone methylations in HSCs are maintained by the balance between histone methyltransferases and demethylases. Dysregulations of histone demethylases may also lead to HSC defects. One isoform of LSD1, a H3K9 and H3K4 demethylase, can maintain HSC quiescence [28]. It also controls the differentiation of erythroid progenitors by regulating the expressions of GATA2 and HOX genes. Interestingly, LSD1 can also interact with MLL1 complex. JARID1B, a Jumonji C domain-containing histone H3K4 demethylase, is not required for steady-state hematopoiesis but essential for long-term HSC self-renewal and maintenance [29]. More significantly, KDM6A and KDM6B, which show similar biochemical activity to remove methyl groups from H3K27, reverse to EZH2 activity, have contrasting functions in hematopoietic system. KDM6B knockout mice are largely normal with minimal phenotype, but it is essential for long-term repopulating ability of HSCs revealed by competitive transplant assay. However, KDM6A is critical for regulating the self-renewal of HSCs, and loss of KDM6A can promote the transformation into both myeloid and lymphoid malignancies through both demethylation activity-dependent and demethylation activity-independent mechanisms [30]. Interestingly, KDM6A is also a component of the MLL3/4-containing COMPASS-like complex. Thus, fine-tuning the histone modifications is essential for normal HSC self-renewal and functions.

8.3 DNA Modifications in Leukemia and LSCs

Abnormal DNA methylation pattern is a common feature of blood malignancies. Profiling the genome-wide methylation of 344 AML sample showed that these patients could be divided into separate subgroups with distinct DNA methylation patterns, which were associated with different driver mutations [31]. Methylations at promoter CpG islands and CpG shores and enhancers had numerous changes during leukemogenesis, though the functional significance of individual sites remains largely unknown. Strikingly, genomic profiling revealed unexpected high frequent mutations in the DNA methylation factors, such as DNMT3A, TET2, and IDH1/2 [9]. Accumulating data suggest that the DNA modifiers play significant roles in leukemia pathology and also opportunities for novel therapy options.

8.3.1 DNMT3A in AML

DNMT3A alters gene expression by de novo methylation of 5′-position of cytosine at CpG dinucleotides, which are among the most frequently mutated epigenetic regulators in hematological malignancies [32]. It was shown that approximately 22% of cytogenetically normal AML cases carry DNMT3A mutations, and the hotspot mutation lies on arginine 882 in the DNMT3A catalytic domain, with around 60% mut cases affecting this codon [33]. This somatic missense mutation is usually associated with poor prognosis in AML patients and, particularly when accompanied by FLT3-ITD positivity, is an extremely poor prognostic factor even after allogeneic hematopoietic stem cell transplantation [34].

Relative studies have identified DNMT3A as a tumor suppressor gene in many hematological malignancies. In mice transplanted with Dnmt3a-KO HSCs, both lymphoid and myeloid malignancies emerged, and the latency was 3–14 months [35, 36]. Mice transplanted with Dnmt3a-R882H represented a disease phenotype mimicking chronic myelomonocytic leukemia (CMML) in humans 1 year after bone marrow transplantation [37]. Loss of Dnmt3a synergizes with the Flt3 internal tandem duplication (ITD) in mice to initiate both AML and T-ALL, with Dnmt3a dosage influencing leukemia type [38]. In another genetic background, Dnmt3a mutation cooperates with NRAS mutation to transform hematopoietic stem/progenitor cells and induce acute leukemia development [39].

DNMT3A functions primarily at the stem cell level. Dnmt3a is highly expressed in mice HSC, where its loss promotes HSC self-renewal and inhibits differentiation [40], the mechanism of which was proposed as the interruption of downstream regulators such as RUNX1 [41]. HSC acquires a DNMT3A mutation prime a preleukemic clone to promote the impact of secondary hits, such as FLT3-ITD, to initiate myeloid or lymphoid leukemia. The underlying mechanism of DNMT3A in leukemogenesis is intricate. Despite the global 5mC level in HSC is unaffected, DNMT3A mutation induces focal epigenetic alteration. CpG hypomethylations are significantly enriched at gene regulatory sites, notably, putative enhancers marked by H3K4me1 as well as binding sites of master hematopoietic transcription factors [39]. DNMT3A R882H directly binds to and potentiates transactivation of stemness genes for leukemogenicity, including a Meis1-Mn1-Hoxa regulatory node, which encodes a set of transcription factors and cofactors crucial for sustaining self-renewal of LSCs [39, 42, 43]. Demethylation increases the representation of transcription factors, such as FLI1, GFI1B, and PU.1, which determine cell fate during stem, myeloid, and lymphoid cell differentiation [38]. In DNMT3A-mutated CMML, disturbed DNA methylation – either hypomethylation or hypermethylation – was detected. Most hematopoiesis-related genes, such as Hoxa3 and Hoxa6, were hypomethylated in the gene body, while genes that regulate lymphocyte development, such as Notch1 and Notch3, have shown more hypermethylated status in gene body regions. Thus, the relative balance of transcriptional controls between proliferation and differentiation of hematopoietic cells may be broken [37]. In addition, DNA methylation-independent manner of mutated DNMT3A is explored as well. Aberrant increase of CDK1 in the presence of the DNMT3A mutant can promote cell-cycle activity [37]. Junji Koya revealed aberrant recruitment of polycomb repressive complex 1 (PRC1) to specific differentiation-associated gene loci, such as PU.1 and Cebpa, resulting in the differentiation block of LSCs [44].

8.3.2 TET2 (Ten-Eleven Translocation 2)

While DNA methylation is carried out by DNMT enzymes, the removal of methyl group and conversion to unmodified state are catalyzed by TET gene family. TET2 is the only mutated gene among TET family members reported in a variety of human hematological malignancies such as leukemia, myelodysplastic syndrome (MDS), and malignant lymphoma, with nearly 8–23% of adult AML patients carrying such mutation.

TET2 mutation alone is not sufficient for disease formation but rather leads to a generation of premalignant clones, setting a critical basis for malignant transformation. Additional mutations are required for developing full-blown diseases. Most Tet2-mutant mice develop CMML-like myeloid disease or T-cell lymphoma after a long latency, usually more than several months [45, 46]. The combination of Tet2 disruption and AML1-ETO expression can induce AML [47].

TET2 is critical for the function of hematopoietic stem cells. Disruption of TET2 results in the expansion of multipotent as well as myeloid progenitors, leading to the accumulation of pre-LSCs [48]. However, the detailed mechanism of how TET2 mutations propagate preleukemic and leukemic states remains poorly understood. Study showed that the catalytic activity of TET2 was essential and critical for its myeloid malignancy-suppressive function in HSC/HPCs [49]. Thus, DNA demethylation processes mediated by the disruption of TET2 suggest a causative role of TET2 in disease initiation. Hypermethylation at non-CpG islands is reported in TET2 mutation in multiple studies [50, 51]. TET2 disruption predominantly affects the epigenetic state of enhancers. Hypermethylation at enhancers can be directly linked to deregulation of specific genes that in turn results in oncogenic transformation, including downregulated putative tumor suppressor genes such as Mtss1, Las2, Lxn, Ctdspl, and Grap2 and upregulated putative oncogenes such as Aff3, Pim2, Nepn, Notch3, and Igf1r [47]. Disruption of TET2 activity confers enhanced self-renewal in HSC resulting in the expansion advantage in the mutated clone to form pre-LSCs population. Moreover, sustained Tet2 deficiency is required to maintain self-renewal of pre-LSCs, since its effects are reversed upon Tet2 restoration [52].

8.3.3 IDH1/2

Isocitrate dehydrogenases (IDHs) are metabolic enzymes that catalyze the oxidative decarboxylation of isocitrate, producing alpha-ketoglutarate, a critical step in the citric acid cycle. However, because alpha-ketoglutarate is the critical cofactor of multiple epigenetic regulatory factors, including TET and Jumonji C domain-containing histone demethylases, IDHs can also affect the epigenetic regulations beyond metabolism. Surprisingly, IDH1 and IDH2 have been found to be mutated in about 20% AML and many other human cancers [9]. IDH mutations are associated with intermediated prognosis and more frequent in normal karyotype AML. Majority of these mutations are on IDH1 R132, IDH2 R140, and R172. These mutations not only disrupt the normal catalytic function of IDHs to generate alpha-ketoglutarate from isocitrate but consume alpha-ketoglutarate to produce a new metabolite 2-hydroxyglutarate [53, 54]. It turns out that this new metabolite greatly impairs the functions of the epigenetic factors TET and histone demethylases which require alpha-ketoglutarate as cofactor. 2-Hydroxyglutarate leads to DNA hypermethylation, abnormal histone methylations, and, subsequently, aberrant gene expressions and HSC dysregulated self-renewal and differentiation block [55, 56]. Thus 2-hydroxyglutarate is an onco-metabolite, and IDH mutations provide a direct unique link between cancer metabolism and epigenetics.

IDH1 R132H knockin mice have increased numbers of HSC and progenitor cells, suggesting this mutation impairs the differentiation of HSC and early progenitor cells. Probably due to the defects of these mutant HSCs and progenitors, these mice also have anemia and extramedullary hematopoiesis [57]. However, IDH mutations themselves can't lead to full-blown AML. IDH mutations frequently occur with NPM1, DNMT3A, and FLT3 mutations. Our work, together with others' work, has shown that IDH2 R140 or R172 mutant, cooperating with FLT3ITD or NRASG12D, can promote leukemogenesis by blocking the differentiation of HSC and progenitors [58]. IDH1 R132 or IDH2 R140 mutations can also cooperate with JAK2 mutations to promote myeloproliferative neoplasm progression [59]. Further, we showed that pharmacologic or genetic inhibition of IDH2 triggers the differentiation and death of AML cells, albeit only with prolonged IDH2 inhibition. Inhibition of the bromodomain-containing protein Brd4 triggers rapid differentiation and death of IDH2 mutant AML. Combined inhibition of JAK2 and IDH2 in JAK2 and IDH2 double mutated NPM can normalize the hematopoiesis and impede MPN progression. Thus, IDH mutations are putative AML drivers and also therapeutic target in diseases with these mutations. Indeed, the first IDH mutant inhibitor has been approved for AML recently.

8.4 Histone Modifications

Despite the great heterogeneity among leukemia, abnormal histone modifications and mutations in histone modifiers are common in these diseases. Recently cancer genomics studies show that more than 30% AML had mutations on chromatin modifiers, such as MLL fusions, ASXL1, EZH2, and KDM6A. MLL fusions are among the most studied histone modifiers in AML [9].

8.4.1 MLL Fusions

MLL1, located at chromosome 11q23, is among the most frequently rearranged genes in both pediatric and adult leukemias, associated with intermediate prognosis. MLL fusions drive leukemia by regulating HOX gene expressions through histone modifications and also direct promoter binding as transcriptional factor. Hox genes, such as HOX B4 and A9, are important regulators of hematopoiesis and act, in part, by promoting stem cell renewal. MEIS1, the common transcriptional factor for HOX genes, is also a direct target of MLL-fusion proteins and involved in AML pathology. The serine-threonine-rich sequence of MLL in the fusion proteins is required for HSC transformation by DNA binding. Most MLL-fusion proteins have gain of functions that the truncated MLL as a transcriptional activator directed by its partners. MLL fusions can have numerous partners with distinct functions. ENL and AF9 can recruit SWI-SNF chromatin-remodeling complexes. And MLL-fusion transcriptional activity to regulate the expressions of HOX genes is SWI-SNF dependent [60].

DOT1L, a histone H3K79 methyltransferase, forms a complex with multiple MLL fusions and is required for the aberrant transcriptional activity of these MLL-fusion proteins. Inhibition of DOT1L can reduce the expressions of HOX genes and further inhibit the transformation capacity of MLL fusions [61, 62]. Thus, DOT1L can be a potential therapeutic target in AML with MLL fusions.

Interestingly, 10% of adult AML with normal cytogenetics have internal tandem duplications of MLL. This duplication can alter the DNA binding capacity of MLL, which is required for its transformation activity. However, MLL PTD doesn't lead to upregulate HOX genes, and the functional targets of MLL PTD are still largely unknown [63]. MLL copy number variations are also observed in some MDS and AML. AML with MLL amplifications shares some dysregulated gene expression signature, including HOX genes, with those with MLL fusions, suggesting similar molecular mechanisms of transformation [64]. But how wild-type MLL in MLL-amplified AML is recruited to its targets is still unclear.

8.4.2 EZH2 (Enhancer of Zeste Homolog 2)

EZH2, the catalytic subunit of polycomb repressive complex (PRC2), trimethylates H3K27 to promote transcriptional repression of target genes, which plays a major role in normal hematopoiesis. Aberrant expression of EZH2 differentially contributes to the tumor initiation in different types of leukemia. EZH2 acts as a double facet factor either as a tumor suppressor or oncogene in acute and chronic myeloid and/or lymphoid leukemias. Overexpressing of EZH2 in many different malignancies and mutations that elevate its trimethyltransferase activity is found in subtypes of lymphoma, suggesting a pro-tumorigenic role. However, EZH2 loss-of-function mutations have also been observed in MDS, suggesting a tumor suppressor function in certain cellular contexts.

Comparing the gene expression profiles of human CML LSC to normal HSC, EZH2 is found to be upregulated in LSC at all three phases of disease. It was reported that the survival of CML LSC relies on the expression of EZH2, and more specifically its enzymatic activity, since inactivation of EZH2 affects the survival of LSC and prevents disease from initiation and maintenance. Expression of the EZH2 homolog EZH1 is reduced in EZH2-deficient CML LSC, creating a context resembling complete loss of PRC2 [65]. The mechanism by which EZH2 regulates stemness of LSC is proposed as EZH2 binds to and trimethylates H3K27 at the promoter of PTEN, thus repressing its gene transcription and, meanwhile, activating its major downstream AKT/mTOR signaling [66].

Leukemia generated from fetal liver (FL) and adult bone marrow (BM) differed dramatically in their LSC activity with FL leukemia characterized by relatively poorer transplantability and more rapid cycling. EZH2 plays a critical role in this phenomenon. FL leukemia possesses robust NOTCH1-driven autocrine IGF1 signaling, which drives cells into cycling and thereby depletes the LSC compartment. However, the IGF1 pathway is restricted in BM leukemia by EZH2-dependent H3K27 trimethylation. Therefore, EZH2-mediated epigenetic downregulation of IGF1 signaling promotes LSC activity [67].

Other mechanisms concerning the requirement of EZH2 in LSC are proposed as well. The following listed pathways involving in the LSC initiation and maintenance are all regulated by the histone-modifying function of EZH2. Koki Ueda et al. and J Shi et al. found that inhibition of EZH2 depletes LSC of MLL-fusion leukemia through upregulation of Cdkn2a and its product p16 [68, 69]. The oncogenic function of Ezh2 in AML indicated that besides the aforementioned pathway, Ezh2 plays a critical role in the maintenance of LSC by reinforcing their differentiation blockage through the regulation of its direct targets including EGR1, which is a positive regulator of differentiation of myeloid cells and a tumor suppressor that transactivates tumor suppressor genes, including TGFβ1, PTEN, and p53 [70].

On the other hand, the tumor suppressor role of EZH2 is analyzed as well. A study performed in a murine model demonstrated that PRC2, along with MLL-AF9, is responsible for blockade of myeloid differentiation in AML [68]. In the model of ETP-ALL generated by deleting EZH2 in p53-null hematopoietic cells, EZH2 loss propagates hypermethylation at T-cell differentiation-regulating genes to promote leukemic transformation [71].

8.5 COMPASS Family (Complex of Proteins Associated with Set1)

In human, COMPASS family consists of six macromolecular complexes which are capable of methylating H3K4. The core components – Set1a/b, MLL1/MLL2, and MLL3/MLL4 – acting as scaffolds of their respective complexes, are critical for the enzymatic activity. MLL1/MLL2 are necessary for gene-specific H3K4 trimethylation, while MLL3/MLL4 have been accredited as key H3K4 monomethyltransferases at enhancers. MLL3/MLL4 associates with WRAD (WDR5, RbBP5, ASH2L, and DPY30), NCOA6, PTIP, PA1, and UTX in its own separate protein complex. MLL1/MLL2/MLL3/MLL4 together with UTX have been documented as the most frequently mutated histone modifiers in human cancers [72].

A model of AML induced by the MLL-AF10 oncogene was established by Wong et al. to interrogate the LSC epigenome. They reported that c-kit+ cells enriched for LSCs were maintained in a global epigenetic state characterized by relative H3K4 hypermethylation and H3K79 hypomethylation. H3K4 hypermethylation is the major epigenetic alteration that sustains the LSC maintenance transcriptional program and the expression of Hoxa and Meis1 genes, which are direct targets of MLL oncoproteins required for AML pathogenesis. Conversely, global erasure of H3K4me3 is associated with LSC terminal differentiation [73].

MLL3 (also known as KMT2C) is validated as a haploinsufficient 7q tumor suppressor in AML in our study. MLL3 is required for efficient differentiation of HSPC from long-term hematopoietic stem cells (LT-HSC) to short-term HSC (ST-HSC), multipotent progenitors (MPP), and more committed myeloid progenitors. Genes differentially expressed in shMll3;p53 −/− HSPC suggest that Mll3 suppression affects a LSC signature that is linked to Myc activity. Mll3 suppression produces a repressive chromatin context on genes linked to differentiation [74].

8.6 Epigenetic Therapy

AML is the most prevalent acute leukemia in adults. With the exception of a few specific AML subtypes, the mainstay of treatment for the majority of AML subtypes has not significantly changed over the last 20 years and continues to be based on standard cytotoxic chemotherapy. As a result, clinical outcome remains poor for the majority of patients, with overall long-term survival in the region of 20–30%, posing the demand for novel therapies [75].

Leukemia stem cells initiate hematopoietic malignancies, and epigenetic alterations are early events during LSC formation. LSCs are resistant to conventional chemotherapy, and persistent LSCs, which are insensitive to upfront regimens, are the root of resistance and relapse. Patients harboring epigenetic mutations are usually associated with poor prognosis. Therefore, therapies aimed at eradicating LSCs are supposed to be imperative for curing leukemia. Nowadays, significant progress has been made in the exploring of LSCs-targeted therapies. Monoclonal antibodies targeting LSC surface markers, agents blocking the interactions between LSCs and their microenvironment, and agents targeting signaling pathways involved in the regulation of survival and self-renewal of LSCs have been either approved or investigated in preclinical and clinical trials [76]. Typically, epigenetic genes have been considered as potentially attractive targets for new therapies, due to the reversible nature of epigenetic signatures.

Two DNMT inhibitors, azacitidine (5-azacytidine) and decitabine (5-aza-2′-deoxycytidine), have already been approved by the Food and Drug Administration for the treatment of patients with MDS and are also being routinely used for AML based on results from several clinical trials. AML patients with DNMT3A mutation are found to be associated with increased response to azanucleoside DNMT inhibitors. One study reported that 75% of patients with DNMT3A mutations achieve complete response (CR), while the CR rate in patients with wild-type DNMT3A was 34% [77]. The mechanisms underlying the inhibition of DNA methylation by Aza and Aza-dC are mainly focused on their apoptotic- and differentiation-inducing effects on LSC [78]. For instance, 5-AZA-dC upregulates the expression of DIF2 through the demethylation of its promoter and thus facilitates TNFα-induced transcriptional pathways leading to differentiation [79]. Similarly, AZA treatment significantly demethylates upstream regulatory element within PU.1 promoter, leading to upregulation of PU.1, followed by derepression of its transcriptional targets and onset of myeloid differentiation [80]. Another study showed that Aza-dC induces expression of OLFM4, of which an overexpression leads to apoptosis and differentiation in HL60 cell line [81].

Apart from the DNA methylome, histone methylome is another major epigenetic modification as the target in epigenetic drugs (Fig. 8.1). HDAC inhibitors were among the first epigenetic drugs developed, and several have been tested in clinical trials for malignancies including AML, such as valproic acid (VPA), panobinostat, and vorinostat. HDAC inhibitors alone have shown limited clinical activity in MDS and AML. Clinical trials evaluating the combination of HDACi with DNMTi are still ongoing. EZH2 was highly overexpressed in primary human CML CD34 + cells. Study showed that targeting EZH2 by either small molecule GSK126 or specific shRNA displayed a remarkable inhibitory effect on LSCs and drastically prolonged the survival of CML mice. EZH2 knockdown resulted in elevation of PTEN and led to impaired recruitment of EZH2 and H3K27me3 on the promoter of PTEN gene [66]. Another study found that EZH2 inhibition not only suppressed MLL-fusion leukemia proliferation but also reduced LSC frequency through upregulation of p16 [69]. Tyrosine kinase inhibitor (TKI) combined with EZH2 inhibitor significantly upregulates H3K27me3 level compared to either TKI or EZH2i alone, resulting in

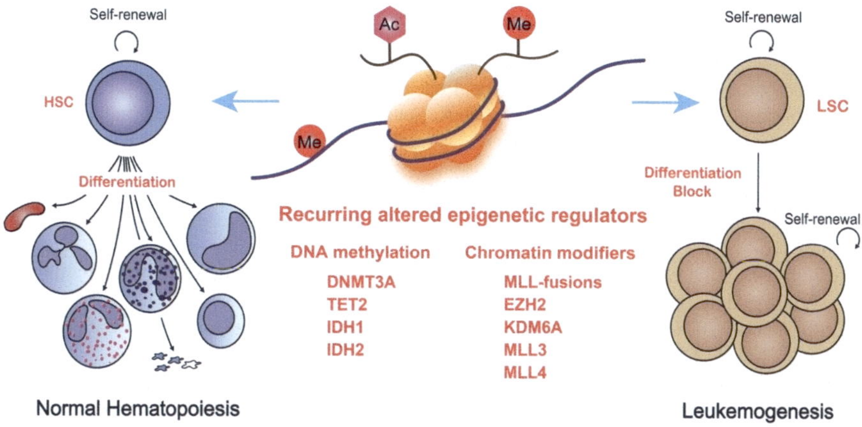

Fig. 8.1 Epigenetic regulation of LSCs. LSCs undergo epigenetic changes during leukemogenesis, providing new therapeutic strategies for leukemias

further loss of LSCs. Combined effects provide an approach to eradicate TKI-persistent LSCs in refractory patients with CML [82].

8.7 Perspective

Since LSCs were firstly found in AML in 1997, extensive studies have been contributed to identification and characterization of such cell populations. During the course of malignant transformation, both genetic and epigenetic abnormalities play an essential role in the pathogenesis. However, despite the genetic programs that maintain LSCs self-renewal have been greatly explored, the comprehensive epigenetic landscape that sustains LSCs' cellular identity and functionality is not clear yet. Epigenetic-targeting therapies raise the possibility of eliminating LSC, but these are still early day for epigenetic therapies.

On the one hand, further studies are needed to shed light on the role of epigenetic modifications in leukemia and leukemia stem cells. Preclinical and basic research are required in order to define exact consequences of epigenetic regulator mutations that have been discovered in AML and their molecular mechanism in leukemia. On the other hand, efficient LSC-based treatments of leukemia remain to be explored which can eventually benefit leukemia patients. The next decade will see unprecedented activity in preclinical and clinical investigation of epigenome-based therapies. More efforts have to be made and the prospects are very exciting.

References

1. Waddington CH (1942) The epigenotype. Endeavour 1:18–20
2. Arrowsmith CH, Bountra C, Fish PV, Lee K, Schapira M (2012) Epigenetic protein families: a new frontier for drug discovery. Nat Rev Drug Discov 11:384–400. https://doi.org/10.1038/nrd3674
3. Chi P, Allis CD, Wang GG (2010) Covalent histone modifications – miswritten, misinterpreted and mis-erased in human cancers. Nat Rev Cancer 10:457–469. https://doi.org/10.1038/nrc2876
4. Strahl BD, Allis CD (2000) The language of covalent histone modifications. Nature 403:41–45. https://doi.org/10.1038/47412
5. Shi Y et al (2004) Histone demethylation mediated by the nuclear amine oxidase homolog LSD1. Cell 119:941–953. https://doi.org/10.1016/j.cell.2004.12.012
6. Lindroth AM et al (2001) Requirement of CHROMOMETHYLASE3 for maintenance of CpXpG methylation. Science (New York, NY) 292:2077–2080. https://doi.org/10.1126/science.1059745
7. Weber M et al (2007) Distribution, silencing potential and evolutionary impact of promoter DNA methylation in the human genome. Nat Genet 39:457–466. https://doi.org/10.1038/ng1990
8. Tamaru H, Selker EU (2001) A histone H3 methyltransferase controls DNA methylation in Neurospora crassa. Nature 414:277–283. https://doi.org/10.1038/35104508
9. Cancer Genome Atlas Research, N et al (2013) Genomic and epigenomic landscapes of adult de novo acute myeloid leukemia. N Engl J Med 368:2059–2074. https://doi.org/10.1056/NEJMoa1301689
10. Zhang J et al (2012) The genetic basis of early T-cell precursor acute lymphoblastic leukaemia. Nature 481:157–163. https://doi.org/10.1038/nature10725
11. Ye H, Wolf RA, Kurz T, Corr PB (1994) Phosphatidic acid increases in response to noradrenaline and endothelin-1 in adult rabbit ventricular myocytes. Cardiovasc Res 28:1828–1834
12. Trowbridge JJ, Snow JW, Kim J, Orkin SH (2009) DNA methyltransferase 1 is essential for and uniquely regulates hematopoietic stem and progenitor cells. Cell Stem Cell 5:442–449. https://doi.org/10.1016/j.stem.2009.08.016
13. Challen GA et al (2011) Dnmt3a is essential for hematopoietic stem cell differentiation. Nat Genet 44:23–31. https://doi.org/10.1038/ng.1009
14. Challen GA et al (2014) Dnmt3a and Dnmt3b have overlapping and distinct functions in hematopoietic stem cells. Cell Stem Cell 15:350–364. https://doi.org/10.1016/j.stem.2014.06.018
15. Moran-Crusio K et al (2011) Tet2 loss leads to increased hematopoietic stem cell self-renewal and myeloid transformation. Cancer Cell 20:11–24. https://doi.org/10.1016/j.ccr.2011.06.001
16. Pan F et al (2017) Tet2 loss leads to hypermutagenicity in haematopoietic stem/progenitor cells. Nat Commun 8:15102. https://doi.org/10.1038/ncomms15102
17. Beerman I et al (2013) Proliferation-dependent alterations of the DNA methylation landscape underlie hematopoietic stem cell aging. Cell Stem Cell 12:413–425. https://doi.org/10.1016/j.stem.2013.01.017
18. Buscarlet M et al (2017) DNMT3A and TET2 dominate clonal hematopoiesis and demonstrate benign phenotypes and different genetic predispositions. Blood 130:753–762. https://doi.org/10.1182/blood-2017-04-777029
19. Busque L et al (2012) Recurrent somatic TET2 mutations in normal elderly individuals with clonal hematopoiesis. Nat Genet 44:1179–1181. https://doi.org/10.1038/ng.2413
20. Xie M et al (2014) Age-related mutations associated with clonal hematopoietic expansion and malignancies. Nat Med 20:1472–1478. https://doi.org/10.1038/nm.3733
21. Park IK et al (2003) Bmi-1 is required for maintenance of adult self-renewing haematopoietic stem cells. Nature 423:302–305. https://doi.org/10.1038/nature01587

22. Iwama A et al (2004) Enhanced self-renewal of hematopoietic stem cells mediated by the polycomb gene product Bmi-1. Immunity 21:843–851. https://doi.org/10.1016/j.immuni.2004.11.004
23. Shao L et al (2011) Reactive oxygen species and hematopoietic stem cell senescence. Int J Hematol 94:24–32. https://doi.org/10.1007/s12185-011-0872-1
24. Vo LT et al (2018) Regulation of embryonic haematopoietic multipotency by EZH1. Nature 553:506–510. https://doi.org/10.1038/nature25435
25. Mochizuki-Kashio M et al (2011) Dependency on the polycomb gene Ezh2 distinguishes fetal from adult hematopoietic stem cells. Blood 118:6553–6561. https://doi.org/10.1182/blood-2011-03-340554
26. Xie H et al (2014) Polycomb repressive complex 2 regulates normal hematopoietic stem cell function in a developmental-stage-specific manner. Cell Stem Cell 14:68–80. https://doi.org/10.1016/j.stem.2013.10.001
27. Jude CD et al (2007) Unique and independent roles for MLL in adult hematopoietic stem cells and progenitors. Cell Stem Cell 1:324–337. https://doi.org/10.1016/j.stem.2007.05.019
28. Wada T, Koyama D, Kikuchi J, Honda H, Furukawa Y (2015) Overexpression of the shortest isoform of histone demethylase LSD1 primes hematopoietic stem cells for malignant transformation. Blood 125:3731–3746. https://doi.org/10.1182/blood-2014-11-610907
29. Stewart MH et al (2015) The histone demethylase Jarid1b is required for hematopoietic stem cell self-renewal in mice. Blood 125:2075–2078. https://doi.org/10.1182/blood-2014-08-596734
30. Ntziachristos P et al (2014) Contrasting roles of histone 3 lysine 27 demethylases in acute lymphoblastic leukaemia. Nature 514:513–517. https://doi.org/10.1038/nature13605
31. Figueroa ME et al (2010) DNA methylation signatures identify biologically distinct subtypes in acute myeloid leukemia. Cancer Cell 17:13–27. https://doi.org/10.1016/j.ccr.2009.11.020
32. Chaudry SF, Chevassut TJ (2017) Epigenetic guardian: a review of the DNA methyltransferase DNMT3A in acute myeloid leukaemia and clonal haematopoiesis. Biomed Res Int 2017:5473197
33. Ley TJ et al (2010) DNMT3A mutations in acute myeloid leukemia. N Engl J Med 363:2424–2433
34. Ahn J-S et al (2016) DNMT3A R882 mutation with FLT3-ITD positivity is an extremely poor prognostic factor in patients with normal-karyotype acute myeloid leukemia after allogeneic hematopoietic cell transplantation. Biol Blood Marrow Transplant 22:61–70
35. Celik H et al (2015) Enforced differentiation of Dnmt3a-null bone marrow leads to failure with c-Kit mutations driving leukemic transformation. Blood 125:619–628
36. Mayle A et al (2015) Dnmt3a loss predisposes murine hematopoietic stem cells to malignant transformation. Blood 125:629–638
37. Xu J et al (2014) DNMT3A Arg882 mutation drives chronic myelomonocytic leukemia through disturbing gene expression/DNA methylation in hematopoietic cells. Proc Natl Acad Sci 111:2620–2625
38. Weiskopf K et al (2016) CD47-blocking immunotherapies stimulate macrophage-mediated destruction of small-cell lung cancer. J Clin Invest 126:2610–2620. https://doi.org/10.1172/JCI81603
39. Lu R et al (2016) Epigenetic perturbations by Arg882-mutated DNMT3A potentiate aberrant stem cell gene-expression program and acute leukemia development. Cancer Cell 30:92–107
40. Jeong M et al (2018) Loss of Dnmt3a immortalizes hematopoietic stem cells in vivo. Cell Rep 23:1–10
41. Challen GA et al (2012) Dnmt3a is essential for hematopoietic stem cell differentiation. Nat Genet 44:23
42. Ferreira HJ et al (2016) DNMT3A mutations mediate the epigenetic reactivation of the leukemogenic factor MEIS1 in acute myeloid leukemia. Oncogene 35:3079
43. Qu Y et al (2014) Differential methylation in CN-AML preferentially targets non-CGI regions and is dictated by DNMT3A mutational status and associated with predominant hypomethylation of HOX genes. Epigenetics 9:1108–1119

44. Koya J et al (2016) DNMT3A R882 mutants interact with polycomb proteins to block haematopoietic stem and leukaemic cell differentiation. Nat Commun 7:10924
45. Moran-Crusio K et al (2011) Tet2 loss leads to increased hematopoietic stem cell self-renewal and myeloid transformation. Cancer Cell 20:11–24
46. Muto H et al (2014) Reduced TET2 function leads to T-cell lymphoma with follicular helper T-cell-like features in mice. Blood Cancer J 4:e264
47. Rasmussen KD et al (2015) Loss of TET2 in hematopoietic cells leads to DNA hypermethylation of active enhancers and induction of leukemogenesis. Genes Dev 29:910–922
48. Nakajima H, Kunimoto H (2014) TET2 as an epigenetic master regulator for normal and malignant hematopoiesis. Cancer Sci 105:1093–1099
49. Zhao Z et al (2016) The catalytic activity of TET2 is essential for its myeloid malignancy-suppressive function in hematopoietic stem/progenitor cells. Leukemia 30:1784
50. Ko M et al (2010) Impaired hydroxylation of 5-methylcytosine in myeloid cancers with mutant TET2. Nature 468:839
51. Yamazaki J et al (2012) Effects of TET2 mutations on DNA methylation in chronic myelomonocytic leukemia. Epigenetics 7:201–207
52. Cimmino L et al (2017) Restoration of TET2 function blocks aberrant self-renewal and leukemia progression. Cell 170:1079–1095. e1020
53. Dang L et al (2009) Cancer-associated IDH1 mutations produce 2-hydroxyglutarate. Nature 462:739–744. https://doi.org/10.1038/nature08617
54. Lu C et al (2012) IDH mutation impairs histone demethylation and results in a block to cell differentiation. Nature 483:474–478. https://doi.org/10.1038/nature10860
55. Ward PS et al (2010) The common feature of leukemia-associated IDH1 and IDH2 mutations is a neomorphic enzyme activity converting alpha-ketoglutarate to 2-hydroxyglutarate. Cancer Cell 17:225–234. https://doi.org/10.1016/j.ccr.2010.01.020
56. Ward PS et al (2012) Identification of additional IDH mutations associated with oncometabolite R(−)-2-hydroxyglutarate production. Oncogene 31:2491–2498. https://doi.org/10.1038/onc.2011.416
57. Sasaki M et al (2012) IDH1(R132H) mutation increases murine haematopoietic progenitors and alters epigenetics. Nature 488:656–659. https://doi.org/10.1038/nature11323
58. Chen C et al (2013) Cancer-associated IDH2 mutants drive an acute myeloid leukemia that is susceptible to Brd4 inhibition. Genes Dev 27:1974–1985. https://doi.org/10.1101/gad.226613.113
59. McKenney AS et al (2018) JAK2/IDH-mutant-driven myeloproliferative neoplasm is sensitive to combined targeted inhibition. J Clin Invest 128:789–804. https://doi.org/10.1172/JCI94516
60. Krivtsov AV, Armstrong SA (2007) MLL translocations, histone modifications and leukaemia stem-cell development. Nat Rev Cancer 7:823–833. https://doi.org/10.1038/nrc2253
61. Bernt KM et al (2011) MLL-rearranged leukemia is dependent on aberrant H3K79 methylation by DOT1L. Cancer Cell 20:66–78. https://doi.org/10.1016/j.ccr.2011.06.010
62. Daigle SR et al (2013) Potent inhibition of DOT1L as treatment of MLL-fusion leukemia. Blood 122:1017–1025. https://doi.org/10.1182/blood-2013-04-497644
63. Dohner K et al (2002) Prognostic significance of partial tandem duplications of the MLL gene in adult patients 16 to 60 years old with acute myeloid leukemia and normal cytogenetics: a study of the acute myeloid leukemia study group Ulm. J Clin Oncol 20:3254–3261. https://doi.org/10.1200/JCO.2002.09.088
64. Pajuelo-Gamez JC et al (2007) MLL amplification in acute myeloid leukemia. Cancer Genet Cytogenet 174:127–131. https://doi.org/10.1016/j.cancergencyto.2006.11.019
65. Xie H et al (2016) Chronic myelogenous leukemia initiating cells require Polycomb group protein EZH2. Cancer Discov 6:1237–1247. CD-15-1439
66. Zhou J et al (2018) PTEN is fundamental for elimination of leukemia stem cells mediated by GSK126 targeting EZH2 in chronic myelogenous leukemia. Clin Cancer Res 24:145–157
67. Giambra V et al (2018) Epigenetic restoration of fetal-like IGF1 signaling inhibits leukemia stem cell activity. Cell Stem Cell 23:714–726. e717

68. Shi J et al (2013) The Polycomb complex PRC2 supports aberrant self-renewal in a mouse model of MLL-AF9; Nras G12D acute myeloid leukemia. Oncogene 32:930
69. Ueda K et al (2014) Inhibition of histone methyltransferase EZH 2 depletes leukemia stem cell of mixed lineage leukemia fusion leukemia through upregulation of p16. Cancer Sci 105:512–519
70. Tanaka S et al (2012) Ezh2 augments leukemogenecity by reinforcing differentiation blockage in acute myeloid leukemia. Blood 120:1107–1117. blood-2011-2011-394932
71. Wang C et al (2018) Ezh2 loss propagates hypermethylation at T cell differentiation–regulating genes to promote leukemic transformation. J Clin Invest 128:3872–3886
72. Kandoth C et al (2013) Mutational landscape and significance across 12 major cancer types. Nature 502:333
73. Wong SH et al (2015) The H3K4-methyl epigenome regulates leukemia stem cell oncogenic potential. Cancer Cell 28:198–209
74. Chen C et al (2014) MLL3 is a haploinsufficient 7q tumor suppressor in acute myeloid leukemia. Cancer Cell 25:652–665
75. Liesveld J (2012) Management of AML: who do we really cure? Leuk Res 36:1475–1480
76. Wang X, Huang S, Chen J-L (2017) Understanding of leukemic stem cells and their clinical implications. Mol Cancer 16:2
77. Metzeler KH et al (2012) DNMT3A mutations and response to the hypomethylating agent decitabine in acute myeloid leukemia. Leukemia 26:1106
78. Pinto A et al (1984) 5-Aza-2′-deoxycytidine induces terminal differentiation of leukemic blasts from patients with acute myeloid leukemias. Blood 64:922–929
79. Laurenzana A et al (2009) Inhibition of DNA methyltransferase activates tumor necrosis factor α–induced monocytic differentiation in acute myeloid leukemia cells. Cancer Res 69:55–64
80. Curik N et al (2012) 5-azacitidine in aggressive myelodysplastic syndromes regulates chromatin structure at PU. 1 gene and cell differentiation capacity. Leukemia 26:1804
81. Liu W, Lee HW, Liu Y, Wang R, Rodgers GP (2010) Olfactomedin 4 is a novel target gene of retinoic acids and 5-aza-2′-deoxycytidine involved in human myeloid leukemia cell growth, differentiation, and apoptosis. Blood 116:4938–4947. blood-2009-2010-246439
82. Scott MT et al (2016) Epigenetic reprogramming sensitizes CML stem cells to combined EZH2 and tyrosine kinase inhibition. Cancer Discov 6:1248–1257

Chapter 9
Leukemia Stem Cells in Chronic Myeloid Leukemia

Yi Shan, Ngoc DeSouza, Qiang Qiu, and Shaoguang Li

Contents

Abstract Chronic myeloid leukemia (CML) is a clonal myeloproliferative disorder characterized by a chromosome translocation that generates the BCR-ABL oncogene encoding a constitutively activated tyrosine kinase. Although BCR-ABL tyrosine kinase inhibitors (TKIs) are highly effective in treating CML at chronic phase, a number of patients develop drug resistance due to the inability of TKIs to kill leukemia stem cells (LSCs). Similar to other types of hematopoietic malignancies, LSCs in CML are believed to be a rare cell population responsible for leukemia initiation, disease progression, and drug resistance. Therefore, a full understanding of the biology of LSCs will help to develop novel therapeutic strategies for effective treatment of CML to possibly reach a cure. In recent years, a significant progress has been made in studying the biology of LSCs in both animal models and human patients at cellular and molecular levels, providing a basis for designing and testing potential molecular targets for eradicating LSCs in CML.

Keywords CML · Leukemia stem cell · TKIs · Therapeutic targets

Y. Shan · N. DeSouza · Q. Qiu · S. Li (✉)
Division of Hematology/Oncology, Department of Medicine, University of Massachusetts Medical School, Worcester, Massachusetts, USA
e-mail: shaoguang.li@umassmed.edu

H. Zhang, S. Li (eds.), *Leukemia Stem Cells in Hematologic Malignancies*, Advances in Experimental Medicine and Biology 1143,
https://doi.org/10.1007/978-981-13-7342-8_9

9.1 Introduction

Chronic myeloid leukemia (CML) is a myeloproliferative neoplasm, characterized by clonal expansion and accumulation of differentiated myeloid cells in the bone marrow and blood [82]. CML usually begins in a chronic phase with leukemia cells in the bone marrow and then, over time, spreads to the blood. If the disease is not controlled, leukemia cells will eventually spread to other areas of the body. Initially, CML cells spreads and grows slowly but can change from a slowly progressing into a rapidly growing form of acute leukemia that can spread and invade almost all major organs in the body. Specifically, CML is a triphasic disease, initiating from a chronic phase (CP) and then progressing to an intermediate accelerated phase (AP) and a terminal blastic phase (BP) (Fig. 9.1a). In CP, the myeloid cell compartment expands, and the cells still retain the capacity to differentiate. In general, CML patients in CP have mild symptoms with many patients remaining asymptomatic. In AP, patients begin to show more immature cells in the blood with frequent constitutional symptoms and a reduced sensitivity to therapy. In BP, immature cells become dominant and survival is poor [83]. The American Cancer Society estimates about 8990 new cases will be diagnosed with CML, and about 1140 people will die of CML in the United States in 2019. Overall, about 15% of all new cases of leukemia are CML, and about 1 person in 526 will get CML in their lifetime in the United States.

A hallmark of CML is the existence of "Philadelphia chromosome" resulting from the reciprocal translocation of the *c-ABL* gene on chromosome 9 and the *BCR* gene on chromosome 22 (Fig. 9.1b). The molecular basis of Philadelphia chromosome is the formation of BCR-ABL, an oncogenic fusion gene encoding a constitutively activated ABL tyrosine kinase [71, 81]. In general, tyrosine kinases are important mediators of signaling cascades in cells and play key roles in diverse biological processes including growth, differentiation, and apoptosis of cells in response to internal and external stimuli. Deregulation of protein kinase activity has been shown to play a central role in the pathogenesis of human cancers [46]. In normal cells, the activity of c-ABL tyrosine kinase is typically regulated in an autoinhibitory fashion in native c-ABL due to the binding of its myristoylated N-terminal "cap" (N-cap) to a deep hydrophobic pocket in the C-terminal lobe of the kinase domain. After the formation of the chimeric BCR-ABL fusion protein occurs, the myristoylated N-cap of c-ABL is replaced with a truncated portion of the BCR protein rendering the C-terminal kinase domain active as a consequence of the conformational change [62]. BCR-ABL is the driving force of leukemogenesis in CML and is responsible for initiating and maintaining leukemia phenotype of CML as well as the phosphorylation, activation, and dysregulation of intracellular signaling pathways that regulate the survival and growth of leukemia cells [93]. CML is considered as a paradigm of a disease to be caused and marked by a single acquired chromosomal translocation.

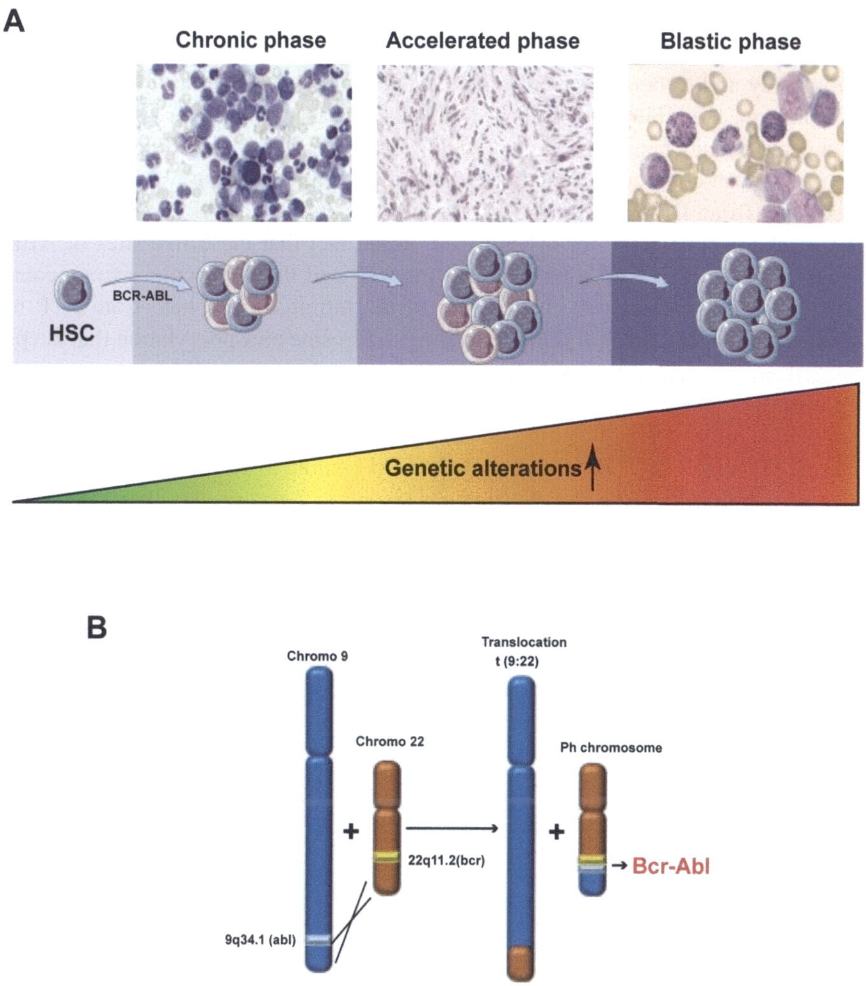

Fig. 9.1 Clinical features of chronic myeloid leukemia. (**a**) Progression of CML from chronic phase (CP) to blastic phase (BP). Biologically, the transition is associated with the accumulation of additional hits in Bcr-Abl itself or in other genes/chromosomes. (**b**) Philadelphia chromosome. A piece of chromosome 9 and a piece of chromosome 22 break off and trade places. The BCR-ABL gene is formed on chromosome 22 where the piece of chromosome 9 attaches. The changed chromosome 22 is called the Philadelphia chromosome

9.2 BCR-ABL Inhibitors: The First-Line Therapy in CML

The standard of care for CML is the BCR-ABL tyrosine kinase inhibitors (TKIs) that are widely used as the first-line agents for CML treatment, because TKIs frequently induce complete hematologic response and dramatically improve patient outcome in chronic phase CML. The advances in biology and therapy of CML have set a gigantic milestone in the history of anticancer precision medicine and targeted therapy. BCR-ABL is one of the first tyrosine kinases to be implicated in a human malignancy and to be successfully targeted. In the late 1990s, the first-generation tyrosine kinase inhibitor, imatinib (IM; also known as "Gleevec" or "Glivec"), was discovered by biochemist Nicholas Lyndon through random screening of a library of thousands of compounds and then was developed into a clinically usable drug by oncologist Brian Druker to treat CML through targeting BCR-ABL kinase activity. The mechanism of action of IM is based on the fact that the active sites of ABL tyrosine kinase each have a binding site for ATP and the enzymatic activity catalyzed by the kinase triggers the transfer of the terminal phosphate from ATP to tyrosine residues on the ABL kinase, leading to tyrosine phosphorylation that keeps the ABL kinase active in signaling and cellular transformation. IM works by binding closely to the ATP binding site, locking the ABL kinase in a closed or self-inhibited conformation, resulting in inhibition of the enzymatic activity of the kinase [24] (Fig. 9.2a). The ultimate result of IM binding to the ABL tyrosine kinase is "switching-off" the downstream signaling pathways that promote leukemogenesis.

The first IM clinical trial took place in 1998 and the drug received FDA approval in 2001. The advent of IM as an effective BCR-ABL kinase inhibitor rapidly and dramatically modified the treatment of CML patients and also led to important changes in the disease management [28]. The initial landmark studies showed high response rates to IM in patients with advanced CML [26] and in a randomized trial involving 1106 CP-CML patients [72]. IM induced complete hematological

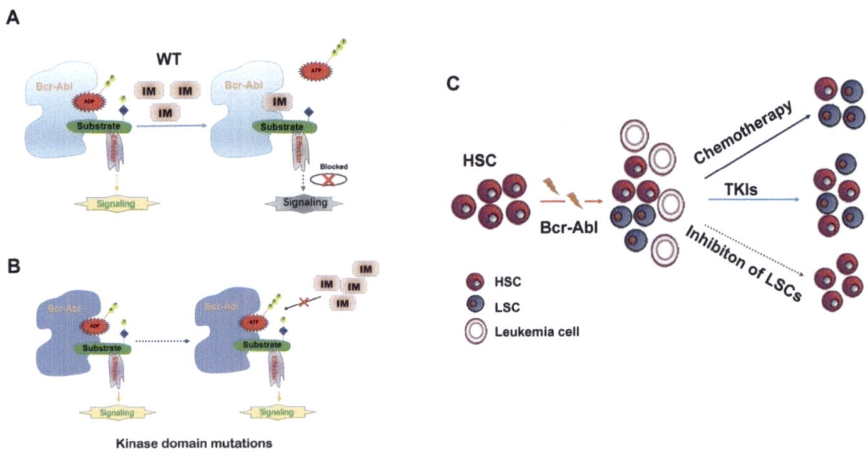

Fig. 9.2 Therapeutic strategies for the treatment of CML. (**a**) Mode of action of the Bcr-Abl kinase inhibitors imatinib (IM). Imatinib blocks the ATP binding to Bcr-Abl, thus inhibiting its phosphorylation and kinase activity. (**b**) When Bcr-Abl kinase domain acquires mutations, imatinib can no longer bind to the kinase domain efficiently to inhibit Bcr-Abl kinase activity, resulting in imatinib resistance. (**c**) Effects of different therapeutic strategies on CML. Chemotherapy and TKIs are effective in killing proliferating Bcr-Abl-positive leukemia cells but fail to eradicate LSCs. Therefore, inhibition of LSCs in CML therapy is the key to curing the disease

response (CHR) in 95.3% of patients and complete cytogenetic response (CCR) in 73.8% of patients. In addition, patients on IM had an improved quality of life [36]. At 6-year follow-up of IRIS trial, IM induced CHR in 98% of patients in chronic phase and CCR in 87% of patients. As a result, IM was described as a "magical bullet" due to its role in converting a fatal cancer into a manageable disease. It is appropriate to say that the use of IM in CML patients revolutionized the treatment of the disease as a consequence of concept breakthrough in targeted therapy of cancer.

Despite a major clinical advance in the treatment of CML through the specific targeting of BCR-ABL kinase activity using TKIs, not all patients experience an optimal response to IM due to a conventional challenging problem associated with the development of drug resistance. In fact, the existence of patients resistant to IM was evident soon after the introduction of the drug into clinical practice. On the other hand, the initial TKI responses were unsatisfactory in patients with advanced-phase (AP) disease, and responses tended to be transient in most responders in AP [84, 92]. Primary resistance is defined as an inability to achieve CHR at 3 months and complete molecular response (CMR) at 6 months. Primary resistance may be caused by differential drug metabolism and/or drug transport, while acquired resistance is defined as progression to advanced disease or loss of response with a five- to tenfold increase in BCR-ABL transcripts. Acquired resistance may be caused by mutations in the BCR-ABL kinase domain, amplification of the BCR-ABL fusion gene, overexpression of drug transporter genes, and overexpression of other tyrosine kinases such as the SRC family kinases [8, 50] (Fig. 9.2b). Consequently, second- and third-generation TKIs, namely, dasatinib, nilotinib, bosutinib, and ponatinib, were rationally designed and approved for clinic use to override the problem of drug resistance. These TKIs are more potent but have been associated with more serious side effects and complications [5]. Patients achieving stable optimal responses to TKI therapy are predicted to have the same life expectancy of the general population. However, it has come to a common recognition that TKIs do not "cure" CML and most CML patients will have to face the perspective of life-long TKI treatment. It is also well established that leukemia stem cells (LSCs) in CML are intrinsically insensitive to TKIs, mainly because their survival does not completely depend on BCR-ABL kinase activity [17] (Fig. 9.2c). LSCs can survive TKI therapy and represent a dangerous reservoir from which TKI resistance and disease relapse may originate. In addition, LSC persistence is thought to be one of the major reasons why treatment-free remission may not be achieved in approximately half of the CML cases [5, 17]. In an attempt to eradicate LSCs, some key molecules/pathways in LSCs have been identified, but the development of new compounds against these molecules/pathways is still ongoing, although we are closer to testing candidate anti-LSC compounds in CML patients.

9.3 LSCs: The Root of CML

The majority of CML patients in CP are highly responsive to TKI therapy that targets BCR-ABL kinase activity. However, a significant number of CML patients treated with TKIs eventually develop drug resistance and relapse. It has been well accepted that LSCs are the root of CML and TKI treatment unlikely leads to a cure of CML due to the ineffectiveness of TKIs in eradicating LSCs. It is necessary to develop new therapies aiming to target LSCs. Typically, stem cells can self-renew to maintain themselves and can also differentiate and proliferate to produce various types of specialized cells [66]. In hematopoietic system, hematopoietic stem cells (HSCs) differentiate and produce multiple lineages of progenitors and mature blood cells [104]. In some cancers, a small population of cells called "cancer stem cells" (CSCs) in bulk of tumor cells is believed to be responsible for maintaining the uncontrolled cell growth. CML is a classic example of a stem cell cancer and arises when the t9;22 translocation occurs in HSCs that function as CSCs or LSCs in CML. This event results in the constitutive expression of the fusion tyrosine kinase BCR-ABL [27, 70, 81], transforming the HSCs into the LSCs, which then gives rise to a clonal myeloproliferative disease (Fig. 9.3). Early evidence regarding the HSC origins of CML came from observations that transfusion of peripheral blood cells from CML patients into severely neutropenic recipients resulted in temporary homologous bone marrow engraftment and Ph1 progeny in the blood. This was later explained by the presence of high numbers of mobilized LSCs in the peripheral blood of chronic phase CML patients. Until very recently, BCR-ABL expression was considered sufficient to cause a CML-like disease in mouse models using retrovirus transduction or transgene insertional mutagenesis to express the oncogene in LSCs [42].

9.4 General Features of the LSCs in CML

9.4.1 Resistance to the Treatment of TKIs

Although it is well accepted that CML is derived from a LSC, biological features of LSCs are largely unclear. For example, LSCs are in general less sensitive or even resistant to inhibition by TKIs, and the underlying mechanisms are poorly understood. CML is caused by the acquisition of the tyrosine kinase BCR-ABL in HSCs, transforming them into LSCs that are able to self-renew, proliferate, and differentiate to give rise to a myeloproliferative disease [75]. Although TKIs that target the kinase activity of BCR-ABL have transformed CML from a once-fatal disease to a manageable one for the vast majority of patients in chronic phase (CP), about only 10% of them can discontinue TKI treatment and maintain a therapy-free remission. Available evidence shows that CML LSCs are resistant to TKIs and persist during long-term therapy, and as a result, LSCs may promote acquired TKI resistance and

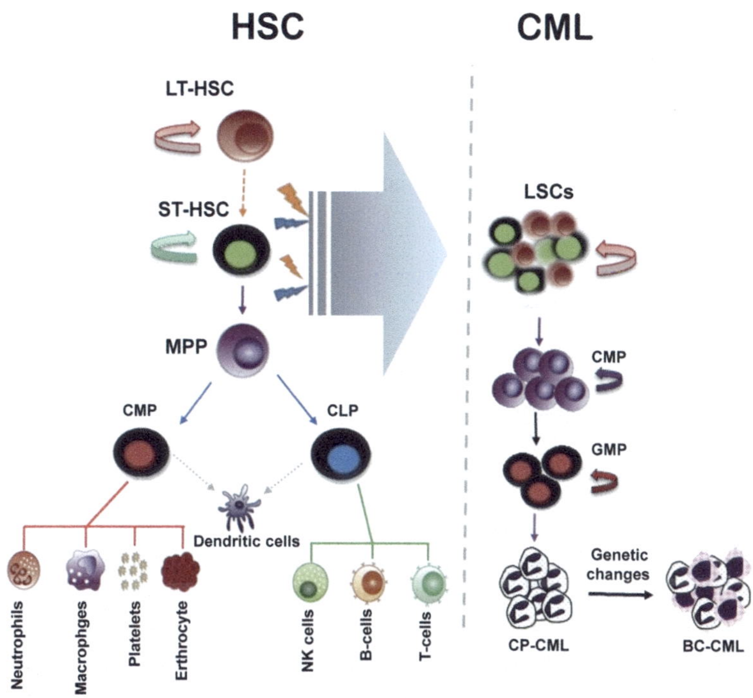

Fig. 9.3 Schematic illustration of the normal and leukemic hematopoietic hierarchies. Hematopoiesis is organized in a hierarchical manner, and rare HSCs give rise to various types of progenitor cells, which proliferate extensively and generate mature blood cells at the bottom of the hierarchy. If HSCs acquire the transforming mutation known as BCR-ABL, the cells gain uncontrolled self-renewal capacity to become LSCs that give rise to CML. Aberrant developmental hierarchy, finally leading high count abnormal white cells in the blood and bone marrow, these cells are determined as LSCs. After the formation of BCR-ABL in CML chronic phase (CP), additional genetic changes can occur, and the disease eventually progresses to fatal CML blastic phase called blastic crisis (BC)

drive relapse or disease progression [5]. The earliest evidence of CML LSCs predated the introduction of TKIs followed by definitive evidence of a deeply, but reversibly, quiescent subpopulation of leukemic cells in patients with CML [41]. In the subsequent years, the consensus view has emerged that virtually all CP patients on TKI therapy and in MMR are not cured of CML and show signs of residual disease burden from the presence of LSCs in the bone marrow (termed "LSC persistence"). In a typical cohort of 100 CP CML patients who undertake TKI therapy over a 5-year period, almost two-thirds will have this "LSC persistence" phenotype. Researchers have consistently detected BCR-ABL primitive cells in the bone marrow of TKI-treated patients in major molecular response (MMR), which are capable of growth in colony-forming cell and stem cell culture assays, even in patients in deep molecular response with no detectable BCR-ABL transcripts by quantitative polymerase chain reaction [17]. The most recent studies have shown that, although

LSCs are not always detectable in cases of deep molecular response, most likely due to technical limitations, some patients with no detectable LSCs can subsequently relapse after TKI discontinuation [16]. Since their discovery in patients almost two decades ago, CML LSCs have become a well-recognized example of CSCs and have been characterized extensively as a significant bottleneck to cure, with the aim of developing new curative therapeutic approaches based on LSC eradication [42].

9.4.2 Quiescence of LSCs

The molecular mechanism of LSC-mediated resistance to current therapies is still largely unknown, but it appears that quiescence of LSCs allows the cells to avoid to be targeted by current therapies [21]. In general, quiescence of CSCs is highly relevant for cancer therapy, because quiescent CSCs are often resistant to both conventional chemotherapy and targeted therapy and contribute to disease relapse following discontinuation of therapies. Therefore, improved understanding of stem cell quiescence helps to develop new strategies for targeting quiescent CSCs [60]. In the case of CML, recent studies have shown that LSCs, particularly those in a quiescent state, are highly resistant to available chemotherapy and targeted therapies, which is closely associated with disease relapse [49, 54]. It is believed that LSC quiescence is controlled by both intrinsic regulatory mechanisms and extrinsic signals from the microenvironment. In addition, signaling molecules involved in cell survival and self-renewal which are the two critical characteristics of quiescent LSCs have been linked to key regulators of cell cycle progression. In CML, BCR-ABL aberrantly phosphorylates and mislocalizes cell cycle proteins which have been implicated in cell cycle deregulation of leukemia cells [21]. Additionally, the interaction of LSCs with bone marrow niche plays a critical role in the long-term maintenance of LSC quiescence. LSCs in the niche are largely dormant and resistant to common chemotherapy. Increasing evidence supports an idea that specific signals from the surrounding stromal cells induce cell cycle arrest of LSCs, allowing them to persist even during TKI treatment. On the other hand, cell-fate decision in HSCs is regulated by several oncogenic transcription factors, including MEIS1 and HOXA9, and is overexpressed in BP of CML [51, 52]. In particular, HOXA9 collaborates with MEIS1 to transform primary bone marrow cells [57], and HOXA9 forms ternary complexes with PBX2 and MEIS1 in myeloid leukemia cells [89]. In addition, *HIF1α*, a master transcriptional regulator of the cellular and systemic hypoxia response, plays a role in cancer progression by activating transcriptional programs for maintaining the ability of self-renewal and multipotency of CSCs in a hypoxic environment [23, 64, 91]. Furthermore, *HIF1α* has been shown to be required for stem cell functions of mouse lymphoma and human acute myeloid leukemia [96]. Some recent studies demonstrate that *HIF1α* is upregulated in BCR-ABL-expressing LSCs and is required for survival maintenance of LSCs in CML. Specifically, deletion of *HIF1α* impairs the cell cycle progression of LSCs with an accumulation of cells in the G0–G1 phases and a concomitant reduction in the S phase of the cell cycle through the upregulation of cyclin-dependent kinase inhibitors p16^{Ink4a}, p19^{Arf}, and p57 [100]. LSCs that survive TKI treatment are sensitive to inhibition by a HIF1 inhibitor acriflavine (ACF), and the reduced survival and growth of CML cells by ACF are mediated through downregulation of c-Myc and stemness-related genes and inhibition of cell cycle progression of LSCs [10].

9.4.3 Critical Signaling Pathways Involved in Self-Renewal and Survival of LSCs

Wnt/β-Catenin Pathway Self-renewal is one of key characteristics of LSCs, and recent studies have reported that a set of critical genes control the self-renewal and survival of LSCs through a wide range of signaling pathways [37]. Among these pathways, the Wnt/β-catenin signaling pathway plays a fundamental role in maintaining the HSC population. Wnt genes encode small secreted proteins existing in all animal genomes, and Wnt signaling is involved in virtually every aspect of embryonic development and also controls homeostatic self-renewal of cells in a number of adult tissues. The activation of the Wnt/β-catenin pathway leads to the translocation of β-catenin into the nucleus, where it induces expression of its target genes such as c-Myc, c-Jun, and cyclin D1 [80] (Fig. 9.4a). A number of studies have demonstrated that Wnt/β-catenin signaling pathway acts as a key regulator in controlling proliferation, survival, and differentiation of hematopoietic cells [19]. Aberrant activation of the Wnt/β-catenin signaling pathway has also been found in CML [43, 102]. Further studies have shown that the Wnt/β-catenin signaling pathway is required for efficient self-renewal of LSCs, indicating that targeting of the Wnt/β-catenin signaling is an attractive therapeutic strategy for CML [43]. It has also been shown that Wnt signaling pathway and the polycomb group protein BMI1 (B lymphoma Mo-MLV insertion region 1 homolog) are involved in the expansion of LSCs [59].

Hedgehog Pathway The hedgehog (Hh) pathway is a highly conserved developmental pathway that regulates the proliferation, migration, and differentiation of cells during development [45]. Three distinct ligands, sonic (Shh), Indian (Ihh), and desert (Dhh) hedgehog, exist in humans. Upon ligand binding to the receptor patched (Ptch), inhibition of smoothened (Smo) receptor is relieved. Smo then activates members of the Gli family of zinc-finger transcription factors, which translocate to the nucleus to regulate the transcription of Hh target genes including Gli1, Gli2, Ptch, and regulators of cell proliferation and survival [94] (Fig. 9.4b). The role of Hh signaling in hematopoietic malignancies has been studied in CML animal models. The Smo$^{-/-}$ CML mice showed a significant decrease in the population of Lin^{-}Sca-1^{+}c-Kit^{+}(LSK) cells, which contains CML LSCs, whereas activated SmoM2 stimulated an increase in LSK cells and acceleration of CML progression [44]. Hh inhibition by cyclopamine, similar to a situation in the Smo$^{-/-}$ cells, caused

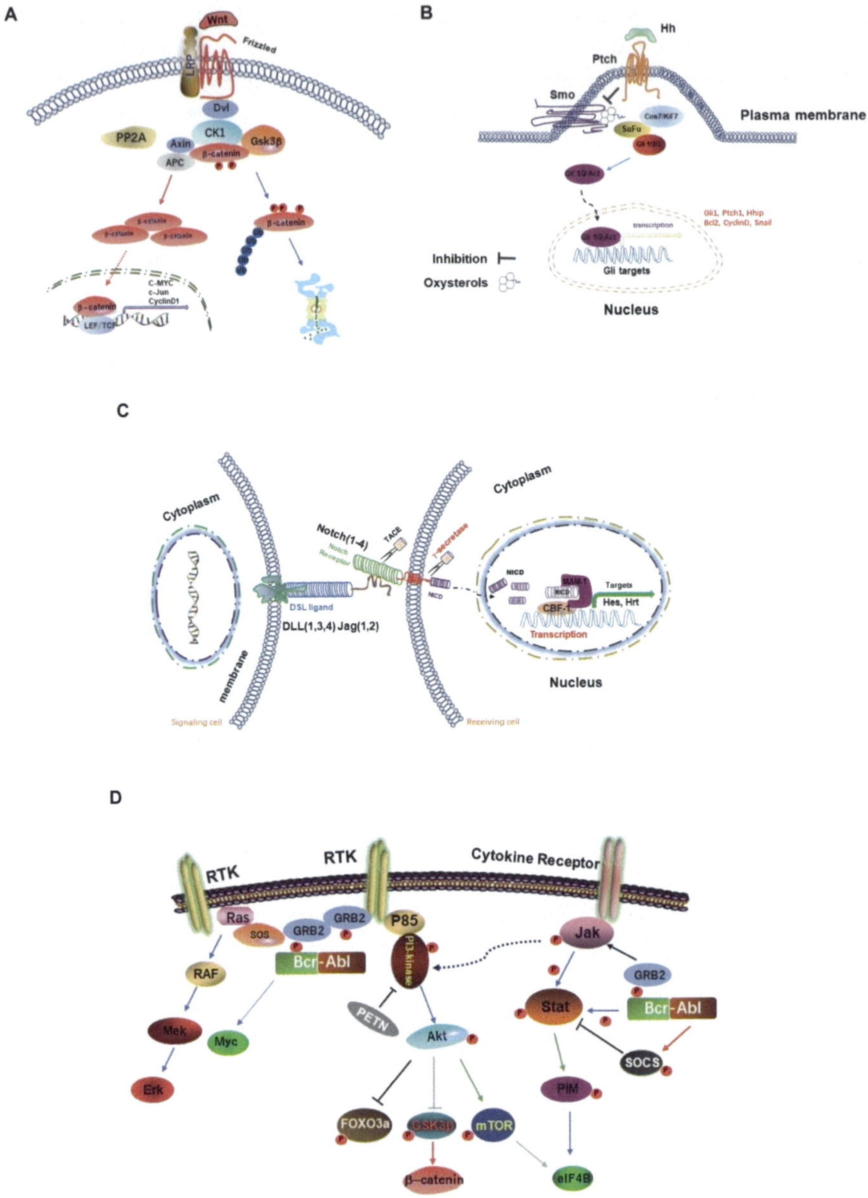

Fig. 9.4 Critical signaling pathways involved in self-renewal and survival of LSCs. (**a**) In the absence of Wnt ligands, APC (adenomatous polyposis coli) and Axin are recruited into the "β-catenin destruction complex," and then Axin is dephosphorylated by the protein phosphatase 2A (PP2A). β-Catenin is phosphorylated by CK1α (casein kinase 1α) and GSK3β (glycogen synthase kinase 3β) and recognized by E3 ubiquitin ligase which subsequently degrades β-catenin via the proteasome. Low cytoplasmic levels of β-catenin ensure activation of TCF/LEF (T-cell factor/lymphoid enhancer factor) transcription factors and transcriptional repression of Wnt target genes. In the accumulation of the extracellular Wnt ligands, the association of Axin with phosphorylated

increased latency of disease and a decrease in the LSK population, even in CML with the imatinib-resistant T315I mutation. In the secondary recipients, no mice receiving $Smo^{-/-}$cells developed disease, even after 9 months of observation [25]. A combination of cyclopamine and the BCR-ABL inhibitor nilotinib had an additive effect in reducing colony-forming units in vitro, and these CML mice took longer to relapse after discontinuation of therapy than those treated with nilotinib alone. A combination of the Smo antagonist NVP-LDE225 and nilotinib reduced replating efficiency of colony formation even further and also improved survival of the treated CML mice after drug discontinuation, similar to the cyclopamine results [47, 98]. These reports support an effect of Hh inhibition on CML stem cells.

Notch Pathway Another pathway that has been shown to have cross talk with other major "stemness pathways" described above in regulation of self-renewal is the Notch signaling pathway. Notch receptors are an evolutionarily conserved family of transmembrane receptors known to be expressed and activated in HSCs [29]. Binding of its physiological ligands of the Delta and Serrata families leads to separation of an intracellular portion of Notch that can enter the nucleus where it binds to the transcriptional factor CBF-1 that interacts with the cofactor MAML-1 (mastermind-like-1) and leads to transcriptional activation of target genes [1] (Fig. 9.4c). There appears to be an interaction between Notch and Wnt signaling-mediated maintenance of HSCs, suggesting a role of Notch in regulation of

Fig. 9.4 (continued) LRP5/6 (lipoprotein receptor-related protein 5/6) and recruitment of phosphorylated DVL (disheveled) to FZD (frizzled) lead to the dissociation of the destruction complex. β-Catenin is stabilized and translocated into the nucleus to form the complex with TCF or LEF and subsequently activates the target genes, including C-MYC, C-Jun, cyclin D1, etc. (**b**) In the absence of Hh, Ptch inhibits the activation of Smo by removing oxysterol from Smo, resulting in Smo inactivation and degradation. In the presence of high level of Hh ligand binding with Ptch, the pump function of Ptch is turned off to allow accumulation of oxysterols around Smo, leading to Smo accumulation at the plasma membrane. In turn, this accumulation of sterols allows SMO to become active or stay on the membrane for a longer period of time. Activated Smo stimulates Hh pathway, ultimately leading to the transcription of Gli target genes, Gli1, Ptch1, Hhip, Bcl2, CyclinD, and Snail. (**c**) There are five Notch ligands in mammals: Delta-like1 (Dll1), Delta-like3 (Dll3), Delta-like4 (Dll4), Jagged1, and Jagged2. All ligands have an extracellular domain called DSL (Delta, Serrate, and Lag-2) that is involved in receptor binding. There are four Notch receptors in mammals: Notch1-4. Binding of the Notch ligand on one cell to the Notch receptor on another cell results in two proteolytic cleavages of the receptor. The ADAM10 or TACE (TNF-α-converting enzyme) metalloprotease catalyzes the S2 cleavage to generate a substrate for S3 cleavage by the γ-secretase complex. This proteolytic processing mediates release of the Notch intracellular domain (NICD), and then NICD translocates into the nucleus and interacts with the transcription factor CBF-1, thereby displacing corepressors (CoR) and recruiting coactivators such as MAM-1. This results in the activation of downstream target genes including Hes and Hrt. (**d**) The cellular effects of BCR-ABL are exerted through interactions with various proteins that transduce the oncogenic signals responsible for the activation or repression of gene transcription. The key pathways implicated so far are those involving RAS, mitogen-activated protein (MAP) kinases, signal transducers and activators of transcription (STAT), phosphatidylinositol 3-kinase (PI3K), and MYC. Most of the interactions are mediated through tyrosine phosphorylation and require the binding of BCR-ABL to adapter proteins such as growth factor receptor-bound protein 2 (GRB-2)

self-renewal capacity [29]. Although Notch-associated self-renewal has not been clearly demonstrated in CML, the Notch target Hes1 was found to be highly expressed in CML-blast crisis [67]. Recently, inactivating mutations of Notch signaling have been described in patients with chronic myelomonocytic leukemia (CMML). Notch inhibition in HSCs leads to accumulation of aberrant myeloid progenitors and a CMML-like disease in mice [56]. It is necessary to point out that the role of Notch signaling in myeloid leukemia stem cells has not yet been examined adequately and seems to be context dependent, and its role in self-renewal of LSCs needs future investigations.

JAK/STAT and PI3K/AKT Pathway It is well established that JAK/STAT and PI3K/AKT are two crucial signaling pathways involved in regulating cellular survival and proliferation in LSCs [30, 78] (Fig. 9.4d). In CML, JAK/STAT/PIM and PI3K/AKT/mTOR pathways are constitutively activated by BCR-ABL, resulting in uncontrolled cellular proliferation [63, 78]. BCR-ABL can also activate tyrosine phosphorylation of suppressors of cytokine signaling 1 and 3 (SOCS-1 and SOCS-3), two potent suppressors of JAK/STAT signaling, leading to a reduction of their inhibitory effects on JAK/STAT activation [77]. In addition, BCR-ABL causes constitutive repression of FoxO3a by continued activation of the PI3K/AKT signaling axis, providing another mechanism of BCR-ABL-mediated apoptotic resistance [55, 87]. Furthermore, AKT inhibits GSK3β activity, which is involved in apoptotic regulation of CML cells. Finally, it was observed that other members of the PI3K/AKT/mTOR pathway, including PTEN (phosphatase and tensin homolog) and mTOR, play a role in the maintenance of LSCs [33, 73], and there is a cross talk between the JAK/STAT/PIM and PI3K/AKT/mTOR pathways that converge on eukaryotic translation initiation factor 4B (eIF4B) to regulate the activity of the ABL oncogene [11, 97].

9.5 Novel Therapeutic Targets in LSCs

The TKIs that inhibit BCR-ABL kinase activity are widely used for treatment of CML and induce a complete hematologic and cytogenetic response in the majority of chronic-phase CML patients. Although these TKIs can readily kill a significant number of leukemia cells, they fail to completely eradicate the primitive LSCs in CML [39]. Also, $CD34^+$ cells from CML patients with complete cytogenetic responses are capable of driving CML to BP [5]. These results demonstrate that LSCs remain in CML patients, even though they do not show evidence of the disease, suggesting that TKIs are unable to eradicate quiescent LSCs that are responsible for disease progression [4, 18, 20, 40, 53]. It is generally accepted that eradication of LSCs is required for curing CML, but the major challenge lies in the identification of differences between LSCs and their normal stem cell counterparts, HSCs. For example, some developmental genes essential for self-renewal and survival of stem cells, including the Wnt/β-catenin and hedgehog signaling pathways described above, are involved in the regulation of both HSCs and LSCs [102, 103]. Success of an anti-stem cell strategy relies on complete inhibition of genes that are required functionally and specifically for the maintenance of LSCs. Because LSCs in CML express markers similar to those on normal stem cells, the major difference between them should be related to the leukemia-initiating genetic changes such as acquiring an oncogene or accumulating a DNA mutation [7, 58]. It is reasonable to hypothesize that these genetic changes (like acquiring BCR-ABL) cause aberrant regulation of signaling pathways, consequently turning a normal stem cell into a LSC. Some studies have shown that although certain genes play roles in both leukemia and normal stem cells, they are functionally more critical for leukemia than for normal stem cell counterpart [25, 103]. In this situation, the difference in the degree of dependence on the same genes for survival between leukemia and normal stem cells provides a therapeutic window for more specifically targeting LSCs [14].

Some novel candidate targets in CML LSCs have been identified, including Wnt, Notch, Hox, Bcl-2 and Bmi-1, etc. It is exciting to see that small molecular inhibitors for some of these targets are currently under development or being tested in clinical trials, aiming to selectively eradicate LSCs while sparing normal HSCs in patients with CML [6, 34, 35, 65, 68]. Three research groups have shown that activators of the peroxisome proliferator-activated receptor γ (PPARγ) have increased antileukemic activities in combination with TKIs [32, 76, 99]. Activators of PPARγ result in transcriptional downregulation of STAT5, whereas TKIs block phosphorylation of STAT5, with the combined effects of both drugs significantly downregulating this pathway and causing LSCs to exit quiescence where they were eradicated by TKIs [69]. Also, the eukaryotic translation initiation factor 4E (eIF4E) is a potent oncogene elevated in an estimated 30% of human cancers [22, 90]. Recent studies showed that targeting of the MNK-eIF4E axis in blast crisis CML inhibits LSC function [61]. It has been demonstrated that RAD51 (Rec A homolog of *E. coli*) recombinase activity plays a critical role for cancer cell survival, proliferation, and drug resistance. IBR2, a novel small molecule RAD51 inactivator, significantly prolonged survival in a murine imatinib-resistant CML model and effectively inhibited the proliferation of $CD34^+$ progenitor cells from CML patients [105].

Hsp90 is an ubiquitously expressed chaperone protein which accounts for about 1–2% of total cellular protein in mammalian cells [86]. It is a chaperone of several oncoproteins, such as Her2, v-Src, and BCR-ABL, and plays a role in regulating survival, proliferation, and apoptosis of cancer cells [2]. Hsp90 is constitutively expressed at a level that is two- to tenfold higher in tumor cells than in their normal stem cell counterparts, suggesting that Hsp90 is a potential therapeutic target in cancer [3, 48]. It has been shown that IPI-504, an inhibitor of Hsp90, efficiently induces dissociation of BCR-ABL and Hsp90 in BCR-ABL-expressing 32D cells as quickly as 30 min after the treatment. In vivo, treatment with IPI-504 alone significantly prolonged the survival of mice with CML induced by BCR-ABL and had a dramatic inhibitory effect on LSCs in CML mice. These studies demonstrate that inhibition of Hsp90 can effectively inhibit the survival and proliferation of LSCs in CML [74].

The arachidonates 5-lipoxygenase (5-LO) and 15-lipoxygenase (15-LO) genes (*Alox5* and *Alox15*) have been shown to regulate numerous physiological and pathological progresses, including inflammation and cancer [9, 12, 79]. The essential role of *Alox5* and *Alox15* in LSCs was originally discovered from a study using a murine model of BCR-ABL-induced CML. These studies showed that BCR-ABL fails to efficiently induce CML in mice in the absence of *Alox5* or *Alox15* due to the impairments of the functions of LSCs [13, 15] (Figs. 9.5 and 9.6). In support of the role of these two genes in the maintenance of LSCs, specific 5-LO and 15-LO inhibitors cause the impairments of the functions of LSCs in a similar manner. Most importantly, the lack of *Alox5* or *Alox15* does not significantly affect the functions of normal HSCs, indicating that *Alox5* and *Alox15* are required specifically for survival regulation of LSCs with a minimal effect on normal HSCs. These findings provide supporting evidence to the notion that it is possible to develop an effective therapeutic strategy for leukemia therapy through specific targeting of LSCs.

HIF1 belongs to the family of basic helix-loop-helix (bHLH) transcription factors and is a heterodimer that consists of a constitutively expressed HIF1β subunit and a HIF1α subunit [88]. Recent studies demonstrated that *HIF1α* plays a role in cancer progression by activating transcriptional programs for maintaining the ability of self-renewal and multipotency of CSCs in a hypoxic environment and that *HIF1α* was required for stem cell functions of mouse lymphoma and human acute myeloid leukemia [23, 64]. *HIF1α* was identified as a critical regulator for survival maintenance of LSCs in CML. Deletion of HIF1α impairs the propagation of CML through impairing cell cycle progression and inducing apoptosis of LSCs [100] (Fig. 9.7). Deletion of HIF1α also results in elevated expression of $p16I^{nk4a}$ and $p19^{Arf}$ in LSCs; knockdown of $p16^{Ink4a}$ and $p19^{Arf}$ rescues the defective colony-forming ability of $HIF1\alpha^{-/-}$ LSCs. Compared with normal HSCs, LSCs appear to be more dependent on the *HIF1α* pathway. Together, these results demonstrate that *HIF1α* represents a critical pathway in LSCs, and inhibition of the *HIF1α* pathway provides a therapeutic strategy for eradicating LSCs in CML.

On the other hand, it has been reported that some tumor suppressor-like genes play a critical role in survival regulation of LSCs in CML. It has been shown that the stearoyl-CoA desaturase 1 (*Scd1*) and B lymphoid kinase (*Blk*) negatively regulate the functions of LSCs in CML mice [99, 101] (Figs. 9.8 and 9.9). Scd1 catalyzes the biosynthesis of monounsaturated fatty acids from saturated fatty acids, which are the most abundant fatty acids present in mammalian organisms [85]. Scd1 expression is detected in almost all tissues, with a predominant expression in the liver, and is involved in regulating metabolic pathways related to preadipocyte differentiation, insulin sensitivity, metabolism, and tumorigenesis [31, 38]. In the BCR-ABL-induced CML mouse model, *Scd1* expression is downregulated in LSCs, and the deletion of *Scd1* accelerated CML development through affecting the function of LSCs with no effect on normal HSCs [99]. In addition, Pten, p53, and Bcl2 are regulated by *Scd1* in LSCs, and the induction of Scd1 expression by a PPARγ agonist suppresses LSCs and delays CML development, suggesting that *Scd1* plays a tumor-suppressive role in functional regulation of LSCs [99]. Blk is a SRC-family protein tyrosine kinase that is typically involved in cell proliferation and differentiation [95].

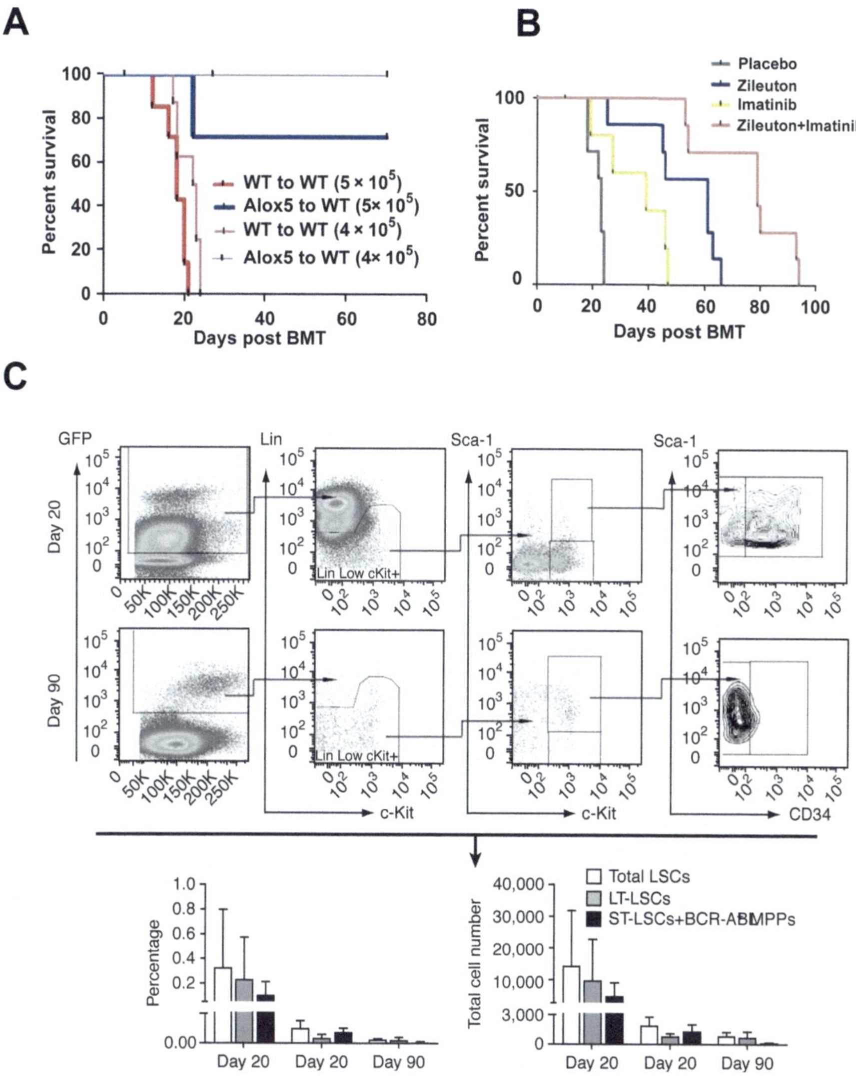

Fig. 9.5 *Alox5* is essential for the functions of LSCs in CML mouse model induced by BCR-ABL. (**a**) Kaplan-Meier survival curves for recipients of *BCR-ABL*-transduced bone marrow cells from wild-type or *Alox5*$^{-/-}$ donor mice (10 mice per group). All recipients of *BCR-ABL*-transduced bone marrow cells from wild-type donor mice developed CML and died within 4 weeks after bone marrow transplantation (days post BMT), whereas recipients of *BCR-ABL*-transduced bone marrow cells from *Alox5*$^{-/-}$ donor mice survived. (**b**) Kaplan-Meier survival curves for CML mice treated with a placebo, zileuton alone, imatinib alone, or both zileuton and imatinib in combination. Zileuton-treated CML mice were much healthier than placebo-treated CML mice, and inhibition of Alox5 by zileuton significantly prolonged survival of CML mice. (**c**) Bone marrow cells were isolated from recipients of *BCR-ABL*-transduced bone marrow cells from wild-type or *Alox5*$^{-/-}$ donor mice, and FACS analysis showed that both percentages and total numbers of LT-LSCs ($GFP^+Lin^-c\text{-}Kit^+Sca\text{-}1^+CD34^-$) and ST-LSCs /BCR-ABL expressing MPP cells ($GFP^+Lin^-c\text{-}Kit^+Sca\text{-}1^+CD34^+$) in recipients of BCR-ABL-transduced *Alox5*$^{-/-}$ (middle and right panels) donor BM cells were much lower than those in recipients of BCR-ABL-transduced wild-type (left panel) donor BM cells ($n = 4$). (Adapted from Chen et al. [13])

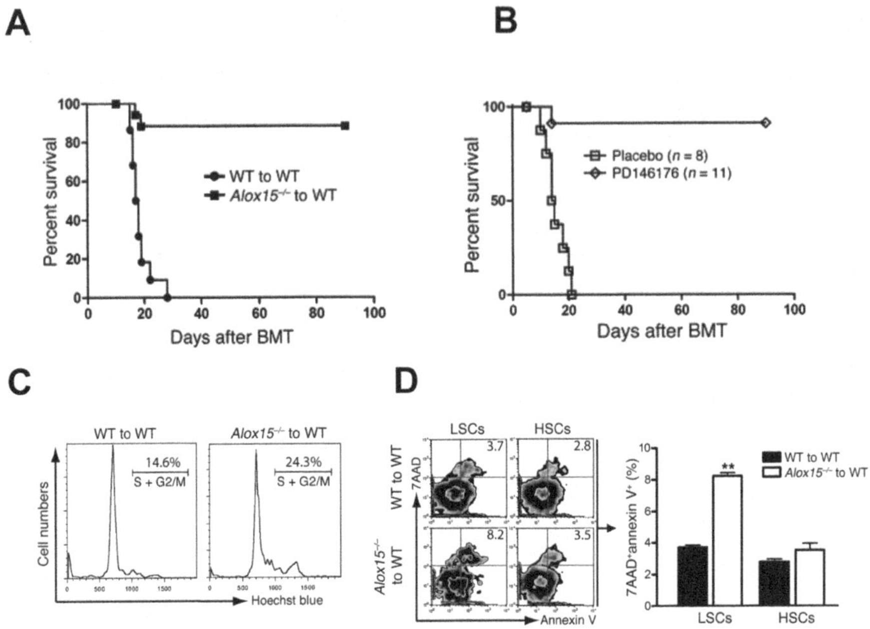

Fig. 9.6 *Alox15* plays a critical role in regulating LSC function and CML development. (**a**) Kaplan-Meier survival curve for recipients of BCR-ABL-transduced bone marrow cells from *Alox15*$^{-/-}$ mice was significantly prolonged compared to recipients of BCR-ABL-transduced bone marrow cells from WT mice (8 mice per group). (**b**) Kaplan-Meier survival curves showed that inhibition of *Alox15* by PD146176 dramatically prolonged survival of the CML mice. (**c**) Cell cycle analysis showed a higher percentage of LSCs in the S + G2/M phase in BM cells of *Alox15*$^{-/-}$ versus WT CML mice. BM cells were stained with Hoechst blue for FACS analysis. The mean percentage for each cell population is shown. (**d**) *Alox15* deficiency caused increased apoptosis of LSCs over normal HSCs by staining the cells with 7-AAD and annexin V. Results represent the mean ± SD. (Adapted from Chen et al. [15])

Inhibition of the Blk pathway accelerates CML development, and overexpression of the Blk delays CML development through suppressing LSC functions via Pax5 and p27, and Blk has no inhibitory effect on normal HSCs [101]. These results indicate that Blk functions as a tumor suppressor specifically in LSCs.

9.6 Perspective

CML is a classic stem cell disease, in which the BCR-ABL oncoprotein is considered essential for abnormal growth and accumulation of neoplastic cells. During the past two decades, despite the BCR-ABL tyrosine kinase inhibitors have successfully been introduced in the treatment of the disease, intrinsic and acquired drug resistance has been recognized as a challenge in clinical practice due to the persistence of CML LSCs that are a small cell population required for the initiation and

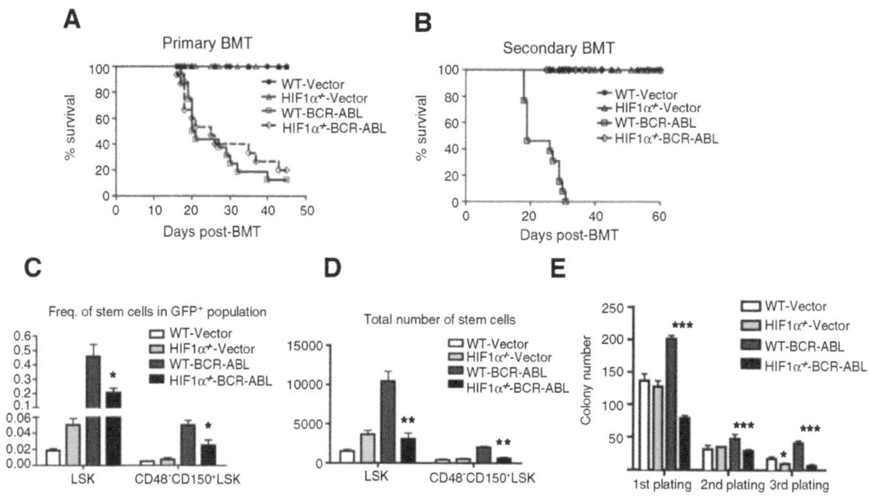

Fig. 9.7 *HIF1α* is required for survival maintenance of CML stem cells. (**a**, **b**) Kaplan-Meier survival curves for primary and secondary recipients of empty vector or *BCR-ABL*-transduced BM cells from WT or $HIF1\alpha^{-/-}$ donor mice. For secondary BM transplantation, BM cells from primary control and CML recipient mice which received empty vector or *BCR-ABL*-transduced WT and $HIF1\alpha^{-/-}$ BM cells were analyzed by FACS, and BM cells containing equal number of WT or $HIF1\alpha^{-/-}$ $GFP^+Lin^-c\text{-}Kit^+Sca\text{-}1^+$ cells along with $2X10^5$ WT BM cells (CD45.1) were transplanted into each lethally irradiated secondary recipient mouse. (**c**, **d**) The percentages and total numbers of stem cells ($GFP^+Lin^-c\text{-}Kit^+Sca\text{-}1^+$ cells and $GFP^+Lin^-c\text{-}Kit^+Sca\text{-}1^+$ $CD150^+CD48^-$ cells) in BM of secondary recipient mice receiving BM cells from WT or $HIF1\alpha^{-/-}$ mice were analyzed by FACS; $*P < 0.05$; $**P < 0.01$. (**e**) The loss of *HIF1α* causes a decrease in the colony-forming ability of BCR-ABL-expressing LSK cells. Sorted normal LSK cells and BCR-ABL-expressing LSK cells ($GFP^+Lin^-c\text{-}Kit^+Sca\text{-}1^+$ cells) were plated into methycellulose medium, colonies were counted, and cells were serially replated. Data show results from representative experiments. Mean values (± SEM) are shown; $*P < 0.05$; $**P < 0.01$. (Adapted from Zhang et al. [100])

maintenance of leukemia. LSCs cannot be eradicated through inhibiting BCR-ABL tyrosine kinase activity by TKIs. Importantly, the insensitivity of LSCs to TKIs is, in some cases, not associated with the development of drug-resistant mutations on BCR-ABL, suggesting that BCR-ABL kinase activity-independent pathways also contribute to the maintenance of survival and self-renewal of LSCs. Therefore, identification of novel genes that play critical role in regulating the function of LSCs will help to develop new therapeutic strategies through targeting LSCs for the treatment of CML, leading to a long-lasting clinical response with the absence of drug resistance and disease relapse. Identification of LSC targets in CML requires a full understanding of genetic changes in BCR-ABL-expressing LSCs. In addition, identification of surface biomarkers and signaling pathways associated with LSCs may provide more LSC targets for eradicating LSCs. On the other hand, in the development of anti-LSC strategies, we should pay much attention to the potential harm of an anti-LSC therapy to normal HSCs.

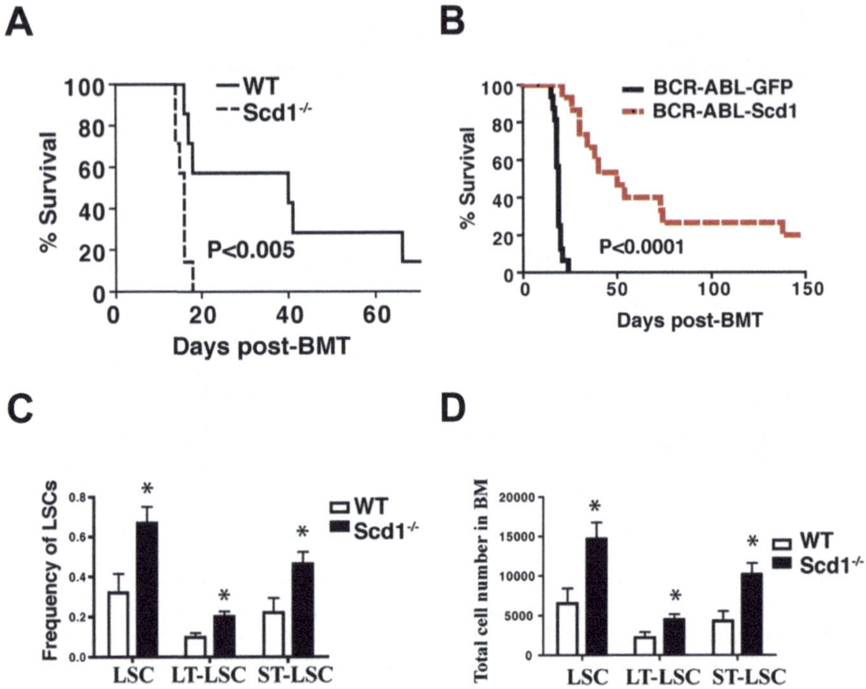

Fig. 9.8 *Scd1* plays a tumor-suppressive role in survival of LSCs in CML. (**a**) Kaplan-Meier survival curves showed that recipients of *BCR-ABL*-transduced *Scd1*$^{-/-}$ bone marrow cells developed and died of CML significantly faster than recipients of *BCR-ABL*-transduced WT mice bone marrow cells ($n = 7$ to 8 mice per group; $P < 0.005$). (**b**) Kaplan-Meier survival curve for secondary recipients of equal number of bone marrow cells from primary CML mice receiving *BCR-ABL*-transduced or *BCR-ABL-Scd1*-transduced WT bone marrow cells. Scd1 overexpression caused a significant delay of CML development in secondary recipients of bone marrow cells from the primary CML mice. ($n = 14$–16 mice per group; $P < 0.0001$). (**c**) Bone marrow cells from recipients of BCR-ABL-transduced WT or *Scd1*$^{-/-}$ bone marrow cells were analyzed by FACS at day 14 after BMT. The percentages and numbers of total LSCs, LT-LSCs, and ST-LSCs in the bone marrow of recipients of *BCR-ABL*-transduced *Scd1*$^{-/-}$ donor bone marrow cells were dramatically higher compared to those in the bone marrow of recipients of *BCR-ABL*-transduced WT donor bone marrow cells ($n = 4$ to 7 mice per group). The data are representative of one of four independent experiments, and mean values (± SEM) are shown ($*$, $P < 0.05$). (Adapted from Zhang et al. [99])

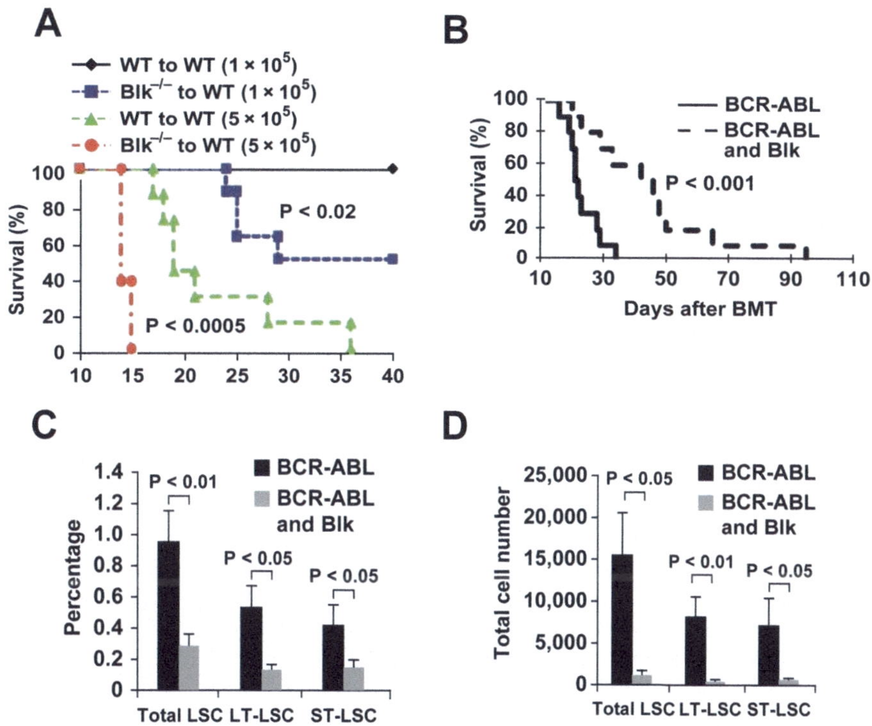

Fig. 9.9 The Blk pathway functions as a tumor suppressor in CML stem cells. (**a**) Kaplan-Meier survival curves for recipients of *BCR-ABL*-transduced bone marrow cells from WT (n = 7) or *Blk*$^{-/-}$ (n = 8) donor mice. Recipients of *BCR-ABL*-transduced bone marrow cells from *Blk*$^{-/-}$donor mice developed CML significantly faster than did recipients of *BCR-ABL*-transduced WT mice bone marrow cells. (**b**) Kaplan-Meier survival curves for recipients of *BCR-ABL* (n = 10) or *BCR-ABL-Blk* (n = 10) transduced bone marrow cells. Overexpression of *Blk* in *BCR-ABL*-transduced bone marrow cells delayed the CML development of recipient mice. (**c**, **d**) The percentages and numbers of total LSCs (GFP$^+$Lin$^-$c-Kit$^+$Sca-1$^+$), LT-LSCs (GFP$^+$Lin$^-$c-Kit$^+$Sca-1$^+$CD34$^-$), and ST-LSCs (GFP$^+$Lin$^-$c-Kit$^+$Sca-1$^+$CD34$^+$) in the bone marrow of recipients of *BCR-ABL*- or *BCR-ABL-Blk*-transduced bone marrow cells were analyzed at day 15 after transplantation. Mean values ± SEM (n = 5). (Adapted from Zhang et al. [101])

References

1. Allman D, Aster JC, Pear WS (2002) Notch signaling in hematopoiesis and early lymphocyte development. Immunol Rev 187:75–86
2. An WG, Schulte TW, Neckers LM (2000) The heat shock protein 90 antagonist geldanamycin alters chaperone association with p210bcr-abl and v-src proteins before their degradation by the proteasome. Cell Growth Differ 11:355–360
3. Banerji U (2009) Heat shock protein 90 as a drug target: some like it hot. Clin Cancer Res 15:9–14
4. Barnes DJ, Melo JV (2006) Primitive, quiescent and difficult to kill: the role of non-proliferating stem cells in chronic myeloid leukemia. Cell Cycle 5:2862–2866
5. Bhatia R, Holtz M, Niu N, Gray R, Snyder DS, Sawyers CL, Arber DA, Slovak ML, Forman SJ (2003) Persistence of malignant hematopoietic progenitors in chronic myelogenous leukemia patients in complete cytogenetic remission following imatinib mesylate treatment. Blood 101:4701–4707
6. Bhattacharyya J, Mihara K, Yasunaga S, Tanaka H, Hoshi M, Takihara Y, Kimura A (2009) BMI-1 expression is enhanced through transcriptional and posttranscriptional regulation during the progression of chronic myeloid leukemia. Ann Hematol 88:333–340
7. Bonnet D, Dick JE (1997) Human acute myeloid leukemia is organized as a hierarchy that originates from a primitive hematopoietic cell. Nat Med 3:730–737
8. Branford S, Cross NC, Hochhaus A, Radich J, Saglio G, Kaeda J, Goldman J, Hughes T (2006) Rationale for the recommendations for harmonizing current methodology for detecting BCR-ABL transcripts in patients with chronic myeloid leukaemia. Leukemia 20:1925–1930
9. Catalano A, Rodilossi S, Caprari P, Coppola V, Procopio A (2005) 5-lipoxygenase regulates senescence-like growth arrest by promoting ROS-dependent p53 activation. EMBO J 24:170–179
10. Cheloni G, Tanturli M, Tusa I, Ho DeSouza N, Shan Y, Gozzini A, Mazurier F, Rovida E, Li S, Dello Sbarba P (2017) Targeting chronic myeloid leukemia stem cells with the hypoxia-inducible factor inhibitor acriflavine. Blood 130:655–665
11. Chen K, Yang J, Li J, Wang X, Chen Y, Huang S, Chen JL (2016) eIF4B is a convergent target and critical effector of oncogenic Pim and PI3K/Akt/mTOR signaling pathways in Abl transformants. Oncotarget 7:10073–10089
12. Chen XS, Sheller JR, Johnson EN, Funk CD (1994) Role of leukotrienes revealed by targeted disruption of the 5-lipoxygenase gene. Nature 372:179–182
13. Chen Y, Hu Y, Zhang H, Peng C, Li S (2009) Loss of the Alox5 gene impairs leukemia stem cells and prevents chronic myeloid leukemia. Nat Genet 41:783–792
14. Chen Y, Li S (2014) Omacetaxine mepesuccinate in the treatment of intractable chronic myeloid leukemia. Oncol Targets Ther 7:177–186
15. Chen Y, Peng C, Abraham SA, Shan Y, Guo Z, Desouza N, Cheloni G, Li D, Holyoake TL, Li S (2014) Arachidonate 15-lipoxygenase is required for chronic myeloid leukemia stem cell survival. J Clin Invest 124:3847–3862
16. Chomel JC, Bonnet ML, Sorel N, Sloma I, Bennaceur-Griscelli A, Rea D, Legros L, Marfaing-Koka A, Bourhis JH, Ame S et al (2016) Leukemic stem cell persistence in chronic myeloid leukemia patients in deep molecular response induced by tyrosine kinase inhibitors and the impact of therapy discontinuation. Oncotarget 7:35293–35301
17. Chu S, McDonald T, Lin A, Chakraborty S, Huang Q, Snyder DS, Bhatia R (2011) Persistence of leukemia stem cells in chronic myelogenous leukemia patients in prolonged remission with imatinib treatment. Blood 118:5565–5572
18. Chu S, Xu H, Shah NP, Snyder DS, Forman SJ, Sawyers CL, Bhatia R (2005) Detection of BCR-ABL kinase mutations in CD34+ cells from chronic myelogenous leukemia patients in complete cytogenetic remission on imatinib mesylate treatment. Blood 105:2093–2098
19. Cobas M, Wilson A, Ernst B, Mancini SJ, MacDonald HR, Kemler R, Radtke F (2004) Beta-catenin is dispensable for hematopoiesis and lymphopoiesis. J Exp Med 199:221–229

20. Copland M, Hamilton A, Elrick LJ, Baird JW, Allan EK, Jordanides N, Barow M, Mountford JC, Holyoake TL (2006) Dasatinib (BMS-354825) targets an earlier progenitor population than imatinib in primary CML but does not eliminate the quiescent fraction. Blood 107:4532–4539
21. Cramer K, Nieborowska-Skorska M, Koptyra M, Slupianek A, Penserga ET, Eaves CJ, Aulitzky W, Skorski T (2008) BCR/ABL and other kinases from chronic myeloproliferative disorders stimulate single-strand annealing, an unfaithful DNA double-strand break repair. Cancer Res 68:6884–6888
22. Culjkovic B, Borden KL (2009) Understanding and targeting the eukaryotic translation initiation factor eIF4E in head and neck cancer. J Oncol 2009:981679
23. Culver C, Melvin A, Mudie S, Rocha S (2011) HIF-1alpha depletion results in SP1-mediated cell cycle disruption and alters the cellular response to chemotherapeutic drugs. Cell Cycle 10:1249–1260
24. Deininger MW, Druker BJ (2003) Specific targeted therapy of chronic myelogenous leukemia with imatinib. Pharmacol Rev 55:401–423
25. Dierks C, Beigi R, Guo GR, Zirlik K, Stegert MR, Manley P, Trussell C, Schmitt-Graeff A, Landwerlin K, Veelken H et al (2008) Expansion of Bcr-Abl-positive leukemic stem cells is dependent on Hedgehog pathway activation. Cancer Cell 14:238–249
26. Druker BJ (2001) Current treatment approaches for chronic myelogenous leukemia. Cancer J 7(Suppl 1):S14–S18
27. Druker BJ (2008) Translation of the Philadelphia chromosome into therapy for CML. Blood 112:4808–4817
28. Druker BJ, Tamura S, Buchdunger E, Ohno S, Segal GM, Fanning S, Zimmermann J, Lydon NB (1996) Effects of a selective inhibitor of the Abl tyrosine kinase on the growth of Bcr-Abl positive cells. Nat Med 2:561–566
29. Duncan AW, Rattis FM, DiMascio LN, Congdon KL, Pazianos G, Zhao C, Yoon K, Cook JM, Willert K, Gaiano N et al (2005) Integration of Notch and Wnt signaling in hematopoietic stem cell maintenance. Nat Immunol 6:314–322
30. Faderl S, Talpaz M, Estrov Z, O'Brien S, Kurzrock R, Kantarjian HM (1999) The biology of chronic myeloid leukemia. N Engl J Med 341:164–172
31. Falvella FS, Pascale RM, Gariboldi M, Manenti G, De Miglio MR, Simile MM, Dragani TA, Feo F (2002) Stearoyl-CoA desaturase 1 (Scd1) gene overexpression is associated with genetic predisposition to hepatocarcinogenesis in mice and rats. Carcinogenesis 23:1933–1936
32. Glodkowska-Mrowka E, Manda-Handzlik A, Stelmaszczyk-Emmel A, Seferynska I, Stoklosa T, Przybylski J, Mrowka P (2016) PPARgamma ligands increase antileukemic activity of second- and third-generation tyrosine kinase inhibitors in chronic myeloid leukemia cells. Blood Cancer J 6:e377
33. Gutierrez A, Sanda T, Grebliunaite R, Carracedo A, Salmena L, Ahn Y, Dahlberg S, Neuberg D, Moreau LA, Winter SS et al (2009) High frequency of PTEN, PI3K, and AKT abnormalities in T-cell acute lymphoblastic leukemia. Blood 114:647–650
34. Guzman ML, Li X, Corbett CA, Rossi RM, Bushnell T, Liesveld JL, Hebert J, Young F, Jordan CT (2007) Rapid and selective death of leukemia stem and progenitor cells induced by the compound 4-benzyl, 2-methyl, 1,2,4-thiadiazolidine, 3,5 dione (TDZD-8). Blood 110:4436–4444
35. Guzman ML, Rossi RM, Karnischky L, Li X, Peterson DR, Howard DS, Jordan CT (2005) The sesquiterpene lactone parthenolide induces apoptosis of human acute myelogenous leukemia stem and progenitor cells. Blood 105:4163–4169
36. Hahn EA, Glendenning GA (2003) Quality of life on imatinib. Semin Hematol 40:31–36
37. Heidel FH, Mar BG, Armstrong SA (2011) Self-renewal related signaling in myeloid leukemia stem cells. Int J Hematol 94:109–117
38. Hess D, Chisholm JW, Igal RA (2010) Inhibition of stearoylCoA desaturase activity blocks cell cycle progression and induces programmed cell death in lung cancer cells. PLoS One 5:e11394

39. Holtz MS, Slovak ML, Zhang F, Sawyers CL, Forman SJ, Bhatia R (2002) Imatinib mesylate (STI571) inhibits growth of primitive malignant progenitors in chronic myelogenous leukemia through reversal of abnormally increased proliferation. Blood 99:3792–3800
40. Holyoake T, Jiang X, Eaves C, Eaves A (1999a) Isolation of a highly quiescent subpopulation of primitive leukemic cells in chronic myeloid leukemia. Blood 94:2056–2064
41. Holyoake TL, Nicolini FE, Eaves CJ (1999b) Functional differences between transplantable human hematopoietic stem cells from fetal liver, cord blood, and adult marrow. Exp Hematol 27:1418–1427
42. Holyoake TL, Vetrie D (2017) The chronic myeloid leukemia stem cell: stemming the tide of persistence. Blood 129:1595–1606
43. Hu Y, Chen Y, Douglas L, Li S (2009) Beta-catenin is essential for survival of leukemic stem cells insensitive to kinase inhibition in mice with BCR-ABL-induced chronic myeloid leukemia. Leukemia 23:109–116
44. Hu Y, Swerdlow S, Duffy TM, Weinmann R, Lee FY, Li S (2006) Targeting multiple kinase pathways in leukemic progenitors and stem cells is essential for improved treatment of Ph+ leukemia in mice. Proc Natl Acad Sci U S A 103:16870–16875
45. Ingham PW et al (2001) Hedgehog signaling: a tale of two lipids. Science 294(5548):1879–1881. PMID:11729305
46. Iqbal N, Iqbal N (2014) Imatinib: a breakthrough of targeted therapy in cancer. Chemother Res Pract 2014:357027
47. Irvine DA, Shepherd JD (2009) Recurrence of lymphoid blast crisis over 20 years after successful sibling allo-SCT for CML: short lived complete cytogenetic response to imatinib. Bone Marrow Transplant 44:267–268
48. Isaacs JS, Xu W, Neckers L (2003) Heat shock protein 90 as a molecular target for cancer therapeutics. Cancer Cell 3:213–217
49. Ito K, Bernardi R, Morotti A, Matsuoka S, Saglio G, Ikeda Y, Rosenblatt J, Avigan DE, Teruya-Feldstein J, Pandolfi PP (2008) PML targeting eradicates quiescent leukaemia-initiating cells. Nature 453:1072–1078
50. Jabbour E, Cortes J, Kantarjian H (2006) Novel tyrosine kinase inhibitors in chronic myelogenous leukemia. Curr Opin Oncol 18:578–583
51. Jamieson CH, Ailles LE, Dylla SJ, Muijtjens M, Jones C, Zehnder JL, Gotlib J, Li K, Manz MG, Keating A et al (2004a) Granulocyte-macrophage progenitors as candidate leukemic stem cells in blast-crisis CML. N Engl J Med 351:657–667
52. Jamieson CH, Weissman IL, Passegue E (2004b) Chronic versus acute myelogenous leukemia: a question of self-renewal. Cancer Cell 6:531–533
53. Jorgensen HG, Allan EK, Jordanides NE, Mountford JC, Holyoake TL (2007) Nilotinib exerts equipotent antiproliferative effects to imatinib and does not induce apoptosis in CD34+ CML cells. Blood 109:4016–4019
54. Kaminska J, Klimczak-Jajor E, Bany-Laszewicz U (2008) Src kinases in the process of maturation megakryocyte progenitors. Postepy Biochem 54:378–383
55. Khorashad JS, de Lavallade H, Apperley JF, Milojkovic D, Reid AG, Bua M, Szydlo R, Olavarria E, Kaeda J, Goldman JM et al (2008) Finding of kinase domain mutations in patients with chronic phase chronic myeloid leukemia responding to imatinib may identify those at high risk of disease progression. J Clin Oncol 26:4806–4813
56. Klinakis A, Lobry C, Abdel-Wahab O, Oh P, Haeno H, Buonamici S, van De Walle I, Cathelin S, Trimarchi T, Araldi E et al (2011) A novel tumour-suppressor function for the Notch pathway in myeloid leukaemia. Nature 473:230–233
57. Kroon E, Krosl J, Thorsteinsdottir U, Baban S, Buchberg AM, Sauvageau G (1998) Hoxa9 transforms primary bone marrow cells through specific collaboration with Meis1a but not Pbx1b. EMBO J 17:3714–3725
58. Lapidot T, Sirard C, Vormoor J, Murdoch B, Hoang T, Caceres-Cortes J, Minden M, Paterson B, Caligiuri MA, Dick JE (1994) A cell initiating human acute myeloid leukaemia after transplantation into SCID mice. Nature 367:645–648

59. Lessard J, Sauvageau G (2003) Bmi-1 determines the proliferative capacity of normal and leukaemic stem cells. Nature 423:255–260
60. Li L, Bhatia R (2011) Stem cell quiescence. Clin Cancer Res 17:4936–4941
61. Lim S, Saw TY, Zhang M, Janes MR, Nacro K, Hill J, Lim AQ, Chang CT, Fruman DA, Rizzieri DA et al (2013) Targeting of the MNK-eIF4E axis in blast crisis chronic myeloid leukemia inhibits leukemia stem cell function. Proc Natl Acad Sci U S A 110:E2298–E2307
62. Lugo TG, Pendergast AM, Muller AJ, Witte ON (1990) Tyrosine kinase activity and transformation potency of bcr-abl oncogene products. Science 247:1079–1082
63. Melo JV, Deininger MW (2004) Biology of chronic myelogenous leukemia – signaling pathways of initiation and transformation. Hematol Oncol Clin North Am 18:545–568. vii–viii
64. Mendez O, Zavadil J, Esencay M, Lukyanov Y, Santovasi D, Wang SC, Newcomb EW, Zagzag D (2010) Knock down of HIF-1alpha in glioma cells reduces migration in vitro and invasion in vivo and impairs their ability to form tumor spheres. Mol Cancer 9:133
65. Mohty M, Szydlo RM, Yong AS, Apperley JF, Goldman JM, Melo JV (2008) Association between BMI-1 expression, acute graft-versus-host disease, and outcome following allogeneic stem cell transplantation from HLA-identical siblings in chronic myeloid leukemia. Blood 112:2163–2166
66. Molofsky AV, He S, Bydon M, Morrison SJ, Pardal R (2005) Bmi-1 promotes neural stem cell self-renewal and neural development but not mouse growth and survival by repressing the p16Ink4a and p19Arf senescence pathways. Genes Dev 19:1432–1437
67. Nakahara F, Sakata-Yanagimoto M, Komeno Y, Kato N, Uchida T, Haraguchi K, Kumano K, Harada Y, Harada H, Kitaura J et al (2010) Hes1 immortalizes committed progenitors and plays a role in blast crisis transition in chronic myelogenous leukemia. Blood 115:2872–2881
68. Neviani P, Santhanam R, Trotta R, Notari M, Blaser BW, Liu S, Mao H, Chang JS, Galietta A, Uttam A et al (2005) The tumor suppressor PP2A is functionally inactivated in blast crisis CML through the inhibitory activity of the BCR/ABL-regulated SET protein. Cancer Cell 8:355–368
69. Nishioka C, Ikezoe T, Yang J, Nobumoto A, Kataoka S, Tsuda M, Udaka K, Yokoyama A (2014) CD82 regulates STAT5/IL-10 and supports survival of acute myelogenous leukemia cells. Int J Cancer 134:55–64
70. Nowell PC, Hungerford DA (1960) Chromosome studies on normal and leukemic human leukocytes. J Natl Cancer Inst 25:85–109
71. Nowell PC, Hungerford DA (1961) Chromosome studies in human leukemia. II. Chronic granulocytic leukemia. J Natl Cancer Inst 27:1013–1035
72. O'Brien SG, Guilhot F, Larson RA, Gathmann I, Baccarani M, Cervantes F, Cornelissen JJ, Fischer T, Hochhaus A, Hughes T et al (2003) Imatinib compared with interferon and low-dose cytarabine for newly diagnosed chronic-phase chronic myeloid leukemia. N Engl J Med 348:994–1004
73. Peng C, Chen Y, Li D, Li S (2010) Role of Pten in leukemia stem cells. Oncotarget 1:156–160
74. Peng C, Li D, Li S (2007) Heat shock protein 90: a potential therapeutic target in leukemic progenitor and stem cells harboring mutant BCR-ABL resistant to kinase inhibitors. Cell Cycle 6:2227–2231
75. Petzer AL, Eaves CJ, Lansdorp PM, Ponchio L, Barnett MJ, Eaves AC (1996) Characterization of primitive subpopulations of normal and leukemic cells present in the blood of patients with newly diagnosed as well as established chronic myeloid leukemia. Blood 88:2162–2171
76. Prost S, Relouzat F, Spentchian M, Ouzegdouh Y, Saliba J, Massonnet G, Beressi JP, Verhoeyen E, Raggueneau V, Maneglier B et al (2015) Erosion of the chronic myeloid leukaemia stem cell pool by PPARgamma agonists. Nature 525:380–383
77. Qiu X, Guo G, Chen K, Kashiwada M, Druker BJ, Rothman PB, Chen JL (2012) A requirement for SOCS-1 and SOCS-3 phosphorylation in Bcr-Abl-induced tumorigenesis. Neoplasia 14:547–558

78. Quintas-Cardama A, Cortes J (2009) Molecular biology of bcr-abl1-positive chronic myeloid leukemia. Blood 113:1619–1630
79. Radmark O, Samuelsson B (2007) 5-lipoxygenase: regulation and possible involvement in atherosclerosis. Prostaglandins Other Lipid Mediat 83:162–174
80. Reya T, Duncan AW, Ailles L, Domen J, Scherer DC, Willert K, Hintz L, Nusse R, Weissman IL (2003) A role for Wnt signalling in self-renewal of haematopoietic stem cells. Nature 423:409–414
81. Rowley JD (1973a) Chromosomal patterns in myelocytic leukemia. N Engl J Med 289:220–221
82. Rowley JD (1973b) Letter: a new consistent chromosomal abnormality in chronic myelogenous leukaemia identified by quinacrine fluorescence and Giemsa staining. Nature 243:290–293
83. Sawyers CL (1999) Chronic myeloid leukemia. N Engl J Med 340:1330–1340
84. Sawyers CL, Hochhaus A, Feldman E, Goldman JM, Miller CB, Ottmann OG, Schiffer CA, Talpaz M, Guilhot F, Deininger MW et al (2002) Imatinib induces hematologic and cytogenetic responses in patients with chronic myelogenous leukemia in myeloid blast crisis: results of a phase II study. Blood 99:3530–3539
85. Scaglia N, Igal RA (2005) Stearoyl-CoA desaturase is involved in the control of proliferation, anchorage-independent growth, and survival in human transformed cells. J Biol Chem 280:25339–25349
86. Scheibel T, Buchner J (1998) The Hsp90 complex – a super-chaperone machine as a novel drug target. Biochem Pharmacol 56:675–682
87. Scheller M, Schonheit J, Zimmermann K, Leser U, Rosenbauer F, Leutz A (2013) Cross talk between Wnt/beta-catenin and Irf8 in leukemia progression and drug resistance. J Exp Med 210:2239–2256
88. Semenza GL (2007) Hypoxia-inducible factor 1 (HIF-1) pathway. Sci STKE 2007:cm8
89. Shen WF, Rozenfeld S, Kwong A, ves Kom LG, Lawrence HJ, Largman C (1999) HOXA9 forms triple complexes with PBX2 and MEIS1 in myeloid cells. Mol Cell Biol 19:3051–3061
90. Siddiqui N, Tempel W, Nedyalkova L, Volpon L, Wernimont AK, Osborne MJ, Park HW, Borden KL (2012) Structural insights into the allosteric effects of 4EBP1 on the eukaryotic translation initiation factor eIF4E. J Mol Biol 415:781–792
91. Soeda A, Park M, Lee D, Mintz A, Androutsellis-Theotokis A, McKay RD, Engh J, Iwama T, Kunisada T, Kassam AB et al (2009) Hypoxia promotes expansion of the CD133-positive glioma stem cells through activation of HIF-1alpha. Oncogene 28:3949–3959
92. Talpaz M, Silver RT, Druker BJ, Goldman JM, Gambacorti-Passerini C, Guilhot F, Schiffer CA, Fischer T, Deininger MW, Lennard AL et al (2002) Imatinib induces durable hematologic and cytogenetic responses in patients with accelerated phase chronic myeloid leukemia: results of a phase 2 study. Blood 99:1928–1937
93. Tauchi T, Boswell HS, Leibowitz D, Broxmeyer HE (1994) Coupling between p210bcr-abl and Shc and Grb2 adaptor proteins in hematopoietic cells permits growth factor receptor-independent link to ras activation pathway. J Exp Med 179:167–175
94. Teglund S, Toftgard R (2010) Hedgehog beyond medulloblastoma and basal cell carcinoma. Biochim Biophys Acta 1805:181–208
95. Thomas SM, Brugge JS (1997) Cellular functions regulated by Src family kinases. Annu Rev Cell Dev Biol 13:513–609
96. Wang Y, Liu Y, Malek SN, Zheng P, Liu Y (2011) Targeting HIF1alpha eliminates cancer stem cells in hematological malignancies. Cell Stem Cell 8:399–411
97. Yang J, Wang J, Chen K, Guo G, Xi R, Rothman PB, Whitten D, Zhang L, Huang S, Chen JL (2013) eIF4B phosphorylation by pim kinases plays a critical role in cellular transformation by Abl oncogenes. Cancer Res 73:4898–4908
98. Zhang B, Strauss AC, Chu S, Li M, Ho Y, Shiang KD, Snyder DS, Huettner CS, Shultz L, Holyoake T et al (2010) Effective targeting of quiescent chronic myelogenous leukemia stem cells by histone deacetylase inhibitors in combination with imatinib mesylate. Cancer Cell 17:427–442

99. Zhang H, Li H, Ho N, Li D, Li S (2012a) Scd1 plays a tumor-suppressive role in survival of leukemia stem cells and the development of chronic myeloid leukemia. Mol Cell Biol 32:1776–1787
100. Zhang H, Li H, Xi HS, Li S (2012b) HIF1alpha is required for survival maintenance of chronic myeloid leukemia stem cells. Blood 119:2595–2607
101. Zhang H, Peng C, Hu Y, Li H, Sheng Z, Chen Y, Sullivan C, Cerny J, Hutchinson L, Higgins A et al (2012c) The Blk pathway functions as a tumor suppressor in chronic myeloid leukemia stem cells. Nat Genet 44:861–871
102. Zhao C, Blum J, Chen A, Kwon HY, Jung SH, Cook JM, Lagoo A, Reya T (2007) Loss of beta-catenin impairs the renewal of normal and CML stem cells in vivo. Cancer Cell 12:528–541
103. Zhao C, Chen A, Jamieson CH, Fereshteh M, Abrahamsson A, Blum J, Kwon HY, Kim J, Chute JP, Rizzieri D et al (2009) Hedgehog signalling is essential for maintenance of cancer stem cells in myeloid leukaemia. Nature 458:776–779
104. Zhou H, Xu R (2015) Leukemia stem cells: the root of chronic myeloid leukemia. Protein Cell 6:403–412
105. Zhu J, Zhou L, Wu G, Konig H, Lin X, Li G, Qiu XL, Chen CF, Hu CM, Goldblatt E et al (2013) A novel small molecule RAD51 inactivator overcomes imatinib-resistance in chronic myeloid leukaemia. EMBO Mol Med 5:353–365

Chapter 10
Cellular Immunotherapy in the Treatment of Hematopoietic Malignancies

Satoko Matsueda, Thinle Chodon, and Richard C. Koya

Contents

Abstract Cancer immunotherapy has been shown to be an efficacious therapeutic approach in the treatment of cancers including hematopoietic malignancies. Induction of T cell cytotoxicity against tumors by adoptive cell therapies (ACT), cancer vaccines, gene therapies, and monoclonal antibody therapies has been intensively studied. In particular, immune checkpoint blockade and chimeric antigen receptor T (CAR-T) cell therapies are the recent clinical successes in cancer immunotherapy. This article introduces the main concepts and addresses the most relevant clinical modalities of cellular immunotherapies for hematological malignancies: antigen non-specific T cell therapy, genetically modified T cell receptor (TCR) T cell therapy, chimeric antigen receptor (CAR) T cell therapy, and CAR-T cell clinical trials in leukemia, lymphoma, and multiple myeloma. Clinical trials have shown encouraging results, but future studies may need to incorporate novel CAR constructs or targets with enhanced safety and efficacy to ensure long-term benefits.

Keywords Adoptive cell therapies · CAR-T · Leukemia

S. Matsueda · T. Chodon · R. C. Koya (✉)
Center for Immunotherapy, Roswell Park Comprehensive Cancer Center, Buffalo, New York, USA
e-mail: richard.koya@roswellpark.org

H. Zhang, S. Li (eds.), *Leukemia Stem Cells in Hematologic Malignancies*, Advances in Experimental Medicine and Biology 1143,
https://doi.org/10.1007/978-981-13-7342-8_10

10.1 Introduction

Much interest has risen for cancer immunotherapy as a new and efficacious therapeutic approach in the last decade. Especially, adoptive cell therapies (ACT), cancer vaccines, gene therapies, and monoclonal antibody therapies have all been used to induce T cell cytotoxicity against tumors.

Immune checkpoint blockade and chimeric antigen receptor T (CAR-T) cell therapies are the recent clinical successes in cancer immunotherapy. In 2017, CAR-T cell therapy (Kymriah, Novartis) for juvenile acute lymphoblastic leukemia (ALL) was approved by the FDA as the first CAR-T cell therapy for the treatment of B-cell ALL patients that does not respond to treatment or relapsed under 25 years old. Thereafter the FDA approved the use of CAR-T cell (Yescarta, Kite) for large B-cell lymphoma, including diffuse large B-cell lymphoma (DLBCL), primary mediastinal large B-cell lymphoma, high-grade B-cell lymphoma, and transformed follicular lymphoma. In 2018, Kymriah was also approved for the treatment of non-Hodgkin lymphoma. These CAR-Ts are genetically modified autologous T cells made to recognize and lyse cancer cells in a very specific fashion. This article introduces the main concepts and addresses the most relevant clinical modalities of cellular immunotherapies for hematological malignancies (see Table 10.1).

10.2 Antigen Non-specific T Cell Therapy

The immune system, especially T lymphocytes cells (T cells), plays an important role to detect and eliminate "nonself" such as pathogens and transformed cells like cancers. Adoptive T cell therapy (ACT) is based on infusion of tumor-specific T cells. Many different cell types can be used as killer cells. Lymphokine-activated killer cells were utilized for the first trial of ACT in human. Patients with metastatic melanoma were treated with the combination of lymphokine-activated killer cells plus interleukin-2 in that study [1]. The survival and sustained activation of the infused lymphokine-activated killer cells were ensured by IL-2. The approach has resulted in marked tumor regression in up to or approximately 30% of the patients with renal cell, melanoma, colorectal, non-Hodgkin's lymphoma, and adenocarcinoma of the lung [2].

ACT with tumor infiltration lymphocytes (TILs) has provided promising results in clinical studies in metastatic melanoma [3]. However, ACT with TILs faces three problems: (1) It remains limited to particular tumor types, (2) it takes 4–6 weeks for TILs to grow, and (3) it requires specialized cell production facilities and staff [4, 5]. There are approaches being developed to attempt overcoming some of these hurdles. Among the most promising ones are the T cell receptor (TCR) engineered therapy and chimeric antigen receptor (CAR) therapy (Fig. 10.1).

Table 10.1 Selected clinical trials for cellular immunotherapy

Immunotherapy type	Cell type	Antigen	Disease	Trial ID	Compound name
Cell therapy	CMV-specific T cells	CMV	CMV-positive hematological malignancies	NCT03067155	N/A
	Mismatched immune cells	CD3/CD28	Leukemia, lymphoma, multiple myeloma	NCT00558675	AlloStim
	NK -92 cell line	Irradiated NK cells	Leukemia, lymphoma, myeloma, Hodgkin's disease	NCT00990717	N/A
	NK cells	NK	Acute myeloid leukemia	NCT02809092	N/A
TCR-T cell therapy	T cells	NY-ESO-1, LAGE-1	Multiple myeloma	NCT01352286 [1]	N/A
	T cells	NY-ESO-1, LAGE-1	Multiple myeloma	NCT01892293	N/A
	T cells	NY-ESO-1	Solid or hematological malignancies	NCT03391778	GSK3377794
	T cells	NY-ESO-1	Multiple myeloma, other metastatic solid cancers	NCT02457650	N/A
	T cells	WT1	Acute myeloid leukemia	NCT00052520 [2]	N/A
	T cells	WT1	Acute myeloid leukemia, chronic myeloid leukemia	NCT01621724	
	Central memory/naive CD8+ T cells	WT-1	Acute myeloid leukemia	NCT02770820	N/A
	T cells	PRAME TCR	Acute myeloid leukemia	NCT02743611	BPX-701
	T cells	HA-1 T TCR	Acute myeloid leukemia	NCT03326921	N/A
CAR therapy	T cells	CD19 CAR	B-cell acute lymphoblastic leukemia, B-cell non-Hodgkin lymphoma	Approved [3]	KYMRIAH
	T cells	CD19 CAR	Large B-cell lymphomas	Approved [4]	YESCARTA
	T cells	CD19 CAR	B-lineage acute lymphoblastic leukemia	NCT02028455 [5]	N/A
	T cells	Muc1/CLL1/CD33/CD38/CD56/CD123	Acute myeloid leukemia	NCT03222674	N/A

(continued)

Table 10.1 (continued)

Immunotherapy type	Cell type	Antigen	Disease	Trial ID	Compound name
	T cells	CD123	Acute myelocytic leukemia	NCT03556982	N/A
	T cells	CD19	CD19+ refractory or relapsed B-ALL patients	NCT03173417	IM19
	T cells	CD22	B-cell malignancies	NCT03262298	N/A
	T cells	CD19/CD22	B-acute lymphoblastic leukemia	NCT03241940	N/A
	T cells	CD20	B-cell malignancies	NCT02710149	N/A
	T cells	BCMA	Multiple myeloma	NCT02658929	bb2121
	T cells	BCMA	Multiple myeloma	NCT02215967 [6]	N/A
	APRIL CAR-T cells	BCMA and TACI	Multiple myeloma	NCT03287804	AUTO2
	T cells	CD133	Acute myeloid and lymphoid leukemias and solid tumor	NCT02541370	N/A
	T cells	CD138	Multiple myeloma	NCT01886976 [7]	N/A
	T cells	CD19	Multiple myeloma	NCT02135406 [8, 9]	N/A
	T cells	CD30	Hodgkin lymphoma and anaplastic large cell lymphoma	NCT01316146 [10]	N/A
	T cells	CD20	Diffuse large B-cell lymphoma	NCT01735604 [11]	N/A
	T cells	CD20	Non-Hodgkin's lymphoma	NCT00012207 [12]	N/A
	T cells	CD5	T cell leukemia or lymphoma	NCT03081910	N/A
	NK cells	CD7	CD7-positive leukemia and lymphoma	NCT02742727	N/A
	NK cells	CD19	CD19-positive leukemia and lymphoma	NCT02892695	PCAR-119
	NK cells	CD19	Relapsed/refractory B-lymphoid malignancies	NCT03056339	N/A

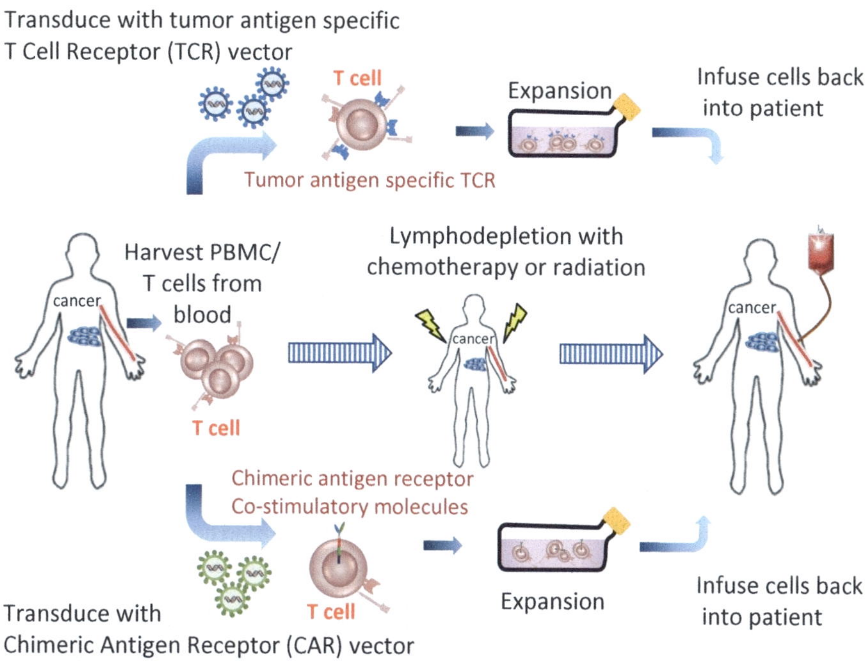

Fig. 10.1 Engineered T cell therapy process. The most promising cellular immunotherapies are the T cell receptor (TCR)-engineered therapy and chimeric antigen receptor (CAR) therapy

10.3 Genetically Modified T Cell Receptor (TCR) T Cell Therapy

In this modality of cellular therapy, T cells are modified to express new T cell receptors that recognize tumor-specific antigens. These receptors can be modified to improve their binding avidity and affinity against their target antigens in an MHC-specific context. PBMC/T cells from patients are harvested and transduced with a viral vector to incorporate the new T cell receptor. TCR therapies have shown tumor regression in some patients with advanced solid tumors, e.g., metastatic melanoma, colorectal cancer, and synovial cell sarcoma [6, 7]. One of the difficulties in TCR therapy is that the TCR must be genetically matched to the patient's MHC type. A possible safety issue is the unpredictable specificities of heterodimer formation of the recombinant α and β TCR chains with the respective chains of the endogenous TCR [8]. Unpredictable and unideal mispairing of TCR chains may not only result in loss of specificity but may also induce potential autoreactivity [9, 10]. Introduction of murine-human hybrid T cell receptor (TCR) [11, 12] and cysteines into the constant region of the α and β chains [13] enforces preferential pairing and improves both the efficacy and safety of TCR therapy to generate tumor-reactive T cells. The first treatment with genetically engineered T cells against MART1 was administrated in metastatic melanoma patients and reported in 2006 [14]. In this study, 2 of 15 patients demonstrated objective regressions of metastatic melanoma lesions, and no toxicities were observed in any patients [14]. A second-generation TCR with greater affinity for MART1 was conducted in a subsequent clinical trial [15]. Six of the 20 patients (30%) treated with MART-1 TCR and 3 of 16 (19%) treated with gp100 TCR experienced an objective antitumor response; however several patients displayed autoimmune T cell-mediated destruction of normal melanocytes [15].

Cancer-testis (CT) antigens are a very attractive class of proteins to be targeted in cancer. These antigens have restricted tissue expression in germ cells and in a significant number of different malignancies. NY-ESO-1 is a CT antigen expressed by many malignancies, including multiple myeloma, melanoma, sarcoma, non-small cell lung cancer, cholangiocarcinoma, and breast and ovarian cancer [16, 17]. A TCR against NY-ESO-1 with single and dual CDR3α and CDR2β amino acid substitutions dramatically enhanced tumor recognition [18]. The response rates of patients with melanoma and synovial cell sarcoma were 45% and 67%, respectively, and no toxicities were observed in a trial [6]. Another CT antigen, MAGE-A3 had been targeted with a high avidity TCR. Five of nine patients demonstrated cancer regression following infusion of anti-MAGE-A3 TCR gene-engineered T cells [19]. However, three of nine patients developed severe neurologic toxicity, probably due to the recognition of the MAGE-A12 protein which is expressed in human brain. Another MAGE-A3-specific TCR led to cardiac shock in two patients and caused death from heart failure when the TCR cross-reacted with Titin, a protein expressed in myocardial tissue [20]. Regarding the use of TCRs in hematologic malignancies, a phase I/II trial for multiple myeloma (MM) patients accessed the safety and efficacy of NY-ESO-1 and LAGE-1-specific TCR-engineered T cells [21]. In this study, 14 out of 20 MM patients showed a near complete response (nCR) or complete response (CR), 2 patients showed a very good partial response, 2 showed a partial response, 1 showed stable disease, and 1 showed progressive disease at day 100 after transplantation [22]. Interestingly, long-term engraftment/persistence of engineered cells was observed up to 2 years after infusion in this study. Another promising target for hematologic malignancies, Wilms' tumor 1 (WT1), is known to be expressed in most acute myeloid leukemia (AML) patients [22, 23]. Since WT1 expression levels correlate with poor clinical prognosis, it would be an attractive target for ACT [22]. WT1-specific TCR-engineered T cell therapy was conducted in a phase I clinical trial for AML and high-risk myelodysplastic syndromes (MDS) [24]. In the study, five out of eight patients showed persistence of engineered cells, and four of these five patients survived over 12 months. These results indicate that targeting WT1 antigen for AML and MDS has the potential to be of clinical benefit [24].

10.4 Chimeric Antigen Receptor (CAR) T Cell Therapy

The chimeric antigen receptor (CAR) is composed of an extracellular domain with a single-chain variable fragment (scFv) from monoclonal antibody for antigen binding and an intracellular part with a TCR/CD3ζ endodomain for the transmission of ζ signal in response to recognition of antigen and provision of T cell activation [25]. Second- and third-generation CARs have been developed for enhanced cytotoxicity, proliferation, cytokine secretion, resistance, and development of memory by including combinations of costimulatory domains, such as CD28, CD137 (4-1BB), and CD134 (OX40) [26]. A fourth-generation CAR construct called "T cells redirected for universal cytokine-mediated killing" (TRUCKs) is equipped with genetic modifications to induce expression of a transgenic product like IL-12 or another cytokine [27].

10.5 CAR-T Cell Clinical Trials in Leukemia

CD19-specific CAR-T cells have been administered in recurrent/refractory CD19+ ALL in children and young adults, and there was 93% minimal residual disease-negative (MRD−) complete remission (CR) rate in treated patients, resulting in 89% overall efficacy [28]. In this study, 45% of patients who achieved an MRD− CR relapsed, and 39% of those relapses showed a CD19-negative escape variant leukemia [28]. In chronic lymphocytic leukemia (CLL), a report of a patient who received ACT with second-generation anti-CD19 CAR-modified T cells demonstrated a complete remission associated with the tumor lysis syndrome [29]. In addition to tumor lysis syndrome, the patient had B-cell depletion and hypogammaglobinemia as a chronic toxic effect [29]. Severe toxic effects with CAR-T cell therapy have been observed from unexpected presence of the target antigens on normal tissues [30]. To address this issue, a novel construct that incorporated the interleukin-15 (IL15) gene and an inducible caspase-9-based suicide gene was developed. It had improved antitumor effects and efficiently eliminated CAR CD19 T lymphocytes [31]. Suicide gene may be useful for treatment of patients, except when acute toxic effects occurred within hours after administration [32, 33]. The most common adverse events, cytokine release syndrome (CRS), and neurologic toxicities (NT) have been observed in CAR-T cell trials. Effective ways to prevent these toxicities remain poorly understood [34]. Modification of T cell composition and dose reduction have been successful in B-cell ALL patients [35], and blocking IL-6 was found to be effective [36, 37]. Myeloid cells, including macrophages and monocytes, were found to be the major cells mediating CRS and NT by releasing IL-1 and IL-6 among other cytokines [38]. Blocking IL-1 receptor effectively abolished CRS as well as NT in mouse model [38]. CD22 targeting CAR-T might be an option to CD19-relapsed patients [39] and is currently under clinical investigation. Another attractive target is CD123. Preclinical studies with CD123-specific CAR-T cells demonstrated tumor eradication and long-term disease-free survival in PDX mice model [40]. A pilot safety study of CD123 CAR-T exhibited a decrease of blasts, and the patient achieved partial remission, although cytokine release syndrome developed [41]. Over ten clinical trials are accruing for leukemia patients with CD123 CAR-T, and further exploration is needed.

10.6 CAR-T Cell Clinical Trials in Multiple Myeloma

Candidate antigens of immunotherapies for multiple myeloma include B-cell maturation antigen (BMCA)/CD269, CD138, and CD19. The expression of BCMA/CD269, a member of the TNF receptor superfamily, has been reported mostly in plasma cells and subsets of mature B cells including multiple myeloma [42–45]. Preclinical studies showed that BCMA-CAR-T exhibits cytotoxicity and in vivo tumor eradication [42]. One clinical trial showed that the overall response rate was 81%, with 63% very good partial response or complete response [46]. In this study, two out of ten patients who received 0.3–3 × 10^6 CAR+ T cells/kg showed partial response (PR) and very good partial response (VGPR). The overall response rate for patients who received 9 × 10^6 CAR+ T cells/kg was 81%, 3 PR, 8 VGPR, 2 stringent complete response (SCR) among 16 patients. Eleven patients were found to be minimal residual disease (MRD) negative, 5 were not evaluated among 16 patients who received 9 × 10^6 CAR+ T cells/kg. The toxicity was mild and no cytokine release syndrome (CRS) at lower dose (0.3–3 × 10^6 CAR+ T cells/kg). However, 6 of 16 patients at the dose of 9 × 10^6 CAR+ T cells/kg had CRS grade 3 or 4 [46]. CD138 (syndecan-1) is a member of the transmembrane heparan sulfate proteoglycan family and is expressed on normal and malignant plasma cells [47, 48]. Five patients were infused CD3+ CD138-CAR-T cells (average of 0.756 × 10^7 cells/kg), and there was no intolerable toxicity observed [49]. Four of five patients had stable disease that lasted over 3 months. CD19 is not typically considered a target in multiple myeloma; however, minor subset of the multiple myeloma cells with drug-resistant, disease-propagating properties has a CD19-positive phenotype [50, 51]. Ten patients received CD19-CAR-T cells following salvage high-dose melphalan and autologous stem cell transplantation (ASCT) with one, two and four patients exhibiting SCR, PR and VGPR, respectively, 100 days after ASCT [52]. Two of ten patients who had VGPR exhibited significantly longer PFS after ASCT + CD19-CAR-T compared to prior ASCT alone (479 vs. 181 days; 249 vs. 127 days). Most of grade 3 or higher adverse events were unrelated to CD19 CAR-T cells. CRS was observed in one patient only and it was grade 1 [52]. This study indicated that CD19-CAR-T may improve duration of response and targeting CD19+ myeloma-propagating cells could be of clinical benefit [52].

10.7 CAR-T Cell Clinical Trial Therapy in Lymphoma

CD30 is a member of the tumor necrosis factor receptor (TNFR) family, and it is a cell surface receptor for activated T cells, Hodgkin lymphoma (HL), anaplastic large cell lymphoma (ALCL), and a few other non-Hodgkin's lymphomas (NHL) [53]. Anti-CD30 monoclonal antibodies in HL and ALCL showed clinical success. This encouraged the development of CD30-CAR-T cell therapy. CD30-CAR-T cell were infused without a conditioning regimen into 7 HL and 2 ALCL patients [54]. Two had CR, one had continued CR (CCR), and three had transient stable disease in this study. One CR lasted over 3 years and another CR lasted for 2 years. These CR responses were mainly achieved in patients infused with CD30-CAR-T cells at the dose of 2×10^8. No severe adverse events were observed, and no patients developed CRS [54]. CD20 is a tetra-transmembrane protein that is expressed in >90% of B-cell lymphomas [55]. Anti-CD20 antibody combined with chemotherapy will produce superior clinical effects in most NHL patients compared to chemotherapy alone. CD20-CAR-T 1×10^7 cells/kg were infused into seven patients, one had CR, and four had PR, and in total five of seven showed tumor regression during the trial with four of seven experiencing grade 3 or 4 delayed adverse events [56]. Seven patients received CD20-CAR-T cells at the dose of 1×10^8 for infusion 1, 1×10^9 for infusion 2, and 3.3×10^9 for infusion 3 in another trial [57]. Two of seven patients achieved CR to cytoreductive chemotherapy administered before the CAR-T cell infusions and remained disease-free 3 months and 13 months after T cell infusions. One patient had PR and lasted for 3 months after treatment with CAR-T cells. Four other had SD. There were no grade 3 or 4 toxicities observed and no adverse events attributable to the T cell infusions [57]. In vitro and in vivo mouse characterization experiments indicated that bispecific CAR-T platform with CD19-CD20 enhanced cytotoxicity by producing more IFNγ, TNFα, and IL-2 compared to original CAR. Bispecific CD19-CD20 CAR-T would trigger robust cytotoxicity against tumor expressing CD19 or CD20 and potentially prevent antigen escape [58].

10.8 Future Prospects for Cellular Immunotherapy

Checkpoint-blocking antibodies and small molecule inhibitors have demonstrated promising potential, but these new treatment modalities are still in their early stages, and persistence of response is short-lived, limiting their overall clinical benefits. Therefore, a combinatorial approach with ACT may offer a number of potential synergies; therefore a number of clinical trials are underway.

Although CAR-T cell therapy has shown great success in hematological malignancies, autologous T cell therapies are costly and labor-intensive, and its safety prediction still remains difficult. Some alternative cellular immunotherapies have been developed to address some of those issues. CAR-NK and CAR-iPSC-NK cells might extend these possibilities since natural killer (NK) cells do not require HLA matching and show cytotoxicity against tumor. The advantages of using CAR-NK are that allogeneic NK cells should not cause significant GVHD; NK cells would have a relatively limited life-span, limiting toxicity. The NK92 cell line was used for a clinical trial as it can easily expand. However, its use requires irradiation prior to infusion to patients due to the potential for tumor engraftment following infusion. And irradiation of CAR-NK can affect its cellular proliferation and persistence in vivo [59, 60]. Cord blood (CB) can be a source of NK cells; however they are phenotypically and functionally immature. Using GMP-grade K562-based artificial antigen-presenting cells (aAPCs) expressing membrane bound IL-21 and 4-1BB ligand, clinically relevant doses of GMP-grade NK cells were generated from a CB [61]. In another approach, NK-CAR-iPSC-NK cells significantly inhibited tumor growth and prolonged survival in an ovarian cancer mouse model [62]. The iPSC-derived NK cells also induced less cytokine release, lymphocyte-related pathological damage, and GVHD [62]. CAR-iPSC-NK cells may be a promising cellular therapy, although further clinical studies will reveal its own safety concerns.

10.9 Conclusions

An ACT, along with the engineering of the TCR or CAR to recognize tumor antigens, and potentially in combination with checkpoint blocking Abs or small molecule inhibitors, has the potential to be very beneficial in assisting in the eradication of various hematological malignancies. However, there are still several gaps in knowledge to be addressed before it becomes a mainstream therapeutic choice. Clinical trials have shown encouraging results, but future studies may need to incorporate novel CAR constructs or targets with enhanced safety and efficacy to ensure long-term benefits.

References

1. Rosenberg SA et al (1985) Observations on the systemic administration of autologous lymphokine-activated killer cells and recombinant interleukin-2 to patients with metastatic cancer. N Engl J Med 313(23):1485–1492
2. Rosenberg SA et al (1987) A progress report on the treatment of 157 patients with advanced cancer using lymphokine-activated killer cells and interleukin-2 or high-dose interleukin-2 alone. N Engl J Med 316(15):889–897
3. Rosenberg SA et al (2011) Durable complete responses in heavily pretreated patients with metastatic melanoma using T-cell transfer immunotherapy. Clin Cancer Res 17(13):4550–4557
4. Humphries C (2013) Adoptive cell therapy: honing that killer instinct. Nature 504(7480):S13–S15
5. Hinrichs CS, Rosenberg SA (2014) Exploiting the curative potential of adoptive T-cell therapy for cancer. Immunol Rev 257(1):56–71

6. Robbins PF et al (2011) Tumor regression in patients with metastatic synovial cell sarcoma and melanoma using genetically engineered lymphocytes reactive with NY-ESO-1. J Clin Oncol 29(7):917–924
7. Parkhurst MR et al (2011) T cells targeting carcinoembryonic antigen can mediate regression of metastatic colorectal cancer but induce severe transient colitis. Mol Ther 19(3):620–626
8. Jennifer Makalowski HA. Chapter13
9. Coccoris M et al (2010) T cell receptor (TCR) gene therapy to treat melanoma: lessons from clinical and preclinical studies. Expert Opin Biol Ther 10(4):547–562
10. Bendle GM et al (2010) Lethal graft-versus-host disease in mouse models of T cell receptor gene therapy. Nat Med 16(5):565–570. 1p following 570
11. Cohen CJ et al (2006) Enhanced antitumor activity of murine-human hybrid T-cell receptor (TCR) in human lymphocytes is associated with improved pairing and TCR/CD3 stability. Cancer Res 66(17):8878–8886
12. Goff SL et al (2010) Enhanced receptor expression and in vitro effector function of a murine-human hybrid MART-1-reactive T cell receptor following a rapid expansion. Cancer Immunol Immunother 59(10):1551–1560
13. Kuball J et al (2007) Facilitating matched pairing and expression of TCR chains introduced into human T cells. Blood 109(6):2331–2338
14. Morgan RA et al (2006) Cancer regression in patients after transfer of genetically engineered lymphocytes. Science 314(5796):126–129
15. Johnson LA et al (2009) Gene therapy with human and mouse T-cell receptors mediates cancer regression and targets normal tissues expressing cognate antigen. Blood 114(3):535–546
16. Hofmann O et al (2008) Genome-wide analysis of cancer/testis gene expression. Proc Natl Acad Sci U S A 105(51):20422–20427
17. Odunsi K et al (2014) Epigenetic potentiation of NY-ESO-1 vaccine therapy in human ovarian cancer. Cancer Immunol Res 2(1):37–49
18. Robbins PF et al (2008) Single and dual amino acid substitutions in TCR CDRs can enhance antigen-specific T cell functions. J Immunol 180(9):6116–6131
19. Morgan RA et al (2013) Cancer regression and neurological toxicity following anti-MAGE-A3 TCR gene therapy. J Immunother 36(2):133–151
20. Linette GP et al (2013) Cardiovascular toxicity and titin cross-reactivity of affinity-enhanced T cells in myeloma and melanoma. Blood 122(6):863–871
21. Rapoport AP et al (2015) NY-ESO-1-specific TCR-engineered T cells mediate sustained antigen-specific antitumor effects in myeloma. Nat Med 21(8):914–921
22. Bergmann L et al (1997) High levels of Wilms' tumor gene (wt1) mRNA in acute myeloid leukemias are associated with a worse long-term outcome. Blood 90(3):1217–1225
23. Greiner J et al (2004) mRNA expression of leukemia-associated antigens in patients with acute myeloid leukemia for the development of specific immunotherapies. Int J Cancer 108(5):704–711
24. Tawara I et al (2017) Safety and persistence of WT1-specific T-cell receptor gene-transduced lymphocytes in patients with AML and MDS. Blood 130(18):1985–1994
25. Eshhar Z et al (1993) Specific activation and targeting of cytotoxic lymphocytes through chimeric single chains consisting of antibody-binding domains and the gamma or zeta subunits of the immunoglobulin and T-cell receptors. Proc Natl Acad Sci U S A 90(2):720–724
26. Urba WJ, Longo DL (2011) Redirecting T cells. N Engl J Med 365(8):754–757
27. Chmielewski M, Abken H (2015) TRUCKs: the fourth generation of CARs. Expert Opin Biol Ther 15(8):1145–1154
28. Gardner RA et al (2017) Intent-to-treat leukemia remission by CD19 CAR T cells of defined formulation and dose in children and young adults. Blood 129(25):3322–3331
29. Porter DL et al (2011) Chimeric antigen receptor-modified T cells in chronic lymphoid leukemia. N Engl J Med 365(8):725–733
30. Heslop HE (2010) Safer CARS. Mol Ther 18(4):661–662

31. Hoyos V et al (2010) Engineering CD19-specific T lymphocytes with interleukin-15 and a suicide gene to enhance their anti-lymphoma/leukemia effects and safety. Leukemia 24(6):1160–1170
32. Brentjens R et al (2010) Treatment of chronic lymphocytic leukemia with genetically targeted autologous T cells: case report of an unforeseen adverse event in a phase I clinical trial. Mol Ther 18(4):666–668
33. Morgan RA et al (2010) Case report of a serious adverse event following the administration of T cells transduced with a chimeric antigen receptor recognizing ERBB2. Mol Ther 18(4):843–851
34. Zheng PP, Kros JM, Li J (2018) Approved CAR T cell therapies: ice bucket challenges on glaring safety risks and long-term impacts. Drug Discov Today 23(6):1175–1182
35. Turtle CJ et al (2016) CD19 CAR-T cells of defined CD4+:CD8+ composition in adult B cell ALL patients. J Clin Invest 126(6):2123–2138
36. Maus MV et al (2014) Adoptive immunotherapy for cancer or viruses. Annu Rev Immunol 32:189–225
37. Davila ML et al (2014) Efficacy and toxicity management of 19-28z CAR T cell therapy in B cell acute lymphoblastic leukemia. Sci Transl Med 6(224):224ra25
38. Liu D, Zhao J (2018) Cytokine release syndrome: grading, modeling, and new therapy. J Hematol Oncol 11(1):121
39. Shalabi H, Angiolillo A, Fry TJ (2015) Beyond CD19: opportunities for future development of targeted immunotherapy in pediatric relapsed-refractory acute leukemia. Front Pediatr 3:80
40. Tianyu Cai RG, Gouble A, Smith J, Cavazos A, Konoplev S, Lane AA, Guzman ML, Kantarjian HM, Pemmaraju N, Konopleva M (2016) Pre-clinical studies of anti-CD123 CAR-T cells for the treatment of Blastic Plasmacytoid Dendritic Cell Neoplasm (BPDCN). Blood 128(22):4039
41. Yi Luo L-JC, Hu Y, Dong L, Wei G, Huang H (2015) First-in-man CD123-specific chimeric antigen receptor-modified T cells for the treatment of refractory acute myeloid leukemia. Blood 126(23):3778
42. Carpenter RO et al (2013) B-cell maturation antigen is a promising target for adoptive T-cell therapy of multiple myeloma. Clin Cancer Res 19(8):2048–2060
43. Ng LG et al (2004) B cell-activating factor belonging to the TNF family (BAFF)-R is the principal BAFF receptor facilitating BAFF costimulation of circulating T and B cells. J Immunol 173(2):807–817
44. O'Connor BP et al (2004) BCMA is essential for the survival of long-lived bone marrow plasma cells. J Exp Med 199(1):91–98
45. Laabi Y et al (1994) The BCMA gene, preferentially expressed during B lymphoid maturation, is bidirectionally transcribed. Nucleic Acids Res 22(7):1147–1154
46. Brudno JN et al (2018) T cells genetically modified to express an anti-B-cell maturation antigen chimeric antigen receptor cause remissions of poor-prognosis relapsed multiple myeloma. J Clin Oncol 36(22):2267–2280
47. Cohen AD (2018) CAR T cells and other cellular therapies for multiple myeloma: 2018 update. Am Soc Clin Oncol Educ Book 38:e6–e15
48. O'Connell FP, Pinkus JL, Pinkus GS (2004) CD138 (syndecan-1), a plasma cell marker immunohistochemical profile in hematopoietic and nonhematopoietic neoplasms. Am J Clin Pathol 121(2):254–263
49. Bo Guo MC, Han Q, Hui F, Dai H, Zhang W, Zhang Y, Wang Y, Zhu H, Han W (2016) CD138-directed adoptive immunotherapy of chimeric antigen receptor (CAR)-modified T cells for multiple myeloma. J Cell Immunother 2(1):28–35
50. Garfall AL et al (2015) Chimeric antigen receptor T cells against CD19 for multiple myeloma. N Engl J Med 373(11):1040–1047
51. Hajek R, Okubote SA, Svachova H (2013) Myeloma stem cell concepts, heterogeneity and plasticity of multiple myeloma. Br J Haematol 163(5):551–564
52. Garfall AL et al (2018) Anti-CD19 CAR T cells with high-dose melphalan and autologous stem cell transplantation for refractory multiple myeloma. JCI Insight 4(4):e127684

53. Schneider C, Hubinger G (2002) Pleiotropic signal transduction mediated by human CD30: a member of the tumor necrosis factor receptor (TNFR) family. Leuk Lymphoma 43(7):1355–1366
54. Ramos CA et al (2017) Clinical and immunological responses after CD30-specific chimeric antigen receptor-redirected lymphocytes. J Clin Invest 127(9):3462–3471
55. Wang Y et al (2014) Effective response and delayed toxicities of refractory advanced diffuse large B-cell lymphoma treated by CD20-directed chimeric antigen receptor-modified T cells. Clin Immunol 155(2):160–175
56. Maloney DG (2012) Anti-CD20 antibody therapy for B-cell lymphomas. N Engl J Med 366(21):2008–2016
57. Till BG et al (2008) Adoptive immunotherapy for indolent non-Hodgkin lymphoma and mantle cell lymphoma using genetically modified autologous CD20-specific T cells. Blood 112(6):2261–2271
58. Zah E et al (2016) T cells expressing CD19/CD20 bispecific chimeric antigen receptors prevent antigen escape by malignant B cells. Cancer Immunol Res 4(6):498–508
59. Tang X et al (2018) First-in-man clinical trial of CAR NK-92 cells: safety test of CD33-CAR NK-92 cells in patients with relapsed and refractory acute myeloid leukemia. Am J Cancer Res 8(6):1083–1089
60. Mehta RS, Rezvani K (2018) Chimeric antigen receptor expressing natural killer cells for the immunotherapy of cancer. Front Immunol 9:283
61. Shah N et al (2013) Antigen presenting cell-mediated expansion of human umbilical cord blood yields log-scale expansion of natural killer cells with anti-myeloma activity. PLoS One 8(10):e76781
62. Li Y et al (2018) Human iPSC-derived natural killer cells engineered with chimeric antigen receptors enhance anti-tumor activity. Cell Stem Cell 23(2):181–192 e5

The manufacturer's authorised representative in the EU is Springer Nature Customer Service Centre GmbH, Europaplatz 3, 69115 Heidelberg, Germany. If you have any concerns regarding our products, please contact ProductSafety@springernature.com

Printed and bound by CPI Group (UK) Ltd, Croydon, CR0 4YY

24/07/2026

02175951-0002